Lecture Notes in Bioengineering

More information about this series at http://www.springer.com/series/11564

Kamran Kiasaleh

Biological Signals Classification and Analysis

Springer

Kamran Kiasaleh
Department of Electrical Engineering
The University of Texas at Dallas
Richardson, TX
USA

ISSN 2195-271X ISSN 2195-2728 (electronic)
Lecture Notes in Bioengineering
ISBN 978-3-662-51203-6 ISBN 978-3-642-54879-6 (eBook)
DOI 10.1007/978-3-642-54879-6

Springer Heidelberg New York Dordrecht London

Softcover reprint of the hardcover 1st edition 2015

Printed on acid-free paper

Springer-Verlag GmbH Berlin Heidelberg is part of Springer Science+Business Media
(www.springer.com)

This book is dedicated to my beloved wife, Tae,
and my children Brandon, Kristen, and Alison.
I am forever indebted to my mother for her endless support
and to my late father for his love of knowledge.

Preface

Biological systems have been known for many decades to produce measurable signals, which often reveal nontrivial information about the underlying processes at work. It is the hope of the scientific community that from the observation and processing of various biologically-generated signals one can draw unambiguous conclusions regarding the state of a biological system. This "state" is of outmost importance to a variety of applications, including disease diagnosis and detection of the onset of many deadly diseases, including heart attack, stroke, cancer, among other illnesses plaguing mankind today. This text will make an attempt to shed light on the concept of signals and systems not only from the "man-made" perspective, but also from the angle of nature-made or biologically-made viewpoint. Ultimately, the goal of this text is to enable the reader to bring to bear the vast knowledge of digital signal processing to tackle the biological signals in order to extract important information of clinical value. To achieve this goal, one has to acquire a healthy knowledge of non-biological signals and systems as well as signal processing techniques before embarking on an endeavor that brings into focus the need for somewhat different types of signal processing mechanisms, concerned with non-stationary signals impaired by nonlinear and even chaotic effects. It is the intention here to enable the reader to examine biological signals and systems from the digital signal processing perspective while keeping an eye on some of the shortcomings of the signal processing systems in use today, which rely heavily on the "linearity" assumption (or approximation) in order to arrive at implementable architectures. It is noteworthy that, throughout this text, we use the term "biological signals" to refer to signals originating from a biological entity. Rather seamlessly, we use the expression "biomedical signal processing" to refer to the signal processing tools that one brings to bear to study the behavior of biological signals.

Although this text is intended for an introductory class on biomedical signal processing for senior-level undergraduate or first-year graduate students, the text can readily serve as a reference textbook for professionals in the field concerned with the development of biomedical devices. To that end, many examples are provided wherever possible to help the practicing engineer who may not be familiar with biological signals and systems or a practicing biologist with solid background in the field of biology who may lack an engineering insight to enhance his or her understanding of key concepts in signal processing.

Contents

Chapter 1
Non-Biological Signals

Although we would like to think of signals as being continuous in time and state, in reality, we observe and process signals in discrete time steps using countable (perhaps, infinitely countable) states. That includes biological signals as well. In this chapter, we focus on signals which can be classified as "non-biological." The biological signals will be tackled in the ensuing chapters. Many of the discussions in this chapter, however, have a direct bearing on understanding non-biological signal processing. Hence, the reader is encouraged to carefully examine the contents of this chapter before embarking on studying biological signal processing.

The above, then, necessitates a "digital" approach to signal processing, even though continuous-time signals are preferred when the subject of signals arises. This preference may be motivated by the false assumption that analog signals somehow provide more information than their digital counterparts. Although such an assertion seems to have some merit when the analog signal's digital counterpart is imperfect or the means of reconstructing an analog signal from its digital counterpart is non-ideal, in theory, one may represent an analog signal using its digital counterpart without loss! We will address this issue in the ensuing sections of this chapter. One is, then, compelled to address signals in their digital format for the sake of bringing to bear the vast array of digital signal processing algorithms and processors that are at our disposal today. Digital signals and digital linear systems (namely, signals and linear systems that only exists in discrete points in time) can, for the most part, be expressed in terms of well-known mathematical functions, enabling one to achieve significant level of abstraction in modeling and analysis of the response of virtually any linear system that is in use today. This convenient approach has its drawbacks. For one, it undermines the inherent nonlinearity that are present in most practical systems. From another perspective, it oversimplifies the process by which the performance of a system must be assessed. That is, the nonlinear behavior may have a nontrivial impact on the performance and behavior of a system. However, some key building blocks of digital signal processing, such as filters, etc., can be accurately modeled as linear. We will have to relax this restriction in later chapters where a large number of biological systems, which do not lend themselves to the linearity assumption, will be examined. Hence, as the first order of approximation, we consider linear signals

K. Kiasaleh, *Biological Signals Classification and Analysis*,
Lecture Notes in Bioengineering, DOI 10.1007/978-3-642-54879-6_1

and systems as a viable mechanism to study signal processing systems. This topic, however, will be discussed in the next chapter.

Although one would expect a detailed coverage of "digital" signaling schemes in this chapter, the focus of the ensuing sections are digital as well as analog signals and systems. Indeed, many signals encountered in practice can be classified as analog, and hence a side-by-side coverage of digital and analog signals and systems will serve the reader well. In view of the fact that information is represented by signals, which, in turn, can be expressed in terms of well-known functions, we begin by examining discrete- and continuous-time signals first. Finally, it is imperative to note that a large body of work on this subject chooses to treat signals as deterministic for the most part. In fact, the concept of "noise" is introduced as a separate subject in most textbooks. A somewhat different approach is taken here; that is, we consider that all signals are impaired by random phenomenon (this observation will be justified in the ensuing discussions), resulting in treating signals as random processes. For this reason, a separate coverage of random signals is not provided here.

1.1 Notation

Before we plunge into a discussion on signals, let us set some ground rules regarding notation. In this discussion, lower case letters refer to signals, such as $x\,(t)$ or $x\,[n]$ (the former describes signals in the continuous time whereas the latter denotes the same signal in the discrete time domain). The bold versions of the same represent row vectors, such as $\mathbf{x}\,(t)$ and $\mathbf{x}\,[n]$. Upper case letters signify the Fourier Transform (FT); such as $X\,[\omega]$ or $X\,(\Omega)$, where the former is the discrete FT of $x\,[n]$ while $X\,[\Omega]$ denotes the continuous-time Fourier Transform of $x\,(t)$. The upper case, bold letters are reserved for representing matrices. If a bold letter is followed by a pair of parenthesis or square brackets, then it denotes a time-varying vector (lower case) or matrix (upper case).

1.2 Discrete- and Continuous-Time Determinist Signals

A signal is considered to be deterministic if its current and past values in time can be used to obtain "all" of its future values precisely (i.e., with infinite accuracy). By the same definition, a random signal is a signal where the future values cannot be estimated precisely using its past and present values. From this simple definition one can conclude that all signals observed in practice, whether man-made or otherwise, should be classified as random. Indeed, a deterministic signal only exists in a mathematical sense and not in a practical sense. For instance, it is a common practice to model signals in the field of electrical engineering as $a \cos\,(\omega t + \theta)$. This signal is deterministic, since its past values can be used to obtain a, ω, and θ. Once such values are known, one will have access to "all" the values of the signal at any point in time. Unfortunately, we cannot generate such a signal in practice (we can get close, but the exact signal cannot be generated) as the oscillators which generate signals such as this exhibit frequency and amplitude uncertainties. So, a more realistic model for an oscillator signal is $a(t)\cos(\omega t + \theta(t))$ where $a(t)$ and $\theta(t)$ are in principle random signals themselves. Note that the phase and frequency uncertainties are lumped together and are represented by $\theta(t)$. In our treatment of this material then, we focus on random signals, with an understanding that a deterministic signal is a special case of a random signal.

In order to provide a consistent notation throughout this text, we define signals in time as $x(t)$ and in discrete time as $x[n]$, given by

$$x[n] = x(nT_s), \tag{1.1}$$

where T_s is the sampling interval. It is assumed that signals are complex in general, with real signals represented by a complex signal with zero imaginary part. That is, in general,

$$x^*[n] \neq x[n]. \tag{1.2}$$

This notation implies that a discrete-time signal may be viewed as the sampled version of a continuous-time signal, although this is not a necessary condition. That is, we can envisage a signal which exists in time only at multiples of T_s and not in the continuous-time domain at all. In the ensuing sections, the conditions that are required to insure a "complete" reconstruction of an analog signal from its samples will be discussed. For the purpose of this section, we assume that T_s is selected in a proper manner to maintain the integrity of the signal. In Fig. 1.1 a familiar signal $\sin(2\pi t)$ and its sampled version with a sampling duration of $T_s = \frac{T}{4}$ where $T = 1$ s (period of the sinusoid) are depicted. It is interesting to note that the "predictable" nature of the signal can be observed both in continuous and discrete time domains. One can clearly observe that an observation of the discrete time version of the signal over a finite period of time (or samples) reveals the underlying structure of the waveform, allowing for a complete reconstruction of the signal from a finite-length observation.

At this juncture, it becomes necessary to relate the samples of a signal in time domain to its continuous time representation. To that end, we need to answer the following question. What must be the sampling rate so that $x[n]$ signifies an "accurate" representation of $x(t)$? More precisely, what are the conditions which must be satisfied in order for us to reconstruct $x(t)$ from $x[n]$, and if such conditions are satisfied, how do we reconstruct $x(t)$ from $x[n]$? To address these questions, we need to consider the Sampling Theorem. However, this theorem is appropriate when deterministic signals are considered, and hence, we first consider the case where $x(t)$ is a deterministic signal. We will then extend this case to the case of random signals at the end of this section.

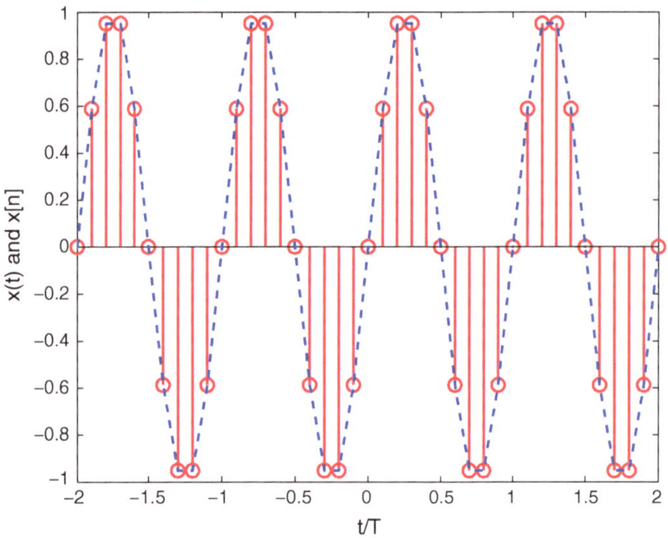

Fig. 1.1 A sinusoid and its sampled version in time

1.2.1 Sampling Theorem

The Sampling Theorem (ST) provides a method for reconstructing a continuous time signal from its samples in time. In order to achieve this goal, ST requires that the signal be sampled at a rate that is at least twice the largest frequency component of the signal. This rate is known as the *Nyquist* rate. This condition is merely imposed to insure that the signal can be reconstructed from its samples without distortion. Therefore, ST answers both questions raised above; that is, ST outlines a prescribe sampling rate in order to reconstruct the original signal perfectly. Furthermore, ST outlines a method for a perfect reconstruction of the signal. The term 'perfect' has been used a few times in the context of signal reconstruction so far. We need to quantify this term. We will do so via an example, which requires that we consider this problem in the Fourier domain. Let us consider a signal $x(t)$, which is sampled at a rate of $\frac{1}{T_s}$ as follows:

$$\begin{aligned} x_s(t) &= \sum_{n=-\infty}^{\infty} x(t)\,\delta(t - nT_s) \\ &= \sum_{n=-\infty}^{\infty} x[n]\,\delta(t - nT_s) \end{aligned} \tag{1.3}$$

where $\delta(t)$ is the delta function. This, obviously, is the ideal case for sampling and is known as 'impulse' sampling. Notice that

$$x_s(t) = \begin{cases} x[n]\,\delta(t - nT_s) & t = nT_s;\ n = 0, \pm 1, \pm 2, \ldots \\ 0 & \text{otherwise} \end{cases}.$$

Now, the use of Fourier Series Expansion (FSE) leads to

$$\sum_{n=-\infty}^{\infty} \delta(t - nT_s) = \frac{1}{T_s} \sum_{m=-\infty}^{\infty} \exp\left(-\frac{2\pi m t}{T_s}\right) \tag{1.4}$$

where $\sum_{n=-\infty}^{\infty} \delta(t - nT_s)$ is considered to be a periodic signal with period T_s. The Fourier coefficients for this periodic signals are all identical to $\frac{1}{T_s}$. Note that the Fourier Transform, denoted here by

$$\Im\{x(t)\} = X(\Omega) = \int_{-\infty}^{\infty} x(t)\, e^{-j\Omega t} dt, \tag{1.5}$$

of

$$\frac{1}{T_s} \sum_{m=-\infty}^{\infty} \exp\left(-\frac{2\pi m t}{T_s}\right)$$

is given by

$$\frac{1}{T_s} \Im\left\{ \sum_{m=-\infty}^{\infty} \exp\left(-\frac{2\pi m t}{T_s}\right) \right\} = \frac{2\pi}{T_s} \sum_{q=-\infty}^{\infty} \delta\left(\Omega - \frac{2\pi q}{T_s}\right). \tag{1.6}$$

Using the definition of FT and (1.3),

$$X_s(\Omega) = \Im\{x_s(t)\}$$

$$= \Im\left\{ \sum_{n=-\infty}^{\infty} x[n]\,\delta(t - nT_s) \right\} \tag{1.7}$$

$$= \sum_{n=-\infty}^{\infty} x[n]\, e^{-jn\Omega T_s} \tag{1.8}$$

However, realizing that multiplication in time results in convolution in frequency,

$$X_s(\Omega) = X(\Omega) \copyright \left\{ \frac{2\pi}{T_s} \sum_{q=-\infty}^{\infty} \delta(\Omega - q\Omega_s) \right\}$$

$$= \frac{1}{T_s} \sum_{n=-\infty}^{\infty} X\left(\Omega - \frac{2\pi n}{T_s}\right) \tag{1.9}$$

where $X_s(\Omega)$ and $X(\Omega)$ are the Fourier Transforms of $x_s(t)$ and $x(t)$, respectively, $\Omega_s = \frac{2\pi}{T_s}$ is the sampling rate in rad/s, and © denotes convolution (in time or frequency).

We can also examine the discrete time Fourier transform (DTFT) of the sequence $x[n]$. That is (we use ω to denote digital frequency and Ω as its analog counterpart),

$$X_s[\omega] = \sum_{n=-\infty}^{\infty} x[n] e^{-jn\omega} \tag{1.10}$$

Examining (1.8) and (1.10) leads one to this identity:

$$X_s[\omega] = X_s\left(\frac{\omega}{T_s}\right) = \frac{1}{T_s} \sum_{n=-\infty}^{\infty} X\left(\frac{\omega - 2n\pi}{T_s}\right) \tag{1.11}$$

Figure 1.2 depicts $X(\Omega)$, $X_s(\Omega)$, and $X_s[\omega]$ for a typical signal (with its FT concentrated about $\Omega = 0$). Note that since $x(t)$ may be complex, its spectrum is depicted as an asymmetrical signal about $\Omega = 0$. Also, $X_s[\omega]$ is periodic with a period of 2π (as expected from its definition). One can make two immediate observations. First, for any signal whose frequency content is confined to $\Omega_{\max} = \max\{|\Omega_2|, |\Omega_1|\}$ (or $\omega_{\max} = \Omega_{\max} T_s = \max\{|\omega_2|, |\omega_1|\}$), the sampling rate must satisfy $\Omega_s > 2\Omega_{\max}$ (or $\omega_s > 2\omega_{\max}$) or an irrecoverable distortion known as "aliasing" will occur. $2\Omega_{\max}$ hence is the Nyquist rate introduced above. The other observation is that, if one uses a sampling rate which is greater than the Nyquist rate, then one can reconstruct the signal perfectly by using an "ideal" low pass filter (see Fig. 1.2), which is confined to $|\Omega| \le \frac{\Omega_s}{2}$ (or $|\omega| \le \pi$ in the discrete domain). That is,

$$H_p(\Omega) = \begin{cases} T_s & |\Omega| \le \frac{\Omega_s}{2} \\ 0 & \text{otherwise} \end{cases}.$$

will recover the original signal perfectly. Such a filter will have an impulse response (in continuous time domain)

$$h_p(t) = \operatorname{sinc}\left(\frac{\pi t}{T_s}\right) \tag{1.12}$$

Now, in time domain, we have

$$\begin{aligned} x(t) &= x_s(t) © h_p(t) \\ &= \sum_{n=-\infty}^{\infty} x[n]\,\delta(t - nT_s) © h_p(t) \end{aligned}$$

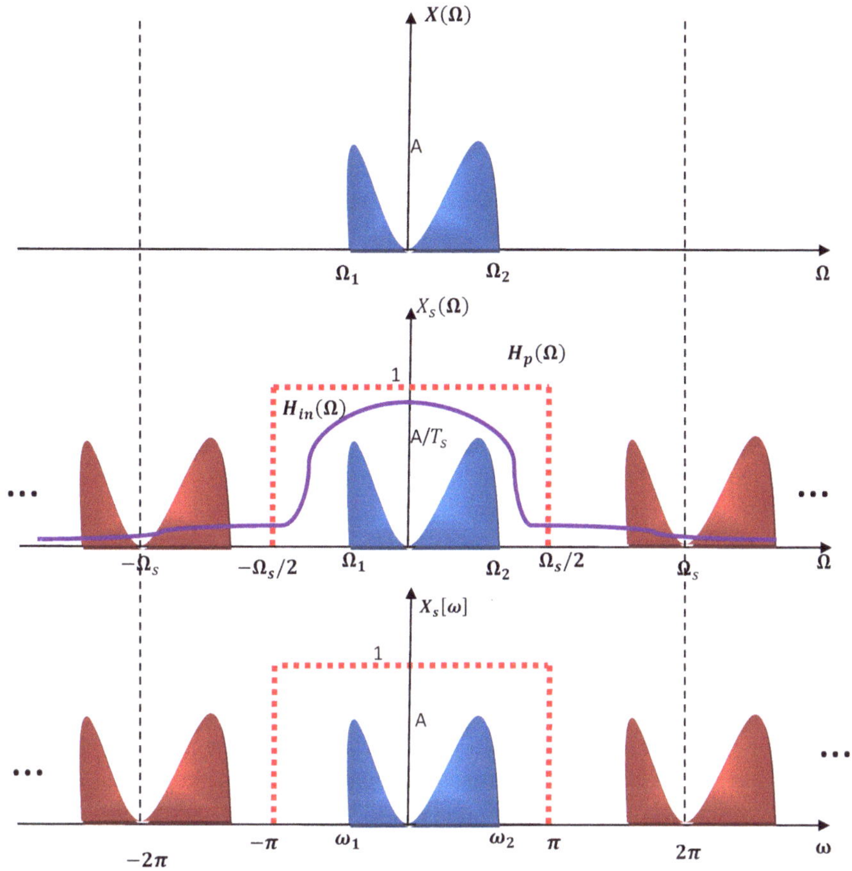

Fig. 1.2 Frequency domain representation of the sampling theorem

$$= \sum_{n=-\infty}^{\infty} x\,[n]\,\mathrm{sinc}\left[\frac{\pi\,(t - nT_s)}{T_s}\right] \tag{1.13}$$

Note that $h_p\,[n]$ is zero at all $n \neq 0$, as one would expect to see for the impulse response of a perfect reconstruction filter (in discrete time domain) and that $x\,(t)\,|_{t=nT_s}$ reduces to $x\,[n]$. The last equation, attributed to Claude Shannon, clearly shows a method which allows for a perfect reconstruction of $x\,(t)$ from its samples. This is a remarkable result in that it allows us to obtain values of a signal at "any" point in time (unaccountably infinite) given that $x\,(t)$ is known to us at countably infinite number of points in time. Unfortunately, the term "perfect" imposes an impossible restriction (as one would expect); note that, in order to reconstruct $x\,(t_1)$ for any $t_1 \neq nT_s$, one requires all samples of the signal ($x\,[n\,]$; for all n) (past, present, and future). This further implies that a perfect reconstruction at time t_1 requires an infinite delay; a condition that is clearly unacceptable in a practical environment. This observation could have been reached by examining $h_p(t) = \operatorname{sinc}\left(\frac{\pi t}{T_s}\right)$; that is, this impulse response corresponds to the impulse response of a non-Causal, and, hence, unrealizable, filter.

Before leaving this section, however, it is instructive to examine the above representation from an entirely different perspective. That is, let us consider a finite summation (and hence a practical solution), instead of an infinite one, to represent the signal as follows:

$$x_N(t) = \sum_{n=-N}^{N} x[n] \operatorname{sinc}\left[\frac{\pi (t - nT_s)}{T_s}\right] \tag{1.14}$$

Note that $x_N(t)$ useds $2N + 1$ samples of $x(t)$ to generate $x_N(t)$. It can be shown (we leave the proof to the reader) that

$$\lim_{N\to\infty} \langle |x(t) - x_N(t)|^2 \rangle \to 0 \tag{1.15}$$

where

$$\langle y(t) \rangle = \lim_{T\to\infty} \frac{1}{T} \int_{-\frac{T}{2}}^{\frac{T}{2}} y(\tau)\, d\tau \tag{1.16}$$

denotes the time average of the enclosed. The condition stated above implies that a convergence in "power" of the truncated representation to that of the original signal as $N \to \infty$. It can be seen from above that, in practice, a reconstruction (or interpolation) can only be asymptotically optimum. Finally, the operation described above is known as interpolation (perfect one in the above case) as the samples $x[n]$ are extended to reconstruct the signal from its samples. There are other (non-optimum) means of interpolation, which are used in practice (we will not comment on them here). One remedy would be to consider imperfect reconstruction (also known as interpolation since this process interpolates the values of the signal at $t = nT_s$ to that at $t \neq nT_s$). There are a number of obvious interpolation mechanisms. For instance, in Fig. 1.2 a possible practical filter is depicted (purple curve), where the system function $H_{in}(\Omega)$ does not vanish to zero (this is a necessary condition for realizable systems), and hence the spectral components of the sampled signal from undesired portion of the spectrum influence the output; note that, although the desired spectrum component about $\Omega = 0$ appears to be preserved perfectly as a result of filtering, the spectrum components about nf_s for all n cannot be fully eliminated, invariably causing distortion. In order to qualify as a true interpolation, the filtering mechanism must guarantee that the output signal values at integer multiple of T_s are indeed $x[n]'s$. Hence, arbitrary filters, such as the one shown in Fig. 1.2, may not meet this requirement. However, any operation that is an approximation to the perfect reconstruction filter can be used to interpolate the signal, bearing in mind that the interpolation will be subject to imperfections and that the interpolation must insure that the interpolated signal values at $t = nT$ is $x[n];\ n = 0, \pm 1, \pm 2, \ldots$. Furthermore, if $x_N(t)$ denotes the interpolated version of $x(t)$ using $2N+1$ samples of $x(t)$, one can use

$$ER_N = \langle |x(t) - x_N(t)|^2 \rangle \tag{1.17}$$

as a metric to establish the effectiveness of the interpolation procedure as a function of N. Note that ER_N may be viewed as the "power" in the error signal, which is computed by finding the difference between the signal and its interpolated version.

Next, we consider upsampling (followed by interpolation) and decimation/down-sampling.

1.2.2 Upsampling

In general, upsampling is primarily used as a first step to the interpolation process. The assumption here is that we are interested in reconstructing the original signal in the discrete domain, which in principle implies that we are interested in samples of the signal at a rate of Ω_{us}, which is restrictedly greater than Ω_s. To that end, let us first consider the signal

$$x_{us}[p] = x(pT_{us})$$

where $T_{us} = \frac{2\pi}{\Omega_{us}}$ with $\Omega_{us} > \Omega_s$. In practice, $\frac{\Omega_{us}}{\Omega_s} = M$ (the upsampling rate is an integer multiple of the original sampling rate). This is the desired output. However, without having access to the actual analog signal, how does one generate $x_{us}[p]$ having access to $x[n]$ only? One can approximate $x_{us}[p]$ by an operation known as interpolation. First, let us define the signal $y[n]$ as follows:

$$y[p] = x\left[\frac{p}{M}\right] \tag{1.18}$$

for some integer M. Note that $y[p] = \begin{cases} x[n] & p = nM \\ 0 & \text{otherwise} \end{cases}$. This signal, although an upsampled version of $x[n]$, assumes a value of 0 at $p \neq nM$, and therefore cannot be labeled as the interpolated version of $x[n]$. This operation is captured for $M = 2$ in Fig. 1.3. In frequency domain, we have

$$Y[\omega] = X_s[M\omega] \tag{1.19}$$

As one can see, $y[p]$ in the frequency domain exhibits significant distortion. However, an ideal low-pass filter ($H_{in}[\omega]$) with a bandwidth of $\frac{\pi}{M}$ can generate a signal whose spectrum is identical to that of $x[n]$ with higher sampling rate. That is,

$$X_{in}[\omega] = Y[\omega]\, H_{in}[\omega] = X_{us}[\omega] \tag{1.20}$$

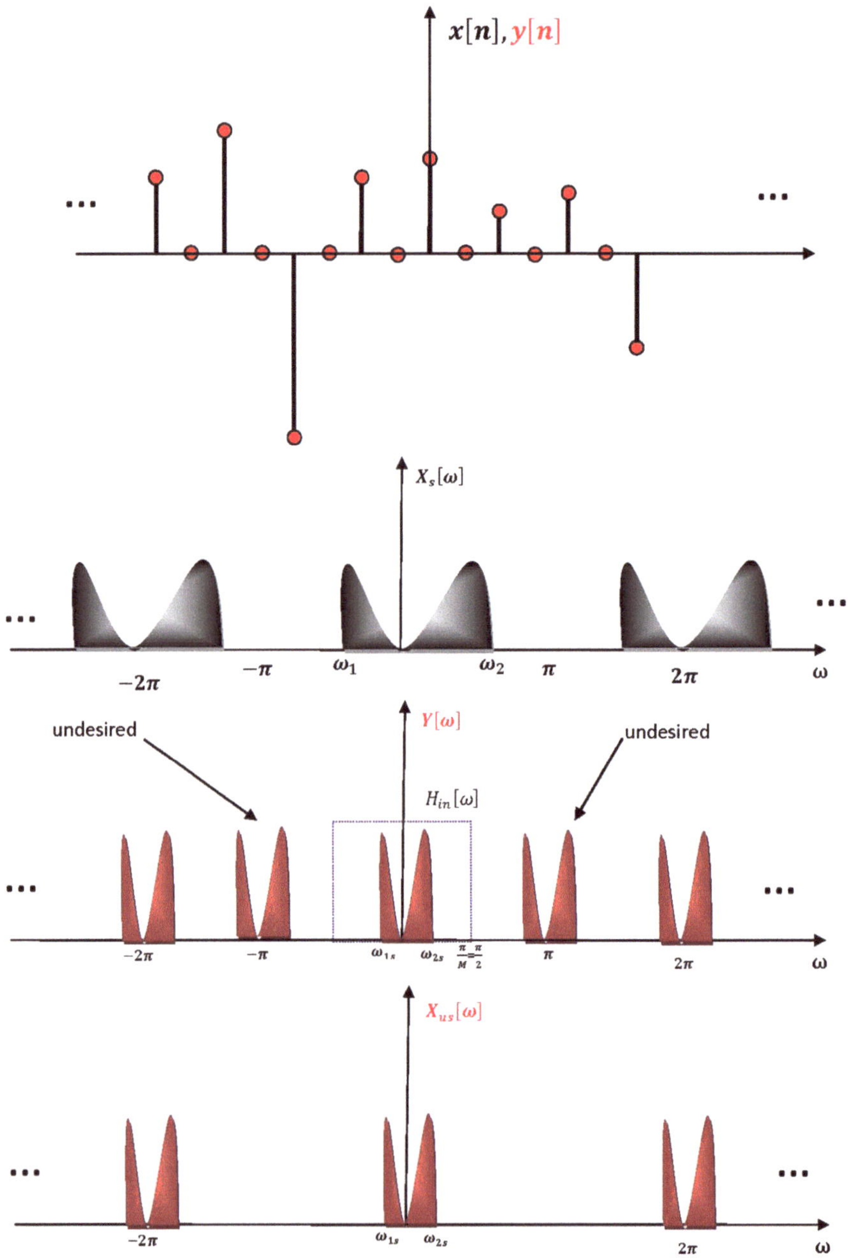

Fig. 1.3 Frequency domain representation of upsampling/interpolation operation

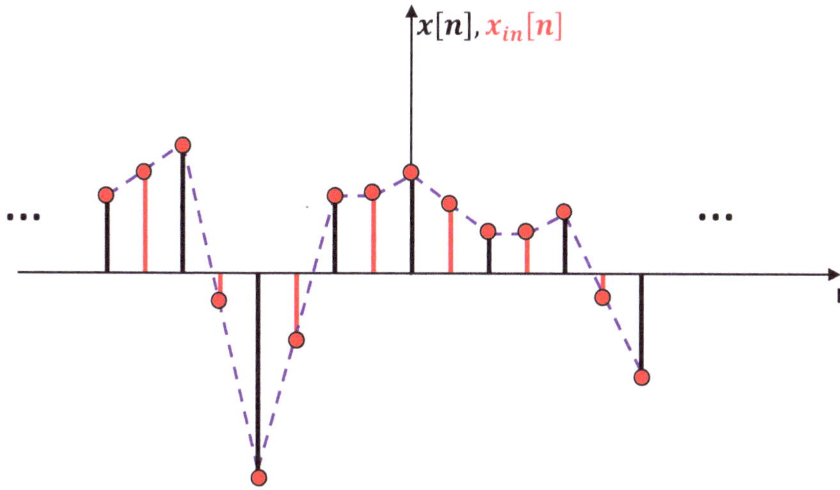

Fig. 1.4 Interpolation of a discrete time signal

or

$$x_{in}[p] = \sum_{m=-\infty}^{\infty} h_{in}[m]\, y[p-m] \tag{1.21}$$

The resulting signal is shown in Fig. 1.4. Hence, the objective of upsampling is to allow a subsequent filtering operation to reconstruct the upsampled version of the analog signal (interpolation). Note that, unlike the intent of the sampling theorem (which was to reconstruct perfectly the analog signal from its samples), we are only interested in generating the samples of $x(t)$ at the sampling rate $\frac{1}{T_{us}}$, which is greater than $\frac{1}{T_s}$. This, all along, was the goal of upsampling; i.e., increasing the sampling rate without causing distortion. One then can conclude that, if a perfect filtering operation can be achieve, $x_{in}[p]$ is indeed identical to $x_{us}[p]$. That is,

$$X_{in}[\omega] = X_{us}[\omega]$$

for a perfect interpolation. However, due to imperfection in the filtering operation, in general,

$$X_{in}[\omega] \neq X_{us}[\omega], \tag{1.22}$$

which leads to

$$x_{in}[p] = \begin{cases} x_{us}[p] & p = nM \\ x_{us}[p] + \Delta_\epsilon[p] & p \neq nM \end{cases}$$

where $\Delta_\epsilon[p]$ denotes the difference between the actual value of $x(t)$ at pT_{us} and that of the interpolated value. As was pointed out earlier, with any form of interpolation, the values of the upsampled signal must coincide with the values of the original signal at $t = nT_s$. That is, $x_{in}[nM] = x_{us}[nM]$.

1.2.3 Downsampling and Decimation

Downsampling is a simple process of reducing the sampling rate of a signal. The objective here is to reduce the computational complexity of digital signal processing system without compromising the integrity of the signal. Figure 1.5 depicts this operation in time and frequency domains. Although from a time domain point of view it appears that this process has caused significant damage to the signal, the frequency domain representation provides us with an important insight; it clearly sets boundaries for this operation. To elaborate, in order for the signal integrity to be maintained (no aliasing to occur), one requires that the downsampling rate $M_d = \frac{\omega_{\max d}}{\omega_{\max}}$ $(M_d \geq 1)$ must be such that $|\omega_{\max d}| < \pi$. In the figure shown below, note that $\omega_{\max d} = \omega_{2d}$. From a mathematical point of view, downsampling can be represented as

$$x_{ds}[n] = x[nM_d] \tag{1.23}$$

which clearly leads to the following:

$$X_{ds}[\omega] = X\left[\frac{\omega}{M_d}\right] \tag{1.24}$$

This transformation is shown in Fig. 1.5. Furthermore, the spectrum of downsampled signal is shown for $|\omega_{\max d}| \geq \pi$, which corresponds to the case where aliasing is introduced as a result of downsampling.

1.2.4 Anti-Aliasing Filter (AAF)

As discussed above, sampling an analog signal results in shifting the spectrum of the signal to all integer multiples of the sampling frequency. This is similar to the spectrum shifting operation, which is observed in the modulation theory. One can see that such an approach is effective if one can limit the spectrum of the signal to a frequency range that is less than half the Nyquist rate (i.e., $2\Omega_{\max}$). However, in reality, a signal of interest may have spectrum that extends beyond the Nyquist rate. In that event, aliasing will occur, which in turn results in irreversible damage to the signal. Furthermore, if the low-pass signal is corrupted by any high frequency components that extends beyond $\Omega_{\max}$, then the resulting sampling at or slightly above Nyquist rate will lead to distortion. Provided that the sampling rate imposes

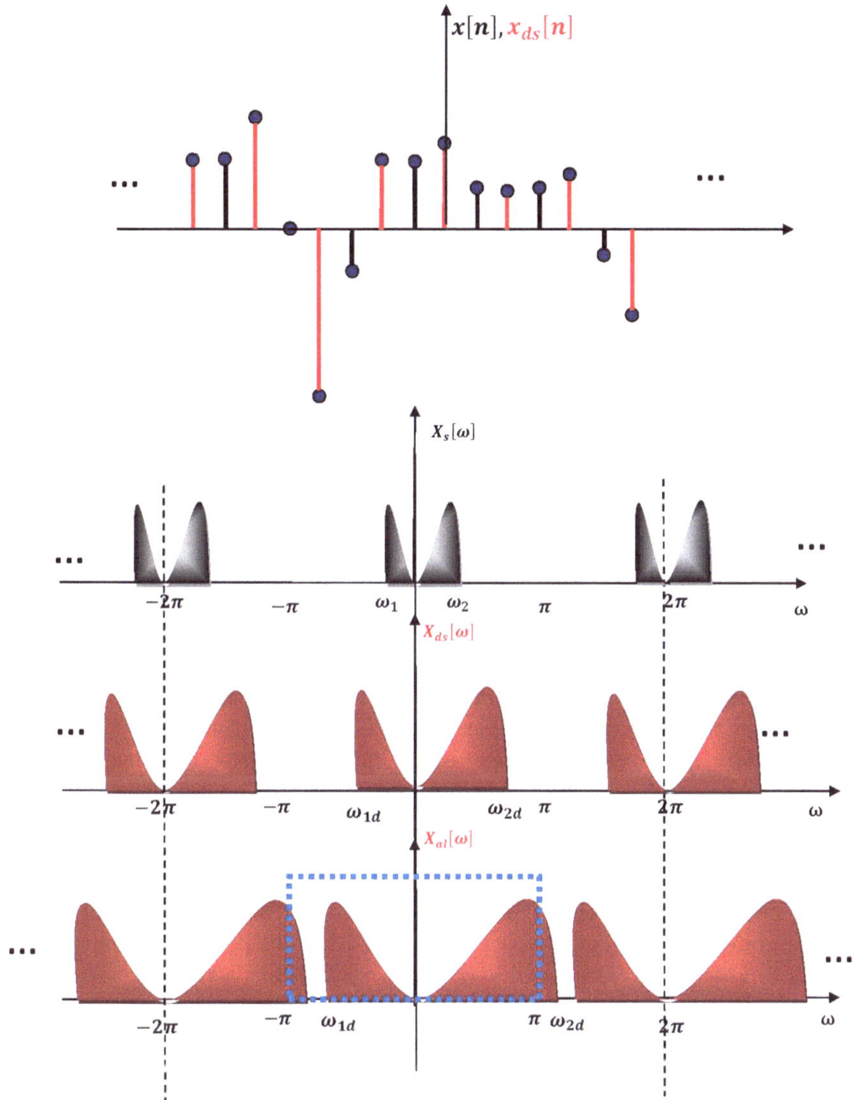

Fig. 1.5 Frequency domain representation of downsampling and decimation

some serious limitations on hardware that follows the digital signal, in practice, there is a desire to reduce the sampling rate to a minimum without distorting the signal. Let $2\Omega_{max}$ (in Hz) denote the maximum sampling rate that we can afford. In that event, a filter whose bandwidth is Ω_{max} must precede the sampling operation. Such a filter is known as AAF and is a common fixture of a digital communication system.

1.2.5 Quantization

As was shown above, an analog signal may be represented by its samples, which are obtained in accordance to the Sampling Theorem. This allows for one to represent the analog signal using samples over a regular time interval, known as the sampling interval. Such a representation allows one to bring to bear digital signal processing tools to store, forward, and process any signal that has an equivalent discrete-time representation. However, before digital signal processing can be utilized, the samples of the signal (which are indeed analog) must be converted to their digital counterparts. This implies that we need to quantize the samples of the signal to one of the q levels, which further implies that any of the samples of the signal is now represented using $\log_2 q$ bits. Such a representation involves processing $\log_2 q$ bits every T_s seconds. This process, however, is not without its drawbacks. For one, we have introduced an error in the representation of the signal. That is,

$$e[n] = x[n] - \tilde{x}[n]$$

may be viewed as the error in the quantization with $\tilde{x}[n]$ denoting the quantized version of $x[n]$. Second, the quantized samples must be converted to their analog part, resulting in an additional level of complexity. Overall, the process of converting $x(t)$ to $\tilde{x}[n]$ may be viewed as analog-to-digital conversion (ADC) in a digital signal processing chain. It is noteworthy that ADC is an important element of any DSP system today. With ever increasing speed of information, ADC remains as an important research topic to date. At the other end, one has to convert $\tilde{x}[n]$ to $x(t)$. Such an operation is labeled as the digital-to-analog conversion (DAC). The following example demonstrates the intricacies of these operations.

Example 1 An analog signal (a frequency -modulated waveform with a center frequency of 60 kHz, modulation index of 3, and the modulating frequency of 3 kHz) is to be converted to a digital signal for transmission via a digital communication system. Give a digital representation of the signal and demonstrate the accuracy of the reconstruction mechanism. Determine the umber of bits per second you need to process this signal.

Solution: *Using a hardware simulation tool (in this case a PSpice model), we can realize an ADC and DAC operation. The ADC used in this example is an 8-bit converter (i.e.,* $q = 2^8 = 256$*).*

The circuit shown in Fig. 1.6 *is used to study the ADC-DAC operation. The PSpice 9.0 model uses an 8-bit ADC, followed by an 8-bit DAC. The sampling rate (DTSM1 block in the figure) is selected as* $1\ MHz\ (T_s = 10^{-6})$. *This sampling rate is high enough to avoid any form of aliasing. An FM signal with a modulation index of* β *and the modulating frequency of* f_m *has a bandwidth of approximately* $2(\beta + 1) f_m = 2(3 + 1) 3 = 24\ kHz$ *(Carson's rule). However, the signal has harmonics and the total spectrum must be considered. That is, although the signal has a bandwidth of 24 kHz, the signal is centered around* 60 *kHz. This gives a bandwidth of ~72 kHz*

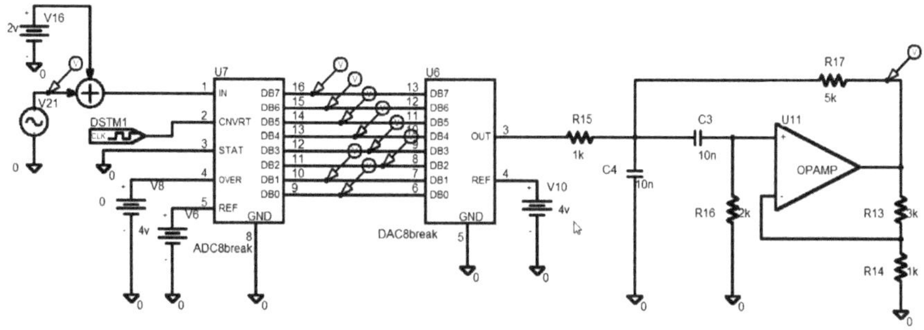

Fig. 1.6 A PSpice model for AD/DA conversion

$(60 + 24/2)$ as far as the sampling theorem is concerned. Hence, using the above sampling rate, which is a little less than 7 times ($\frac{10^6}{2\times 72\times 10^3} = 6.9$) the Nyquist rate is justifiable. The output of the DAC conversion circuit is filtered using a bandpass filter, centered at the center frequency of the input signal (60 kHz) and a bandwidth that accommodates the FM signal. The filter acts as an interpolation filter. In Fig. 1.7, the digital output of ADC along with input and output signals are depicted. Note that, after the initial transient state, the output (green curve) follows the input closely with an exception of a slight amplitude attenutation (a phase shift is introduced due to the delay in the circuit). To establish how well the circuit has performed, the Fourier transforms of the input and output signals are depicted in Fig. 1.8. Interestingly, the reconstructed signal's spectrum overlaps that of the input signal, signifying a satisfactory performance. Note that the Fourier transform confirms the approximate bandwidth of 24 kHz (71 − 47 kHz) about the center frequency of 60 kHz.

In the previous section, we discussed how one can map an analog signal onto its digital counterpart by the use of the Sampling Theorem. However, such a representation requires the exact values of the digital samples, which can be a daunting task for high-bandwidth signals or where memory and speed is of critical importance. As noted above, one has to quantize the digital samples, with a quantization level that does not result in severe degradation in the signal quality while maintaining a reasonable demand on resources. In the above example, we dealt with a 4 volt peak-to-peak signal and a quantization level of 256 ($2^8 = 256$). This was a sufficiently large quantization level for such a well-behaved signal. The result was a minimal level of distortion in the reconstructed signal. The question that is now before us is the impact of the quantization level on the required resources. To answer this question, let us consider the following example.

Example 2 In the previous example, determine the required signal processing resources that will be needed to process a 10 min observation of this signal.

Solution: *In the previous example, we established that a quantization level of 256 resulted in an acceptable performance. Since the samples are taken at the rate of 1 MHz, the resulting bits/s (bps) that needs to be processed is 8 mega bps (Mbps). For*

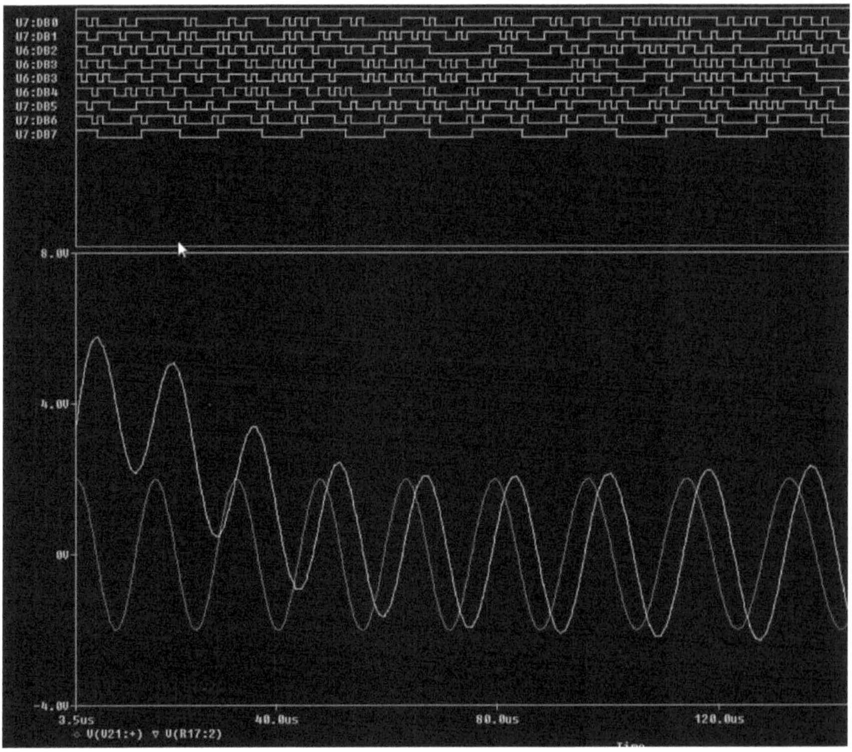

Fig. 1.7 The binary representation of the analog signal along with the reconstructed analog signal

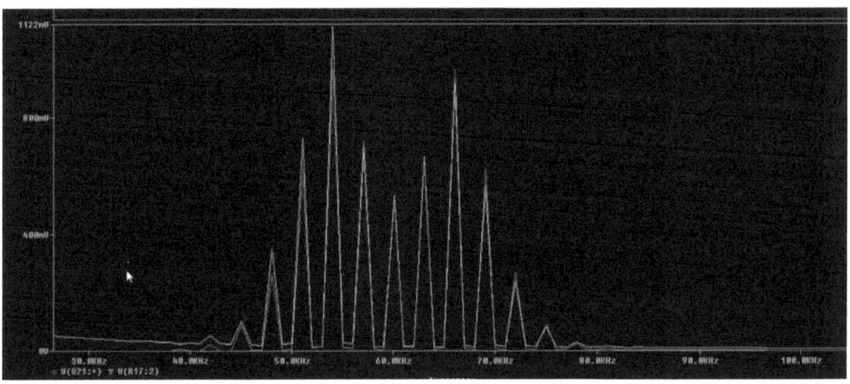

Fig. 1.8 Frequency domain representation of the original and reconstructed analog signal using AD/DA conversion

observing 10 *minutes of the signal then, we need* 4.8 *Gb or* 600 *Mb of storage! This is not a trivial amount of storage for a special purpose computing architecture even by today's standard. Furthermore, the DSP must be able to handle* 8 *Mbps of data rate for any further processing.*

Depending on the problem, the resulting digital system must be able to handle the data rate that is required by the ADC operation. A key question that has not been answered is the following. How does one establishes whether the chosen quantization level is sufficient to represent the signal with a given level of accuracy? This is a bit more complicated question to answer since in addition to the quantization level the reconstruction technique will have an impact on the overall accuracy of the process. Furthermore, in cases where signal accuracy is not well-defined (such as speech signal), a quantitative analysis of accuracy is difficult as the measure of the quality of reconstruction is somewhat subjective (an acceptable reconstructed speech quality to one person may not be unacceptable to another). The second question is the following: How can we remove the need for large quantization levels without sacrificing accuracy? To answer the second question, we first discuss a technique known as delta modulation, which is used in practice to encode/quantize the samples of an analog signal at lower data rates without compromising performance.

1.2.6 Delta Modulation (DM)

Given that the quantization level needed to accurately represent a digital signal may be quite large, an alternate scheme must be devised to deal with this problem. One approach is to increase the sampling rate to several times (typically, 4 times the Nyquist rate) the Nyquist rate in order to create samples that do not change substantially from sample to sample. This approach is feasible in today's system as the clock speed for most digital signal processing platforms exceeds the Nyquist rate of most analog signals by a sizeable margin. In that scenario, one can use only 1 bit to represent the change from sample to sample (1 bit can be interpreted as an indicator of a step up or down) instead of encoding each sample separately. If we let the change be Δ, then the increment can only take values from the set $\{-\Delta, +\Delta\}$. In other words,

$$w[n] = x[n] - x[n-1] \tag{1.25}$$

is quantized to either $+\Delta$ or $-\ \Delta$. Let $\tilde{w}[n]$ denote the quantized version of $w[n]$. Now, it can be shown that

$$\begin{aligned} x[n] &= w[n] + x[n-1] \\ x[n] &= w[n] + w[n-1] + x[n-2] \\ &\cdots\cdot \\ x[n] &= \sum_{q=1}^{n} w[q] - x[0] \end{aligned} \tag{1.26}$$

Assuming $x[0] = 0$, then

$$x[n] = \sum_{q=1}^{n} w[q] \tag{1.27}$$

This implies that a simple adder can be used to recover the signal from $w[q]'s$. Since we have access to only the quantized values of $w[n]$, an estimate for $x[n-1]$ can be obtained as follows:

$$\hat{x}[n-1] = \sum_{q=1}^{n-1} \tilde{w}[q] \tag{1.28}$$

The key assumption is that the quantization error, i.e., $\tilde{w}[n] - w[n]$, is not significant (this is insured by the high sampling rate). Otherwise, a propagation error will occur. Now, let us consider

$$\begin{aligned} x[n] - \hat{x}[n-1] &= x[n] - \sum_{q=1}^{n-1} \tilde{w}[q] \\ &= x[n] - \sum_{q=1}^{n} \tilde{w}[q] + \tilde{w}[n] \\ &= x[n] - \hat{x}[n] + \tilde{w}[n] \end{aligned} \tag{1.29}$$

Hence, if the estimation is close enough (i.e., $x[n] - \hat{x}[n] \approx 0$), $x[n] - \hat{x}[n-1] \approx \tilde{w}[n]$. This was the original assertion that identified $\tilde{w}[n]$ as the quantized version of the change in the signal from sample to sample (i.e., $x[n] - x[n-1]$). This method of digital representation of analog signals is depicted in Fig. 1.9.

Note that the resulting "bit" stream of identical positive and negative levels can be transmitted to the "receiver" and one would only need a simple adder to reconstruct the quantized version of the analog signal. This method of representation is amenable to digital implementation as one does not need to know the frame timing (i.e., when a bit stream representing a quantized level ends and when the next one begins). In fact, a simple running sum is sufficient for the recovery of the signal provided that bit clock is recovered. Finally, the analog samples $\hat{x}[n]$ can be lowpass filtered to obtain an estimate of $x(t)$.

This method of quantization has been used successfully in digital telephony. In the standard digital representation of voice signal, which is sampled at 8 kHz (the Nyquist rate for a speech signal, which is confined to 4 kHz of bandwidth), 256 level quantization has been shown to produce acceptable quality, known as toll quality. This results in a bit rate of 8×8 kHz or 64 kilo bps (kbps). By increasing the sampling rate to 32 kHz (4 times the Nyquist), and using 1 bit quantizer, DM is shown to produce an acceptable voice quality. This results in 32 kbps of data rate, a reduction of 50 % in data rate. Since some non-stationary signals (such as speech) experience wide fluctuations in their amplitudes, it may be instructive to use an

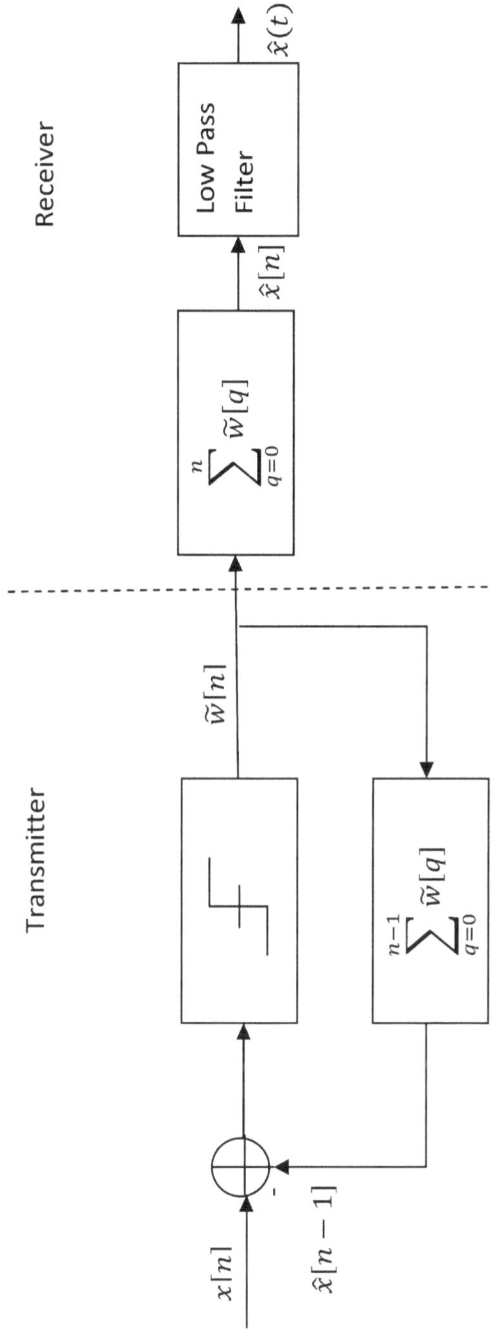

Fig. 1.9 Delta modulation/demodulation

adaptive means to change the number of quantization level (hence, bit per samples). This is also referred to as contiguous variable slope delta (CVSD) modulation in which the quantization step (this is identical to increasing the number of bits in the quantization process) is changed continuously in response to the signal levels rate of change. With adaptive means, one can even reduce the actual bit rate of the speech to 20 kbps. A discussion on adaptive DM and other speech processing techniques is beyond the scope of this text. Instead, we present a useful technique, known as sigma-delta modulation, that has been shown to be useful in many signal encoding scenarios.

1.2.7 Sigma-Delta ($\Sigma - \Delta$) Modulation

In view of the above formulation, we recognize that $x[n-1] = \sum_{q=1}^{n-1} w[q]$. Hence, if one were to quantize $x[n-1]$ and then generate $w[n]$ as $x[n] - \hat{x}[n-1]$, a simplified architecture emerges (see Fig. 1.10). As can be seen, in this arrangement, one does not require a summing device at the receiver. Because of summation (sigma) followed by a differentiation (1-bit quantization may be viewed as a delta operation), this form of digital encoding is known as $\Sigma - \Delta$. Due to its simplicity, this method is used extensively in practice where oversampling is available. Note that in this case $\hat{x}[n-1]$ is the quantized version of $\sum_{q=1}^{n-1} w[q]$, while in the original DM case, $\hat{x}[n-1]$ is the sum of the quantized $w[n]$, i.e., $\hat{x}[n-1] = \sum_{q=1}^{n-1} \tilde{w}[q]$. In both scenarios, if sampling rate is sufficiently large, $w[q]'s$ remain small, and hence the impact of quantization error is somewhat negligible. One drawback of this technique is that the quantizer is now needs to deal with $\sum_{q=1}^{n-1} w[q]$ instead of $w[n]$, which requires a larger dynamic range. This limits the usage of $\Sigma - \Delta$ modulation to applications where the dynamic range of $x(t)$ is small, such as scenarios where $x(t)$ represents a small error signal in a control loop. One remedy is to use a larger quantization level for this scheme to remedy the dynamic range issue. To illustrate this point, let us consider the following example.

Example 3 We are to use DM or $\Sigma - \Delta$ modulation to represent a noise-corrupted sinusoid. The sinusoid is a 5 kHz sinusoid whose amplitude is 1 volt. The noise is a zero-mean Gaussian noise with a standard deviation (STD) of 0.1. Determine the sampling rate and Δ to represent this signal with small distortion. For the $\Sigma - \Delta$ case, determine whether a binary quantization is sufficient to represent the signal assuming that the amplitude of the sinusoid is reduced to 0.1 and the noise STD is reduced to 0.01.

Solution: *Given that a large sampling rate allows DM or $\Sigma - \Delta$ to work properly, we use* 8 *times the Nyquist rate. Using matlab, we plot the* 5 *kHz,* 1 *volt noise-corrupted sinusoid along with its DM representation for* $\Delta = 0.3$ *in Fig.* 1.11 *(note that* $q = 2$ *for DM). Provided the large sampling rate (8 times the Nyquist), DM can be used effectively in this case. The value of* 0.3 *for* Δ *was found by some trial and error*

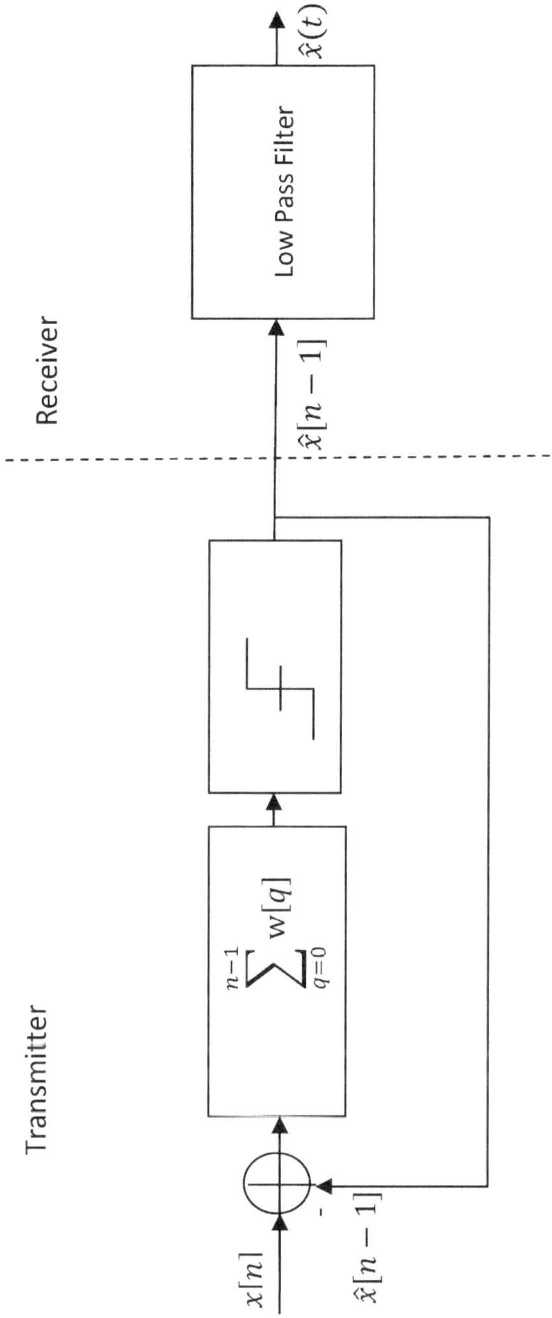

Fig. 1.10 $\Sigma - \Delta$ modulator/demodulator

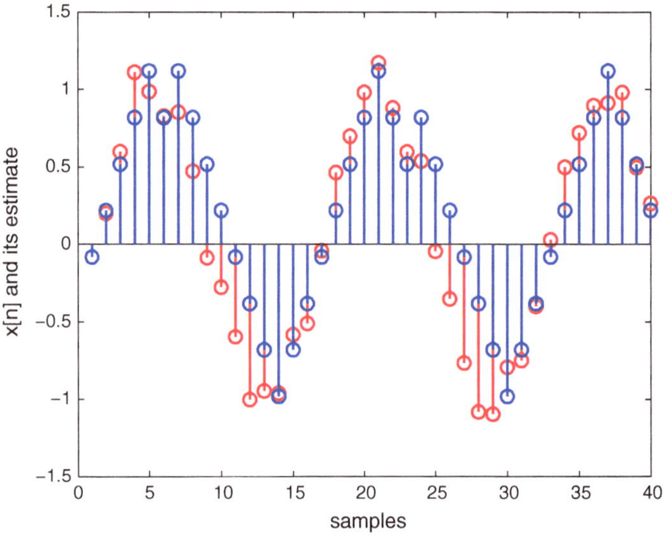

Fig. 1.11 $x[n]$ (*red*) and $\hat{x}[n]$ (*blue*) for a noise-corrputed sinusoidal signal when delta modulation is used to digitize the signal. In this example, $\Delta = 0.3$ and the amplitude of the sinusoid is 1

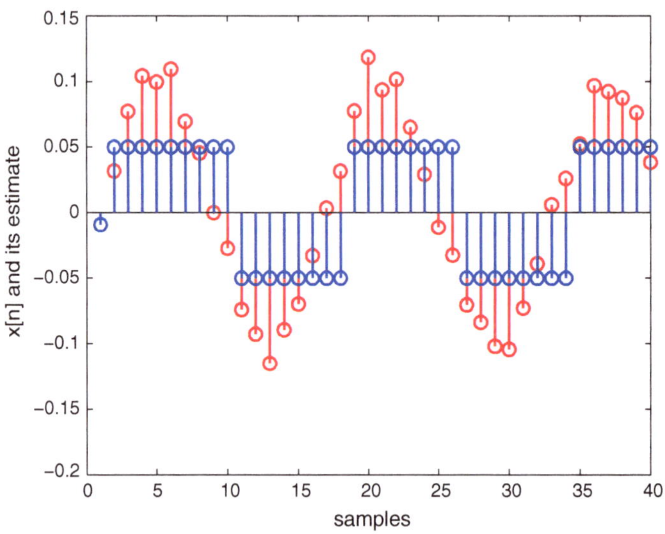

Fig. 1.12 $x[n]$ (*red*) and $\hat{x}[n]$ (*blue*) for a noise-corrputed sinusoidal signal when $\Sigma - \Delta$ modulation is used to digitize the signal. In this example, $\Delta = 0.05$ and the amplitude of the sinusoid is 0.1. In this case, only 2 level quantization is utilized

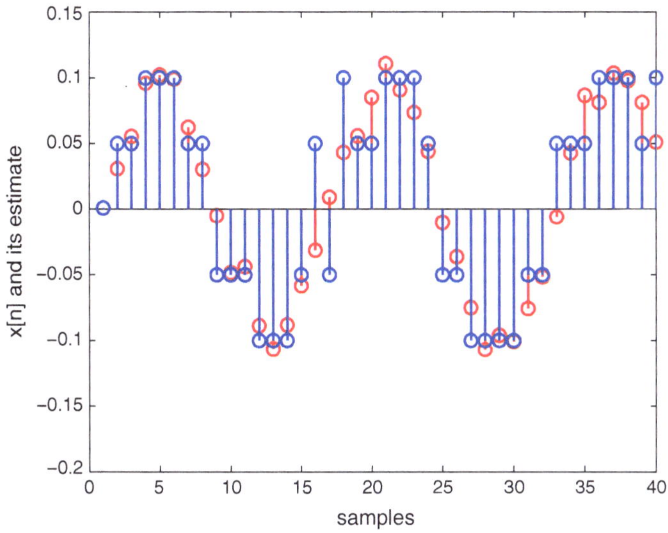

Fig. 1.13 $x[n]$ (*red*) and $\hat{x}[n]$ (*blue*) for a noise-corrputed sinusoidal signal when $\Sigma - \Delta$ modulation is used to digitize the signal. In this example, $\Delta = 0.05$ and the amplitude of the sinusoid is 0.1. In this case, 4-level quantization is utilized

to yield the best performance. As can be seen, although not perfect, this technique provides a reasonably good approximation of the sampled signal. Now, if we attempt $\Sigma - \Delta$ *modulation with* $\Delta = 0.05$ *and only 2-level quantization* ($q = 2$), *large errors are observed (note that the dynamic range of the signal is reduced to allow* $\Sigma - \Delta$ *to function properly), see Fig.* 1.12. *If one increases the quantization level from* $q = 2$ *to* $q = 4$, *a much better better performance can be expected from the* $\Sigma - \Delta$ *modulator, see Fig.* 1.13. *This suggests that* $\Sigma - \Delta$ *is only effective when the dynamic range of the signal is small.*

1.3 Discrete- and Continuous-Time Random Signals

Provided that, in general, signals in communication systems possess random characteristics (as noted earlier), we treat all signals as random processes in this text. In contrast to the example provided above, Fig. 1.14 depicts a random signal, where an observation of the signal in time (discrete or continuous) does not reveal any discernible pattern. The third class of signals, which are of interest to us, is the class of random signals which are information-bearing. This information must be extracted in order to render meaningful decisions regarding the underlying principles. For biomedical signals, this may be an event that has given rise to the signals random attribute. In that event, although the signal appears as deterministic, in reality, the signal is random in nature. Figure 1.15 depicts one such signal. Note that, although

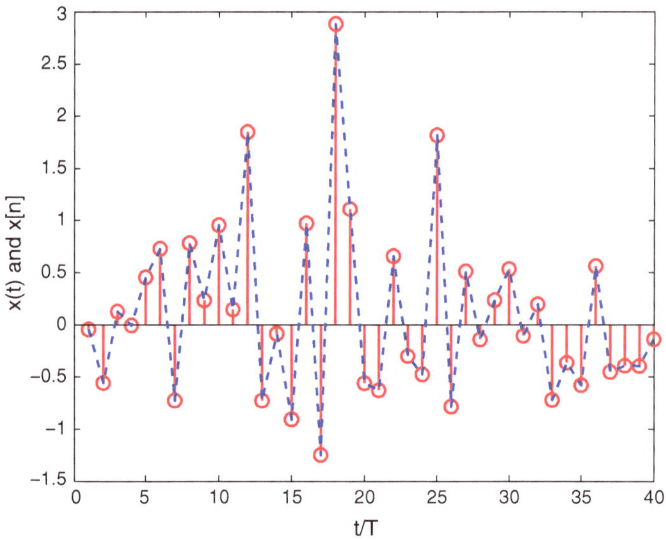

Fig. 1.14 A random signal and its sampled version in time

the levels seem to be random, the signal stays constant (and hence predictable) over a duration of 5 samples. This is the trademark of digitally modulated signals. There is yet another important distinction, which is less obvious and perhaps hidden to the reader. The random waveform in Fig. 1.14 assumes values (at least it appears so at a first glance) from an infinite (countable or uncountable) set of values, whereas the signal depicted in Fig. 1.15 assumes values from a finite (known as the alphabet size) set of values. Incidentally, this is the same distinction which can be made when analog and quantized digital signals are compared. We will elaborate on this issue in the ensuing sections.

As noted earlier, deterministic signals are special cases of random signals, and hence one may eliminate a parallel treatment of deterministic signals. This approach allows one to take a unified approach toward signal and system analysis, resulting in treating the continuous-time version of a signal as a random process and the samples of a random signal as a random sequence. An immediate ramification of this approach is that one needs to have access (or identify) the statistical behavior of the signal. In this text, unless otherwise stated, all vectors of are of length l_v. To address the random nature of signals considered here, one needs to define the joint probability density function (pdf) of $x[n]$, $x[n+1]$, ..., $x[n+l_v-1]$. Before doing so, we introduce a more concise vector notation. That is, let

$$\mathbf{x}[n] = [x[n], x[n+1], \ldots, x[n+l_v-1]]^{\mathsf{T}} \tag{1.30}$$

with its joint pdf defined as

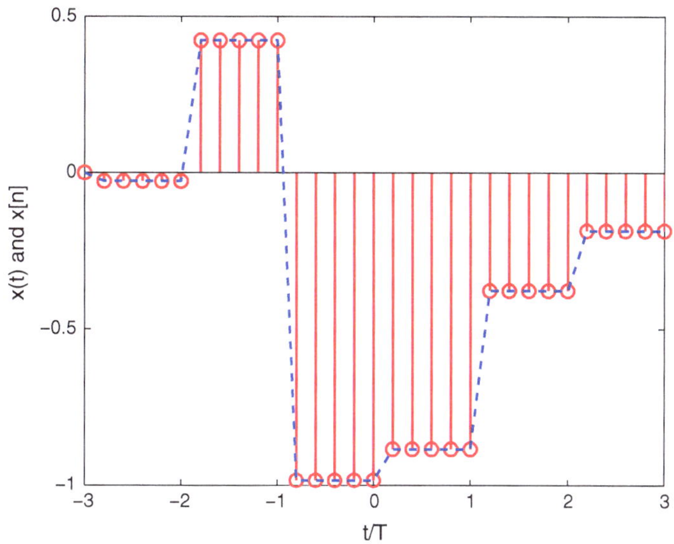

Fig. 1.15 A randomly modulated signal (in amplitude)

$$f_{\mathbf{x}[n]}(\mathbf{x}) = P\left(x[n] \le x_n, x[n+1] \le x_{n+1}, \ldots x[n+l_v-1] \le x_{n+l_v-1}\right) \tag{1.31}$$

be an observed column random vector with $\mathbf{x} = \left[x_n, x_{n+1}, \ldots, x_{n+l_v-1}\right]^{\mathsf{T}}$ denoting a particular realization of the random vector. Similarly, in continuous time, let

$$\mathbf{x}(t_n) = \left[x(t_n), x(t_{n+1}), \ldots, x\left(t_{n+l_v-1}\right)\right]^{\mathsf{T}} \tag{1.32}$$

with

$$f_{\mathbf{x}(t_n)}(\mathbf{x}) = P\left(x(t_n) \le x_n, x(t_{n+1}) \le x_{n+1}, \ldots x\left(t_{n+l_v-1}\right) \le x_{n-l_v+1}\right). \tag{1.33}$$

Furthermore, we define the mean and correlation functions as follows:

$$m_x(t) = E\{x(t)\} \tag{1.34}$$

and

$$R_x(t_1, t_2) = E\left\{x^*(t_1)\, x(t_2)\right\}, \tag{1.35}$$

respectively. $E\{.\}$ denotes the ensemble average of the enclosed. For the discrete time case,

$$m_x[n] = E\{x[n]\} \tag{1.36}$$

and

$$R_x[n, m] = E\left\{x^*[n]\, x[m]\right\}. \tag{1.37}$$

The above general definition is somewhat cumbersome for $l_\upsilon > 2$. An example of $f_{\mathbf{x}[n]}(\mathbf{x})$ for $l_\upsilon = 1$ is depicted in Fig. 1.16. In this case, it is assumed that the process is Gaussian with $R_x[n,n] = 1$ (variance of 1) while $m_x[n] = \sin[n]$. That is,

$$f_{x[n]}(x) = \frac{1}{\sqrt{2\pi R[n,n]}} \exp\left\{-\frac{(x - \sin(n))^2}{2R_x[n,n]}\right\}. \tag{1.38}$$

Gaussian random variables describe a large class of random events in nature, but are not suitable for modeling biomedical signals. We will elaborate on this topic in later chapters. It is, however, important to study this type of random variable as the additive noise in observed biomedical signals, which are often collected using electronic devices, can accurately be modeled as Gaussian. Hence, for those interested in processing biomedical signals, this topic becomes important.

Figure 1.16 clearly indicates the full statistics of $x[n]$ via $f_{\mathbf{x}[n]}(\mathbf{x})$. That is, one can answer the following question. What is $p\{a \le x[n] \le b\}$ for any real a, b, and n (time)? Since

$$p\{a \le x[n] \le b\} = \int_b^b f_{x[n]}(x)\,dx, \tag{1.39}$$

the knowledge of $f_{x[n]}(x)$ provides the desired answer to the above question.

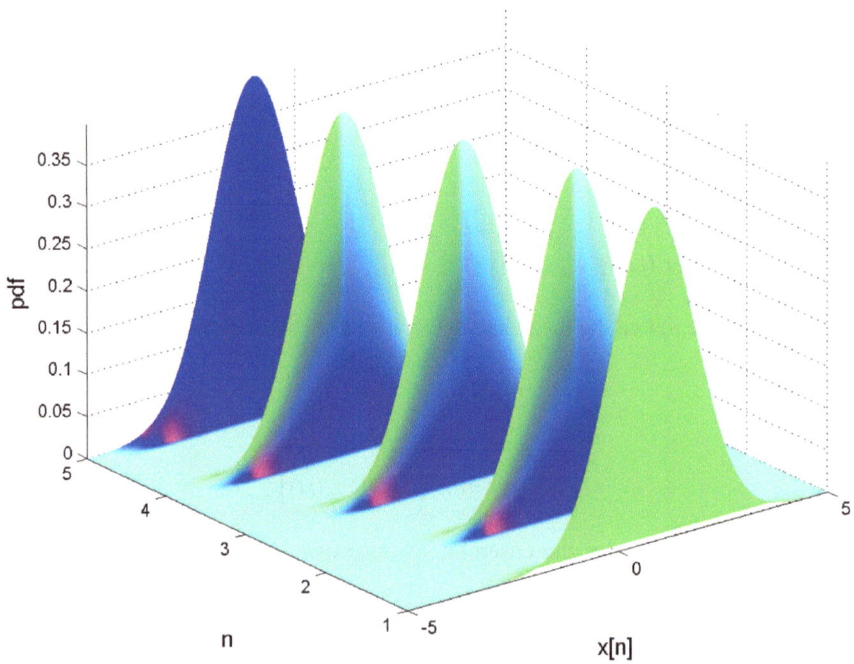

Fig. 1.16 The pdf of $x[n]$

One question remains. What can be said about the statistics of $\mathbf{x}[n]$ when $l_v \neq 1$? To answer this question, one needs the joint statistics of the elements of a random vector. We will address this question later in this section.

Before we can address the distribution function of $\mathbf{x}[n]$ for $l_v > 1$, we need to consider the moments of $\mathbf{x}[n]$. For many applications the first two moments of a random signal have physical significance (i.e., "average" and "variance"), which allow one to draw (albeit inconclusive) conclusions about the behavior of a random signal. Hence, we proceed to define the first two moments of the random signal as

$$\mathbf{m}_x[n] = E\{\mathbf{x}[n]\} \tag{1.40}$$

and

$$\mathbf{R}_x[n] = E\left\{\mathbf{x}[n]\mathbf{x}^H[n]\right\} \tag{1.41}$$

where $E\{x\} = \int_{-\infty}^{\infty} x f_X(x)\,dx$ denotes the ensemble average of the enclosed and the superscript H denotes Hermitian operator (or transpose conjugate). Similarly,

$$\mathbf{m}_x(t_n) = E\{\mathbf{x}(t_n)\} \tag{1.42}$$

and

$$\mathbf{R}_x(t_n) = E\left\{\mathbf{x}(t_n)\mathbf{x}^H(t_n)\right\} \tag{1.43}$$

for the continuous time signals. Gaussian signals, as noted earlier, are of particular interest in digital signal processing as they describe somewhat accurately the signals observed in various electronic devices. Therefore, we will dedicated the remainder of this section to this process. To that end, let us consider the pdf of a Gaussian vector. We limit the discussion to the discrete time case, the reader can readily extend the results to the case of continuous time. We, therefore, are concerned with the pdf of $\mathbf{x}[n]$. First, let us introduce the covariance matrix of complex $\mathbf{x}[n]$, which is given by

$$\mathbf{K}_x[n] = E\left\{(\mathbf{x}[n] - \mathbf{m}_x[n])(\mathbf{x}[n] - \mathbf{m}_x[n])^H\right\} = \mathbf{R}_x[n] - \mathbf{m}_x[n]\mathbf{m}_x^H[n] \tag{1.44}$$

Then,

$$f_{\mathbf{x}[n]}(\mathbf{x}) = \frac{1}{\pi^{l_v}\det\mathbf{K}_x[n]} \times \exp\left\{-(\mathbf{x}-\mathbf{m}_x[n])^H\mathbf{K}_x^{-1}[n](\mathbf{x}-\mathbf{m}_x[n])\right\} \tag{1.45}$$

when $\mathbf{x}[n]$ is a complex vector and $\det\mathbf{K}_x[n]$ denotes the determinant of the matrix $\mathbf{K}_x[n]$. For the case when $\mathbf{x}[n]$ is a real vector ($\mathbf{x}$ is a particular realization of $\mathbf{x}[n]$), we have a slightly different expression for the pdf, which is given by

$$f_{\mathbf{x}[n]}(\mathbf{x}) = \frac{1}{(2\pi)^{l_v/2}\{\det \mathbf{K}_x[n]\}^{1/2}} \times \exp\left\{-\frac{1}{2}(\mathbf{x}-\mathbf{m}_X[n])^{\mathsf{T}}\mathbf{K}_x^{-1}[n](\mathbf{x}-\mathbf{m}_x[n])\right\} \quad (1.46)$$

A remarkable fact about Gaussian random signals is underscored in the above. That is, the equation for pdf clearly shows that the complete statistical characteristics of a Gaussian signal can be found using only the first and second order statistics ($\mathbf{m}_x[n]$ and $\mathbf{K}_x[n]$). For this reason, this model has been used widely to study a host of real-world problems when the characteristics of the underlying process can be safely assumed to be Gaussian. For instance, for a typical radio frequency channel, the additive noise has been shown to be Gaussian, which allows for modeling the received signal for a large class of communication systems using a Gaussian pdf. The additive noise generated by electronic noise can also be modeled as Gaussian. This, as noted earlier, is importance observation since biomedical signals are often collected using electronic devices.

Example 4 Let us consider $x[n] = [x[n], x[n+1]]$ ($l_v = 2$), where $x[n]$ is a Gaussian random sequence. Assume that the correlation function $x[n]$ is given by $R_x[n] = \begin{bmatrix} 1 & 0.4 \\ 0.4 & 1 \end{bmatrix}$. Also, assume that $m_x[n] = [1, -2]$. Plot $f_{\mathbf{x}[n]}(\underline{x})$.

Solution: *This pdf is depicted in Fig.* 1.17*. To gain a better perspective, let us consider contours of the pdf (see Fig.* 1.18*). Note that the contours of the joint pdf of a pair of correlated Gaussian random variables is elliptical in shape, centered at the mean of the vector, which is* $(1, -2)$*. The elliptical shape with minor and major axes at some arbitrary directions underscores dependency between the two random variables. For the case where the correlation matrix is diagonal (the off-diagonal elements are zero), the two random variables become uncorrelated. For Gaussian case, when the random variables become uncorrelated they also become independent. That is,* $f_{\mathbf{x}[n]}(\underline{x}) = f_{x[n]}(x_1) f_{x[n+1]}(x_2)$*. This is underscored by the resulting contours, which will become elliptical with the ellipse's major and minor axes coincide with axes for* $x[n]$ *and* $x[n+1]$ *shown in the Fig.* 1.18.

1.3.1 Stationarity

A large number of signals which are encountered in digital signal processing domain exhibit some form of "stationarity" over observation length of interest. For biomedical signals, however, the stationarity condition is often not satisfied. However, in order to understand non-stationarity, one has to understand the concept of stationarity first. Specifically, we consider two class of stationarity: wide-sense stationarity (WSS) and wide-sense cyclostationarity (WSC).

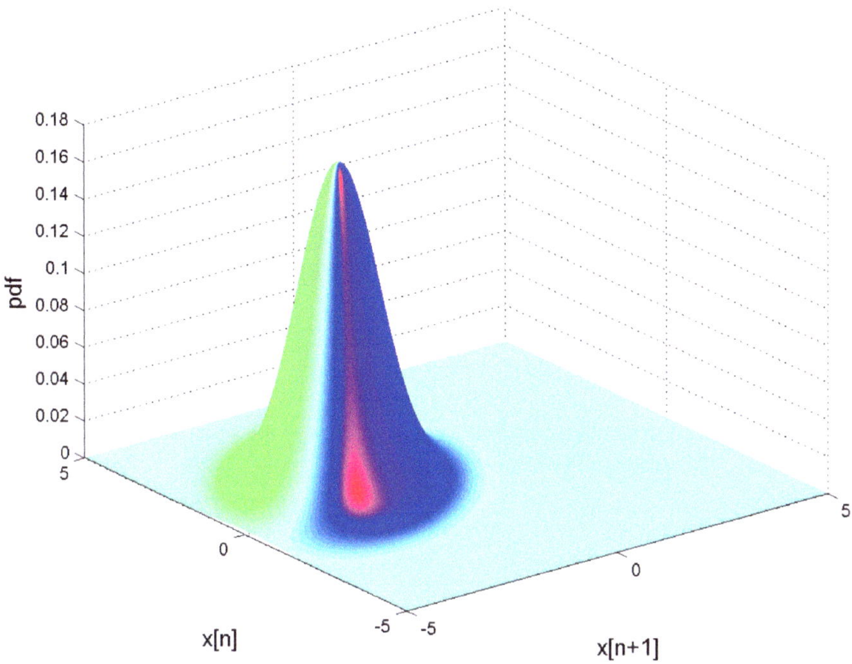

Fig. 1.17 The joint pdf of $x[n]$ and $x[n+1]$

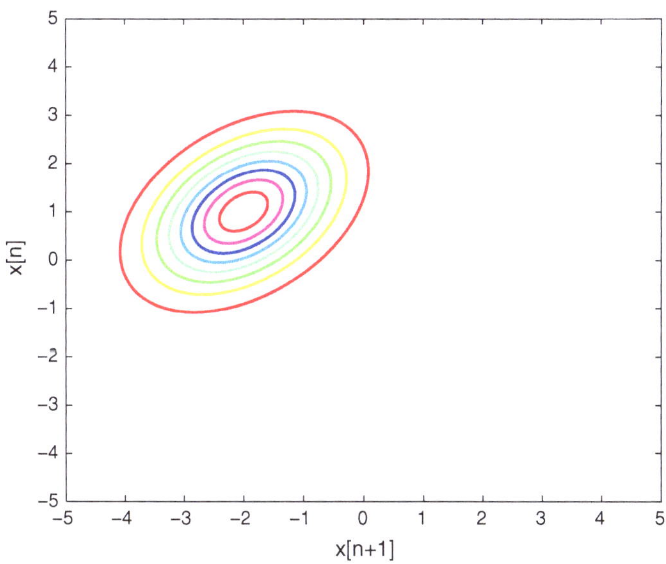

Fig. 1.18 The contours of the joint pdf of $x[n]$ and $x[n+1]$

For the wide-sense stationarity, the ensemble average of the signal is constant and the correlation among a pair of samples is only a function of the separation of samples in time. That is, $E\{x[n]\} = E\{x[m]\}$ for all n and m, and he correlation function, given by

$$R_x[n, n+m] = E\left\{x^*[n]\,x[n+m]\right\} \tag{1.47}$$

is independent of n. That is,

$$R_x[n, n+m] = R_x[m] \tag{1.48}$$

or

$$R_x(t_1, t_2) = R_x(t_1 - t_2) \tag{1.49}$$

for the continuous time case. Furthermore, m_{X_n} or m_{X_t} are independent of n and t, respectively. This leads to the following:

$$\mathbf{m}_x[n] = \mathbf{m}_x \tag{1.50}$$

and

$$\mathbf{R}_x[n] = \mathbf{R}_x \tag{1.51}$$

for all n for the discrete case. Equivalently, for the continuos time case,

$$\mathbf{m}_x(t_n) = \mathbf{m}_x \tag{1.52}$$

and

$$\mathbf{R}_x(t_n) = \mathbf{R}_x(t_n + \tau) \tag{1.53}$$

for any τ. For wide-sense cyclostationarity, we expect the ensemble average of the signal to be a periodic function of time while the correlation between a pair of samples exhibit periodicity. More precisely,

$$E\{x[n]\} = E\left\{x\left[n + kN_p\right]\right\} \tag{1.54}$$

and

$$E\left\{x^*[n]\,x[m]\right\} = E\left\{x^*\left[n + kN_p\right]x\left[m + kN_p\right]\right\} \tag{1.55}$$

for some N_p and any integer k. Furthermore, for the continuos time case,

$$R_x(t_1, t_2) = R_x\left(t_1 + kT_p, t_2 + kT_p\right) \tag{1.56}$$

for some T_p and all k. Furthermore, the mean of the process, if dependent on time, must be a periodic signal of n (discrete time case) or t (for continuos time case). Similar to the above case, these identities lead to the following:

$$\mathbf{m}_x[n] = \mathbf{m}_x\left[n + kN_p\right] \tag{1.57}$$

and

$$\mathbf{R}_x[n] = \mathbf{R}_x\left[n + kN_p\right] \tag{1.58}$$

for some N_p and any integer k. For the continuos time,

$$\mathbf{m}_x(t_n) = \mathbf{m}_x\left(t_n + kT_p\right) \tag{1.59}$$

and

$$\mathbf{R}_x(t_n) = \mathbf{R}_x\left(t_n + kT_p\right) \tag{1.60}$$

for some T_p and all k. Reader is cautioned that the strict-sense stationarity may not be present when a signal is classified as wide-sense stationary. To elaborate, a process is strict-sense stationary (S.S.S.) if

$$f_{\mathbf{x}[n]}(\mathbf{x}) = f_{\mathbf{x}[n+p]}(\mathbf{x}) \tag{1.61}$$

or

$$f_{\mathbf{x}(t_n)}(\mathbf{x}) = f_{\mathbf{x}(t_n+\tau)}(\mathbf{x}) \tag{1.62}$$

for all p, τ, and l_v (length of vectors). In the event that the signal is strict-sense cyclostationary (S.S.C),

$$f_{\mathbf{x}[n]}(\mathbf{x}) = f_{\mathbf{x}[n+kN_p]}(\mathbf{x}) \tag{1.63}$$

or

$$f_{\mathbf{x}(t_n)}(\mathbf{x}) = f_{\mathbf{x}(t_n+kT_p)}(\mathbf{x}) \tag{1.64}$$

for all p, integer k, and some N_p and T_p. In view of the above, one may conclude that strict-sense stationary signals are rarities, specially in the field of biomedical signal processing. It is noteworthy, however, that wide-sense stationary or wide-sense cyclostationary Gaussian signals, which are of particular interest in the field of digital signal processing, are also strict-sense stationary or strict-sense cyclostationary, respectively. Hence, a large class of signals which are encountered in electronic circuits may be classified as strict-sense stationary.

Example 5 Let $x(t) = A\sin(\omega_0 t + \theta)$ be the signal of interest. This signal is used widely in electronic systems, and hence, a treatment of this signals will be of special interest here. Assume that A and θ are a pair of independent random variables with pdf's $f_A(a)$ and $f_\Theta(\theta)$ (not necessarily a uniform distribution), respectively, and $\omega_0 = \frac{2\pi}{T_0}$. We also assume that $f_\Theta(\theta)$ is defined over $[-\pi, \pi]$. We are interested in characterizing a sample of size 4 of the signal over its period. Assume that the samples are taken at 0, $\frac{T_0}{4}$, $\frac{T_0}{2}$, and $\frac{3T_0}{4}$, corresponding to $n = 0$ and $l_v = 4$.

Solution: *In this case,*

$$\mathbf{x}[0] = [A\sin(\theta), A\cos(\theta), -A\sin(\theta), -A\cos(\theta)].$$

This signal is shown in Fig. 1.19. *Due to low number of samples, the signal seems to suffer from distortion. In this case, we have assumed a fixed amplitude and phase (although they are chosen randomly). Note that due to the periodic nature of the signal, we have*

$$\mathbf{x}[0] = \mathbf{x}[4] = \mathbf{x}[8]\ldots$$

Once can also verify that

$$\mathbf{m}[0] = m_a[m_s, m_c, -m_s, -m_c]$$

where

$$m_a = E\{A\} = \int_{-\infty}^{\infty} a f_A(a)\, da,$$

$$m_c = E\{\cos(\theta)\} = \int_{-\pi}^{\pi} \cos(\theta)\, f_\Theta(\theta)\, d\theta,$$

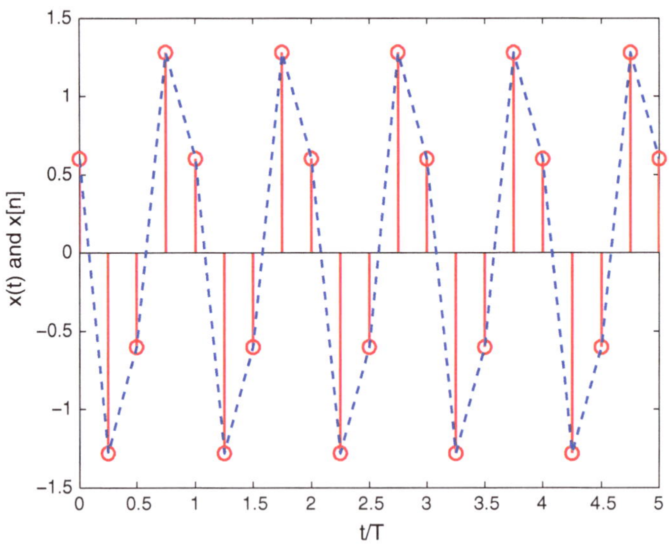

Fig. 1.19 The sampled version of a sinusoidal signal with random phase and amplitude

and

$$m_s = E\{\sin(\theta)\} = \int_{-\pi}^{\pi} \sin(\theta)\, f_\Theta(\theta)\, d\theta.$$

It is rather easy to observe that an arbitrary shift in sampling time will lead to a different outcome. That is, $m_x[0] \neq m_x[1]$ *(assuming that we continue to take samples in the next period). Hence, we cannot establish wide-sense stationarity. However, we note that*

$$\mathbf{m}_x[0] = \mathbf{m}_x[4] = \mathbf{m}_x[8] = \cdots = \mathbf{m}_x[4k].$$

Consequently, we suspect that the signal may exhibit wide-sense cyclostationarity. To prove this point, we must examine the correlation matrix of $\underline{X}_0$*. With little effort, one can show that*

$$\mathbf{R}_x[0] = c \begin{bmatrix} 1-s_2 & s_1 & -1+s_2 & -s_1 \\ s_1 & s_2 & -s_1 & -s_2 \\ -1+s_2 & -s_1 & 1-s_2 & s_1 \\ -s_1 & -s_2 & s_1 & s_2 \end{bmatrix}$$

where $s_1 = \frac{1}{2}E\{\sin(2\theta)\}$, $s_2 = E\left\{\cos^2(\theta)\right\}$, *and* $c = E\left\{A^2\right\}$*. Since*

$$\mathbf{x}[0] = \mathbf{x}[4] = \mathbf{x}[8] \ldots,$$

we can see that

$$\mathbf{R}_x[0] = \mathbf{R}_x[4] = \mathbf{R}_x[8] = \ldots$$

However, any other shift will result in a different outcome. That is, for

$$\mathbf{x}[1] = [A\cos(\theta), -A\sin(\theta), -A\cos(\theta), A\sin(\theta)]$$

(samples are taken at $\frac{T_0}{4}, \frac{T_0}{2}, \frac{3T_0}{4}, T_0$*), we have*

$$\mathbf{R}_x[1] = c \begin{bmatrix} s_2 & -s_1 & -s_2 & s_1 \\ s_1 & 1 \quad s_2 & s_1 & 1 \mid s_2 \\ -s_2 & s_1 & s_2 & -s_1 \\ s_1 & -1+s_2 & -s_1 & 1-s_2 \end{bmatrix}$$

which is indeed not identical to $R_x[0]$*. The above observations confirm the wide-sense cyclostationarity of the signal.*

In order to simulate the above problem, we need to know the statistics of A and θ. Furthermore, we need to generate the samples of these random variable. Without loss of generality, we consider a Rayleigh distributed amplitude (this phenomenon is often observed in practice when narrow-band signals travel through a wireless channel).

This random variable is also observed when one is observing the magnitude of a 2-dimensional vector whose coordinates are zero-mean Gaussian distributed. In the biomedical signal processing field, we often observe signals which are corrupted by the additive Gaussian noise, which is common to all electronic devices. Further, we are often concerned with multiple channel signals. For the case of observing a pair of signal where signals correspond to the dimensions of a 2-dimensional observation, the length of the observed vector is Rayleigh distributed.

In this case, we have $f_A(a) = \frac{1}{\sigma_a^2} \exp\left(-\frac{a^2}{2\sigma_a^2}\right)$ where σ_a^2 is assumed to be the variance of the projection of the vector or either coordinates. Furthermore, we assume that θ is uniformly distributed on $[-\pi, \pi]$. To generate samples of A, we use Monte-Carlo method. This method needs more explanation.

1.3.2 Monte Carlo Simulation

In this method, the samples of A can be generated using $A = F_A^{-1}(U)$ where $F_A(a)$ denotes the Probability Distribution Function (PDF) of A and U is a uniform distributed random variable defined over [0, 1]. This can be readily shown by noting that

$$P\{A \le a\} = P\left\{F_A^{-1}(U) \le a\right\}. \tag{1.65}$$

Now, for a uniform distributed random variable, $P\{U \le u\} = u$. Hence,

$$P\left\{F_A^{-1}(U) \le a\right\} = P\{U \le F_A(a)\} = F_A(a), \tag{1.66}$$

which was our assertion above. This method can be used to generate samples of any random variable for which the inverse PDF is available.

Note that since $m_c = m_s = 0$ (uniformly distributed θ), the mean vectors are zero. Furthermore, $s_1 = 0$ and $s_2 = \frac{1}{2}$. One can also verify that $c = E\left(A^2\right) = 2\sigma_a^2$.

Example 6 Let us consider a zero-mean, real Gaussian random signal whose samples exhibit the following correlation property:

$$E\{x[n]x[m]\} = \begin{cases} \sigma_1^2 & n = m \\ \sigma_2^2 & n = m \pm 1 \\ 0 & \text{otherwise} \end{cases}$$

for any n and m. Is this random process wide-sense stationary or wide-sense cyclo-stationary?

Solution: *Since the mean is zero, one can argue that the mean satisfies the basic condition for wide-sense stationarity as well as for wide-sense cyclostationarity. We need to examine the correlation matrix. To that end, it can readily be shown that the correlation matrix for any l_v is given by*

$$\mathbf{R}_x[N] = \begin{bmatrix} \sigma_1^2 & \sigma_2^2 & 0 & 0 & \cdots \\ \sigma_2^2 & \sigma_1^2 & \sigma_2^2 & 0 & \cdots \\ 0 & \sigma_2^2 & \sigma_1^2 & \sigma_2^2 & 0 \\ 0 & 0 & \sigma_2^2 & \sigma_1^2 & \cdots \\ \vdots & \vdots & \vdots & \vdots & \ddots \end{bmatrix}$$

which is independent of n. *This confirms the wide-sense stationarity of the random signal. Since it is WSS, then it is also WSC.*

To use simulation to study this problem, we consider a vector of size n_sim. We then generate a white Gaussian sequence $y[n]$ *using randn function in matlab. Finally, using the operation* $x[n] = y[n] - 0.1y[n-1]$ *we generate a correlated random vector. Note that*

$$E\{x[n]x[m]\} = \begin{cases} 1 + (0.1)^2 = 1.01 & n = m \\ -0.1 & n = m \pm 1, \\ 0 & \text{otherwise} \end{cases}$$

which satisfies the condition stated in this example.

Example 7 Let us consider the following signal:

$$x(t) = \sum_{m=-\infty}^{\infty} a_m p(t - mT) \tag{1.67}$$

where $\{a_m\}$ is an infinite sequence of independent and identically distributed (i.i.d) random variables. This representation may be considered as a signal for studying the response of a biological system to modulated pulses. It also describes the linearly modulated baseband signals in signal processing. Let $f_A(a)$ denote the pdf of a_n. Also, we assume that $E\{a_n\} = 0$ for all n, and the signal is sampled at a rate of $\frac{4}{T}$ samples/s. Show that the resulting process is wide-sense cyclostationary.

Solution: *In later sections, we will encounter this form of signaling, which is known as linear "modulation" with T denoting the 'symbol' time and* a_n *as digital symbols. At this stage, we do not elaborate on the properties of* $p(t)$. *It suffices to point out that* $p(t)$ *will have a significant impact on the spectrum of the above random signal. This form of signaling is widely used in a variety of scenarios in digital communications world, but it may also be used in biomedical signal processing if the intention is to study the dispersion of a pulse train through a dispersive medium, such as tissue. That is, dispersive media, such as tissue, may be viewed as a bandlimited channel whose bandwidth is smaller than that of the* $p(t)$, *which results in the distortion of the input pulse by stretching it in time. This, in turn, causes the adjacent pulses to overlap, which results in distortion in the information bearing part of the signal. To gain an insight, however,* $x(t)$ *is depicted for a case where* a_n *is randomly selected from the set* $\{-3, -2, -1, 1, 2, 3\}$ *and* $p(t)$ *is a pulse shape known as raised-cosine in Fig.* 1.20, *and is given by*

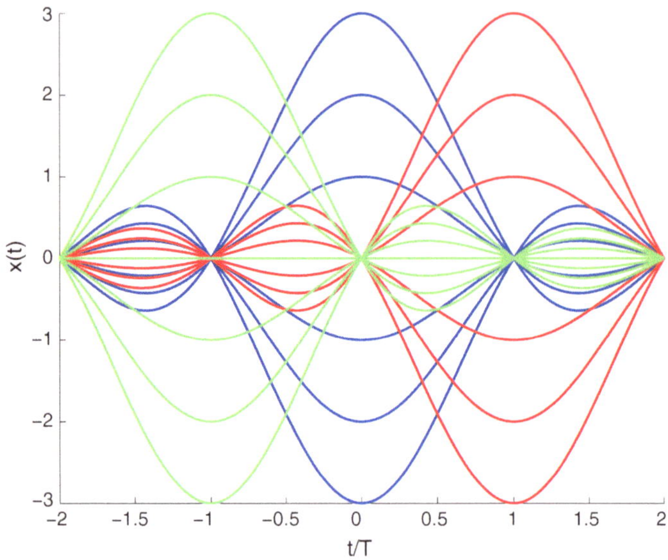

Fig. 1.20 The *eye-diagram* for a 6-level raised-cosine waveform

$$p(t) = \sin c\left(\frac{t}{T}\right)\frac{\cos\left(\frac{\pi\beta t}{T}\right)}{1-\frac{4\beta^2t^2}{T^2}} \text{ where } \sin c(x) = \frac{\sin(\pi x)}{\pi x}$$

These types of plots which are overlaps of all possible waveforms generated by different values of $a_n = a_n$ is known as the eye-diagram. Note that the pulse shape is not confined to T seconds, and hence there are self-interference generated for the adjacent symbol intervals. That is, a_n impacts the signal during the intervals of time associated with $a_{n-1}, a_{n-2}, \ldots$and $a_{n+1}, a_{n+2}, \ldots$. The sampled version of this signal (4 samples per T) is depicted in Fig. 1.21*. One interesting fact that can be obtained from this figure is that the samples taken at nT are only a function of a_n. That is, the symbols in the adjacent interval cannot impact the value of the samples at nT. This fact is also apparent from Fig.* 1.20*, where samples at $-T, 0, T$ are only functions of a_{-1} (green curve), a_o (blue curve), and a_1 (red curve).*

First, we see that

$$\begin{aligned} x[n] &= \sum_{m=-\infty}^{\infty} a_m p(nT_s - 4mT_s) \\ &= \sum_{m=-\infty}^{\infty} a_m p((n-4m)T_s) \end{aligned}$$

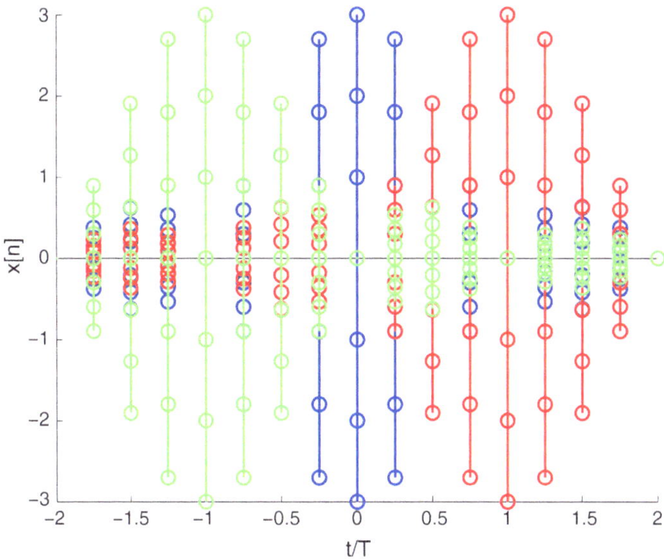

Fig. 1.21 The *eye-diagram* in the discrete domain for a 6-level raised-cosine waveform

$$= \sum_{m=-\infty}^{\infty} a_m p[n-4m] \tag{1.68}$$

where $P[q] = P(qT_s)$. *Next, we observe that*

$$E\{x[n]\} = \bar{a} \sum_{m=-\infty}^{\infty} p[n-4m]$$

where $\bar{a} = E\{a_m\}$. *Since*

$$\sum_{m=-\infty}^{\infty} p[n+4k-4m] = \sum_{m=-\infty}^{\infty} p[n-4m] \longrightarrow E\{x[n]\} = E\{x[n+4k]\} \tag{1.69}$$

for any integer k. *Since we have assumed that* $\bar{a} = 0$, $E\{X[n]\} = 0$ *for all* n. *Next, we need to examine the autocorrelation function of the signal. To that end,*

$$E\left\{x^*[n_1]\,x[n_2]\right\} = \sum_{m_2=-\infty}^{\infty} \sum_{m_1=-\infty}^{\infty} E\left\{a_{m_1} a_{m_2}\right\} p[n_1-4m_1]\, p[n_2-4m_2]$$

$$= \sum_{\substack{m_2=-\infty \\ m_1 \neq m_2}}^{\infty} \sum_{m_1=-\infty}^{\infty} E\left\{a_{m_1} a_{m_2}\right\} p\left[n_1 - 4m_1\right] p\left[n_2 - 4m_2\right] + \sum_{m=-\infty}^{\infty} E\left\{|a_m|^2\right\} p\left[n_1 - 4m\right] p\left[n_2 - 4m\right] \tag{1.70}$$

Since the sequence $\{a_m\}$ *is an uncorrelated sequence,* $E\left\{a_{m_1} a_{m_2}\right\} = E\left\{a_{m_1}\right\} E\left\{a_{m_2}\right\}$. *Hence,*

$$E\left\{x^*\left[n_1\right] x\left[n_2\right]\right\} = \sum_{\substack{m_2=-\infty \\ m_1 \neq m_2}}^{\infty} \sum_{m_1=-\infty}^{\infty} E\left\{a_{m_1}\right\} E\left\{a_{m_2}\right\} p\left[n_1 - 4m_1\right] p\left[n_2 - 4m_2\right] + \sum_{m=-\infty}^{\infty} E\left\{|a_m|^2\right\} p\left[n_1 - 4m\right] p\left[n_2 - 4m\right] \tag{1.71}$$

Now, since $E\{a_m\} = 0$,

$$E\left\{x^*\left[n_1\right] x\left[n_2\right]\right\} = \sum_{m=-\infty}^{\infty} E\left\{|a_m|^2\right\} p\left[n_1 - 4m\right] p\left[n_2 - 4m\right] = \sigma_a^2 \sum_{m=-\infty}^{\infty} p\left[n_1 - 4m\right] p\left[n_2 - 4m\right] \tag{1.72}$$

Note that

$$\sum_{m=-\infty}^{\infty} p\left[n_1 + 4k - 4m\right] p\left[n_2 + 4k - 4m\right] = \sum_{m'=-\infty}^{\infty} p\left[n_1 - 4m'\right] p\left[n_2 - 4m'\right]$$

for any n_1, n_2, *and* k. *Hence,*

$$E\left\{x^*\left[n_1\right] x\left[n_2\right]\right\} = E\left\{x^*\left[n_1 + 4k\right] x\left[n_2 + 4k\right]\right\} \rightarrow \mathbf{R}_x[n] = \mathbf{R}_x[n + 4k].$$

The above observations confirm the cyclostationarity of the process (recall that 4 samples are taken per symbol).

In order to study this problem using matlab, we first generate signal $x[n]$ *for a data sequence of length 10,000. The pulse shape is considered to be a Gaussian pulse. The use of Gaussian pulse, although not of interest in communications systems, is of interest in the biomedical field as it describes accurately the pulsed signals generated by a laser system in tissue optics.*

Three correlation functions are generated. That is, $R_1[jj] = \frac{1}{L} \sum_{n=1}^{4\times10^4} x[n] x[n + jj]$ *is computed for different values of* jj *(*L *is the length of the*

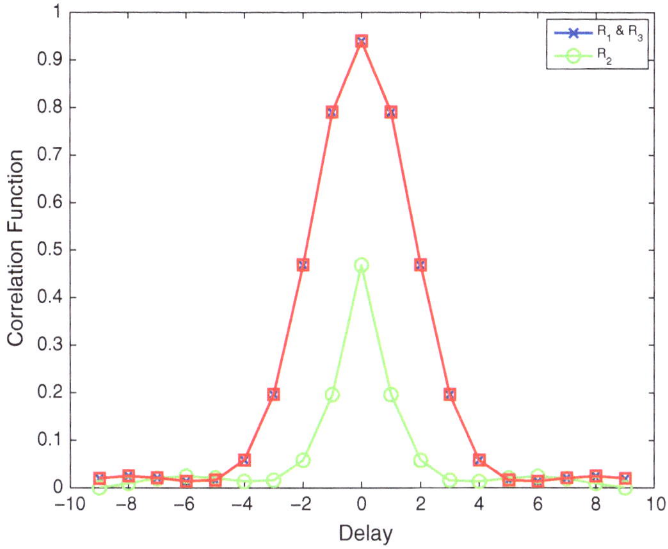

Fig. 1.22 Correlation function of a linearly modulated signal with Gaussian pulses

vector and there are 4×10^4 *samples taken for a data sequence of length* 10^4*). The vector sizes are slightly adjusted to account for vector multiplication. Then,* $R_2[jj] = \frac{1}{L}\sum_{n=1}^{4\times10^4} x[n+2]\,x[n+jj+2]$ *and* $R_3[jj] = \frac{1}{L}\sum_{n=1}^{4\times10^4} x[n+4]\,x[n+jj+4]$ *are computed. Figure 1.22, depicts the result. As expected* $R_1[jj] = R_3[jj]$ *(the red and blue curve overlap). Note that a shift of 4 samples is identical to a delay of one symbol, indicating that the process is cyclostationary. However, as can be seen,* $R_1[jj] \neq R_2[jj]$*, signifying that, although the process is wide-sense cyclostationary, it is not wide-sense stationary.*

Example 8 Let us consider a zero-mean, complex Gaussian signal $x(t) = x_r(t) + jx_i(t)$, where $j = \sqrt{-1}$. We sample this signal at the rate of $\frac{1}{T_s}$ samples per second. Give an expression for the pdf of a sample taken at time nT_s. Assume that the signal has a correlation function $R_x(t_1, t_2) = E\{x^*(t_1)\,x(t_2)\}$, and that $E\{x(t_1)\,x(t_2)\} = 0$ for all t_1 and t_2 (such a complex random process is known as 'proper').

Solution: *Since* $E\{x(t_1)\,x(t_2)\} = 0 \rightarrow$

$$\begin{aligned} E\{x(t_1)\,x(t_2)\} &= E\{\{x_r(t_1) - jx_i(t_1)\}\{x_r(t_2) - jx_i(t_2)\}\} \\ &= E\{x_r(t_1)\,x_r(t_2)\} - E\{x_i(t_1)\,x_i(t_2)\} \\ &\quad + j\{E\{x_r(t_1)\,x_i(t_2)\} + E\{x_r(t_2)\,x_i(t_1)\}\} \\ &= 0 \end{aligned} \tag{1.73}$$

This leads to

$$E\{x_r(t_1)\,x_i(t_2)\} = -E\{x_r(t_2)\,x_i(t_1)\} \tag{1.74}$$

and

$$E\left\{x_i(t_1)\,x_i(t_2)\right\} = E\left\{x_r(t_1)\,x_r(t_2)\right\} = R_r(t_1, t_2) \tag{1.75}$$

Hence,

$$\begin{aligned} R_x(t_1, t_2) &= E\left\{\left\{x_r(t_1) - jx_i(t_1)\right\}\left\{x_r(t_2) + jx_i(t_2)\right\}\right\} \\ &= E\left\{x_i(t_1)\,x_i(t_2)\right\} + E\left\{x_r(t_1)\,x_r(t_2)\right\} \\ &\quad + jE\left\{x_r(t_1)\,x_i(t_2)\right\} + jE\left\{x_r(t_2)\,x_i(t_1)\right\} \end{aligned} \tag{1.76}$$

Considering (1.74) *and* (1.75), *one is led to the following:*

$$R_x(t_1, t_2) = 2E\left\{x_r(t_1)\,x_r(t_2)\right\} = 2R_r(t_1, t_2)\,. \tag{1.77}$$

Clearly, $l_v = 1$, *and*

$$x[n] = x(nT_s) = x_r(nT_s) + jx_i(nT_s)\,. \tag{1.78}$$

Furthermore,

$$\begin{aligned} \mathbf{K}_x[n] &= E\left\{x^*[n]\,x[n]\right\} = R_x(nT_s, nT_s) \\ &= 2R_r(nT_s, nT_s)\,, \end{aligned} \tag{1.79}$$

which leads to

$$\begin{aligned} f_{\mathbf{x}[n]}(x) &= \frac{1}{\pi R_x(nT_s, nT_s)} \exp\left\{-\frac{(x_r - jx_i)(x_r + jx_i)}{R_x(nT_s, nT_s)}\right\} \\ &= \frac{1}{2\pi R_r(nT_s, nT_s)} \exp\left\{-\frac{|x|^2}{2R_r(nT_s, nT_s)}\right\}. \end{aligned} \tag{1.80}$$

In this equation, x_r *and* x_i *are the real and imaginary parts of* x, *respectively. Note the "circular" symmetry of the pdf (namely,* $|x|^2$*) and the fact that the above expression is identical to the product of the pdf's of a pair of independent Gaussian random variables. That is,*

$$\begin{aligned} f_{\mathbf{x}[n]}(x) &= f_{X_r}(x_r)\,f_{X_i}(x_i) \\ &= \frac{1}{\sqrt{2\pi R_r(nT_s, nT_s)}} \exp\left\{-\frac{x_r^2}{2R_r(nT_s, nT_s)}\right\} \\ &\quad \times \frac{1}{\sqrt{2\pi R_r(nT_s, nT_s)}} \exp\left\{-\frac{x_i^2}{2R_r(nT_s, nT_s)}\right\} \end{aligned} \tag{1.81}$$

This is an expected result since (1.74) *for* $t_1 = t_2 = nT_s$ *reduces to*

$$E\left\{x_r\left(nT_s\right)x_i\left(nT_s\right)\right\} = -E\left\{x_i\left(nT_s\right)x_r\left(nT_s\right)\right\}, \tag{1.82}$$

which implies that $E\left\{x_r\left(nT_s\right)x_i\left(nT_s\right)\right\} = 0$. *Therefore, the real and imaginary parts of the signal* $\mathbf{x}[n]$ *are uncorrelated (hence, independent).*

Another desirable aspect of Gaussian signals resides in the fact that its higher moments can be found from the first two moments of the signal. This property comes particularly handy when nonlinear operations are encountered. For this reason, in a variety of fields, there is an interest in modeling observed phenomena/signals using a Gaussian assumption. As we will see in later chapters, such an assumption is invalid when biomedical signals are considered.

Nonlinear operations are often used in digital signal processing to extract periodic structure for clock recovery or non-coherent detection. In biomedical field, nonlinearity is an inherent nature of the underlying system, as a large class of biological systems may be modeled as non-stationary, nonlinear systems with chaotic responses. Nonetheless, we often process the observed signals with Gaussian noise component using nonlinear operations. For instance, we may be interested in energy or power of a signal. This results in nonlinear processing of Gaussian signals, which are present in all systems, including biomedical signals collected using electronic means. In the case of Gaussian signals, we have tractable results for some important cases. To illustrate this point, let us consider the following example.

Example 9 We are interested in modeling the output of a squaring device when the input is a real, zero mean, WSS Gaussian signal. Assume that $R_x(\tau) = E\{x(t)x(t+\tau)\}$ describes the correlation function of the input signal. Let $y(t) = x^2(t)$ denote the output process. Find the correlation function of a vector of size 2 that is obtained by sampling $y(t)$ at regular interval of T_s seconds. Give an expression for the pdf of the $y(t)$.

Solution: *In this case,* $l_v = 2$. *Let* $y[n]$ *denote the output of the nonlinear device. Then, the vector* $\mathbf{y}[n]$ *is given by*

$$\mathbf{y}[n] = \left[x^2[n], x^2[n+1]\right]$$

where $x[n] = x(nT_s)$. *Clearly,* $\mathbf{y}[n]$ *is no longer a Gaussian vector. We, then, have*

$$\mathbf{R}_y[n] = E\left\{\mathbf{y}^T[n]\,\mathbf{y}[n]\right\} = \begin{bmatrix} E\left\{x^4[n]\right\} & E\left\{x^2[n]x^2[n+1]\right\} \\ E\left\{x^2[n]x^2[n+1]\right\} & E\left\{x^4[n+1]\right\} \end{bmatrix}$$

This expression requires the 4th order statistics of the Gaussian signal. This, however, can be found using the following identity for Gaussian signals (one can verify this property by finding the moment generating function of the Gaussian vector $\mathbf{x} = [x(t_1), x(t_2), x(t_3), x(t_4)]$*):*

$$
\begin{aligned}
E\{x(t_1)x(t_2)x(t_3)x(t_4)\} = {} & E\{x(t_1)x(t_2)\}E\{x(t_3)x(t_4)\} \\
& + E\{x(t_1)x(t_3)\}E\{x(t_2)x(t_4)\} \\
& + E\{x(t_1)x(t_4)\}E\{x(t_2)x(t_3)\} \qquad (1.83)
\end{aligned}
$$

Hence,

$$
\mathbf{R}_y[n] = \begin{bmatrix} 3R_x^2(0) & 2R_x^2(T_s) + R_x^2(0) \\ 2R_x^2(T_s) + R_x^2(0) & 3R_x^2(0) \end{bmatrix}
$$

The pdf of $y(t)$ can be found using the function of a random variable transformation, given by

$$
f_y(y) = \sum_{i=1}^{N} \frac{f(x_i)}{\left|\frac{\partial g(x)}{\partial x}|_{x=x_i}\right|} \qquad (1.84)
$$

where $y = g(x)$ and x_i is the ith solution of $x = g^{-1}(y)$ (assuming that this equation leads to N solutions). For the problem at hand, $y = x^2$, and hence $x_1 = \sqrt{y}$ and $x_2 = -\sqrt{y}$ are the two solutions of $x = g^{-1}(y)$. Furthermore, $\frac{\partial g(x)}{\partial x} = 2x$. Note that the pdf of y is zero for $y < 0$. Given that

$$
f_x(x) = \frac{1}{\sqrt{2\pi R_x(0)}} \exp\left(-\frac{x^2}{2R_x(0)}\right), \qquad (1.85)
$$

we have

$$
\begin{aligned}
f_y(y) &= \frac{1}{2\sqrt{2\pi R_x(0)\,y}} \exp\left(-\frac{y}{2R_x(0)}\right) \\
&\quad + \frac{1}{2\sqrt{2\pi R_x(0)\,y}} \exp\left(-\frac{y}{2R_x(0)}\right); \; y \geq 0 \\
&= \begin{cases} \frac{1}{\sqrt{2\pi R_x(0)y}} \exp\left(-\frac{y}{2R_x(0)}\right) & y \geq 0 \\ 0 & \text{otherwise} \end{cases} \qquad (1.86)
\end{aligned}
$$

This is known as the Chi-squared distribution of degree 1. Simulation and theoretical Gaussian random variable and its square are depicted in Figs. 1.23 *and* 1.24. *In this simulation,* 10^5 *samples of a zero-mean, unit variance Gaussian random variable was generated using the randn function in matlab. The samples of y were generated by simply squaring the samples of the Gaussian random variable.*

Another situation which is encountered in practice is when a pair of independent Gaussian signals (signifying 2 independent coordinates of a 2-dimensional signal) are observed and we are interested in the amplitude (energy) of the signal. Such a distribution is subject of the next example.

Example 10 Let $z(t) = \sqrt{x^2(t) + y^2(t)}$ where $x(t)$ and $y(t)$ are a pair of independent, WSS Gaussian random signals with correlation function $R_x(\tau)$. Assume

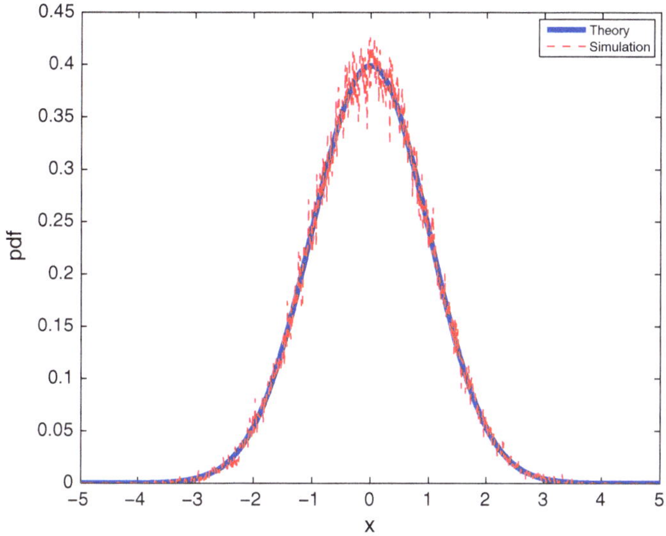

Fig. 1.23 The pdf of a Gaussian random variable and its simulated histogram

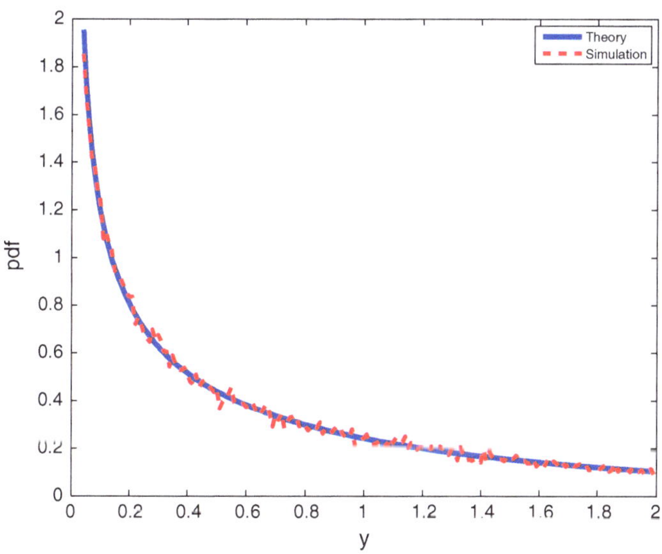

Fig. 1.24 The pdf of the square of a Gaussian random variable and its simulated histogram

that $x(t)$ thinspace has a mean of m_x while $y(t)$ is a zero-mean signal. Find the pdf of $z[n] = z(nT_s)$.

Solution: *It is obvious that* $z[n] = \sqrt{x^2[n] + y^2[n]}$. *Furthermore, the joint pdf of* $x[n]$ *and* $y[n]$ *is given by*

$$f_{x,y}(x,y) = \frac{1}{2\pi R_x(0)} \exp\left\{-\frac{(x-m_x)^2}{2R_x(0)} - \frac{y^2}{2R_x(0)}\right\}$$

Making the change of variable $x[n] = z[n]\cos(\theta)$ *and* $y[n] = z[n]\sin(\theta)$, *and using the general expression for transformation of random variables, results in the following pdf*

$$\begin{aligned} f_{z[n],\theta}(z,\theta) &= \left|\det \tilde{J}\right| f_{x[n],y[n]}(z\cos(\theta), z\sin(\theta)) \\ &= \frac{\left|\det \tilde{J}\right|}{2\pi R_x(0)} \exp\left\{-\frac{(z\cos(\theta)-m_x)^2}{2R_x(0)} - \frac{z^2\sin^2(\theta)}{2R_x(0)}\right\} \\ &= \frac{\left|\det \tilde{J}\right|}{2\pi R_x(0)} \exp\left\{-\frac{z^2+m_x^2}{2R_x(0)}\right\} \exp\left\{\frac{zm_X\cos(\theta)}{R_x(0)}\right\}; z \geq 0 \end{aligned}$$

where $\tilde{J} = \begin{vmatrix} \frac{\partial x[n]}{\partial z_n} & \frac{\partial y[n]}{\partial z_n} \\ \frac{\partial x[n]}{\partial \Theta} & \frac{\partial y[n]}{\partial \Theta} \end{vmatrix}$ *is the Jacobian.* $\left|\det \tilde{J}\right| = z$ *(note that* z *can only be positive and* $\theta \in [0, 2\pi]$*).*

We then have

$$f_{z[n],\Theta}(z,\theta) = \frac{z}{2\pi R_x(0)} \exp\left\{-\frac{z^2+m_x^2}{2R_x(0)}\right\} \exp\left\{\frac{zm_x\cos(\theta)}{R_x(0)}\right\}; z \geq 0$$

Finally, integrating over θ *over the range* $[0, 2\pi]$, *we arrive at*

$$f_{z[n]}(z) = \frac{z}{R_x(0)} \exp\left\{-\frac{z^2+m_x^2}{2R_x(0)}\right\} I_0\left(\frac{zm_x}{R_x(0)}\right); z \geq 0 \quad (1.87)$$

where $I_0(x) = \frac{1}{2\pi}\int_0^{2\pi} e^{x\cos(\theta)}d\theta$ *is the modified Bessel function of the first kind and 0 order. The above pdf is known as the Rician pdf due to Rice. This pdf for two distinct values of* m_x *along with simulated pdf are shown in Fig. 1.25. The sample size is considered to be* 10^5. *The impact of* m_x *on the shape of the pdf is quite profound as it not only shifts the center of the peak of the pdf, but also impacts the overall shape of the pdf, including its tails. This pdf is often used to model the wireless channel when, in addition to the diffused component (modeled as Gaussian signals), a strong signal component (known as specular component) is present. This situation is encountered when the receiver in a wireless communication system is in line-of-sight (LOS) of the transmitter. In biomedical field, this may be viewed as the energy of a 2-dimensional signal which has a dominant signal presence in one dimension (unknown to us) with Gaussian noise present in both dimensions.*

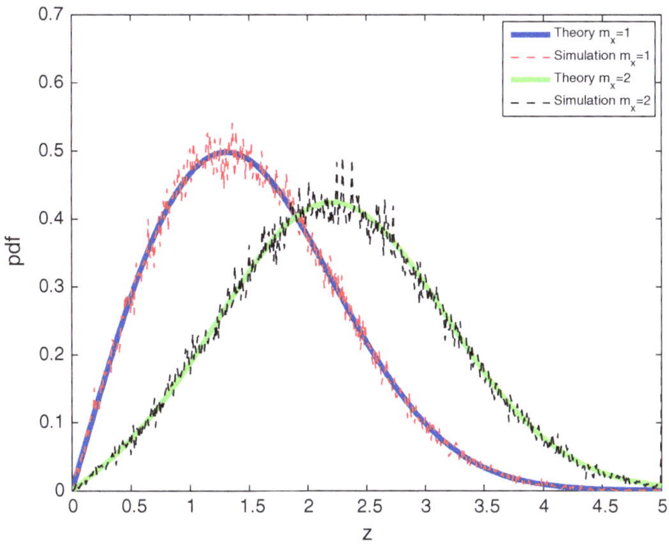

Fig. 1.25 The pdfs of a pair of Rician random variables and their corresponding simulated histograms

A special case of the Rice pdf is when $m_x = 0$, which is the case of the energy of a 2-dimensional signal whose components are purely Gaussian noise. In that case, the amplitude of the received signal can be shown to be the square of the sum of the squares of a pair of zero-mean, independent Gaussian signals. Hence, the Rice pdf reduces to the Rayleigh pdf, given by

$$f_{z[n]}(z) = \frac{z}{R_x(0)} \exp\left\{-\frac{z^2}{2R_x(0)}\right\}; z \geq 0 \tag{1.88}$$

This pdf is identical to that obtained in the previous example (Chi-Squared). In the event of biomedical signal processing where one is interested in determining the presence of an unknown signal with non-zero energy in a pair of noise observations, the Rice and Rayleigh pdfs may be used to model the statistical behavior of the two possible scenarios. As we will discuss in the ensuing chapters, this allows one to arrive at a decision metric, based on Bayesian decision making process, to determine whether a known signal is present in an observed waveform. This is of particular importance when the biomedical signal processing tool is designed to provide a decision in favor of the presence of a phenomenon in the face of noisy observations. So far, we have alluded to the term "energy" several times without providing a formal definition. We will do so next. Furthermore, we will introduce the concept of "spectrum" when dealing with random signals.

1.3.3 Energy and Power

A key feature of any signal, including biomedical signals, is its power or energy. Very often, we observe signals over a limited period of time, which justifies the use of "energy" rather than "power" in signal processing. Such quantities, then, may be of importance for a variety of reasons. Hence, it becomes necessary to have proper definitions for these important parameters. For the signals described in the previous section, we define "average" power and energy as follows:

$$\text{Average Power} = \lim_{N\to\infty} \frac{1}{2N+1} E\left\{\sum_{n=-N}^{N} |x[n]|^2\right\} \tag{1.89}$$

and

$$\text{Energy} = \lim_{N\to\infty} E\left\{\sum_{n=-N}^{N} |x[n]|^2\right\}, \tag{1.90}$$

respectively. Note that if a process if WSS, then

$$E\left\{\sum_{n=-N}^{N} |x[n]|^2\right\} = R_x[0] \tag{1.91}$$

and

$$\text{Average Power} = R_x[0]. \tag{1.92}$$

Otherwise,

$$\text{Average Power} = \lim_{N\to\infty} \frac{1}{2N+1} \sum_{n=-N}^{N} R_x[n]. \tag{1.93}$$

For continuous-time random signals, we have

$$\text{Average Power} = \lim_{T\to\infty} \frac{1}{T} E\left\{\int_{-\frac{T}{2}}^{\frac{T}{2}} |x(t)|^2\, dt\right\} \tag{1.94}$$

and

$$\text{Energy} = \lim_{T\to\infty} E\left\{\int_{-\frac{T}{2}}^{\frac{T}{2}} |x(t)|^2\, dt\right\}. \tag{1.95}$$

It is obvious that a signal with finite energy (energy signal) has zero average power, while a power signal (a signal with finite average power) has infinite energy. Furthermore, note that the ensemble average ($E\{\}$) is needed when random signals are considered (which is the case for almost all signals encountered in biomedical signal processing or otherwise). This is consistent with our approach of considering deterministic signals as a special case of random signals. In the event that signal is deterministic, the above expressions reduce to

$$\text{Average Power} = \lim_{N\to\infty} \frac{1}{2N+1} \sum_{n=-N}^{N} |x[n]|^2 \tag{1.96}$$

and

$$\text{Energy} = \lim_{N\to\infty} \sum_{n=-N}^{N} |x[n]|^2 \tag{1.97}$$

For continuous-time deterministic signals, we have

$$\text{Average Power} = \lim_{T\to\infty} \frac{1}{T} \int_{-\frac{T}{2}}^{\frac{T}{2}} |x(t)|^2 \, dt \tag{1.98}$$

and

$$\text{Energy} = \lim_{T\to\infty} \int_{-\frac{T}{2}}^{\frac{T}{2}} |x(t)|^2 \, dt. \tag{1.99}$$

It is noteworthy that there are classes of signals where the time averages may be used to replace the ensemble averages. In other cases, such a substitution is unwarranted and indeed unjustifiable. However, replacing the ensemble averages with time averages is quite desirable in practice where statistical behavior of signals are often unknown. Although this appears as a viable mechanism to compute power and energy (or for that matter, other statistical characteristics), as will be shown shortly, the time domain analysis may lead to substantial errors in computing the true ensemble averages.

To see the two scenarios, first let us consider the case where time domain averaging leads to correct ensemble averages. For instance, in Fig. 1.26 have depicted the mean and variance of a Gaussian random variable with zero mean and variance of 1 (generated by *randn* in matlab) using ensemble average and time average. It is interesting to note that the estimation of the mean-square requires a number of samples which is substantially larger than that required for the estimation of the mean. One can then conclude that in order to estimate higher order statistics (power, correlation, etc.) one requires larger number of samples than that required for the estimation of the lower order statistics (such as mean). The other observation is that we can estimate the mean as well as the mean-square of a random process in this case using time averaging without resorting to the computation of the ensemble averages. The reason for this convenience is that the program generates ensemble samples of a Gaussian random variable each "time" the *randn* function is called, allowing for the time observation to be an observation of the sample space of the random variable. This may not be the case for observations made in practice.

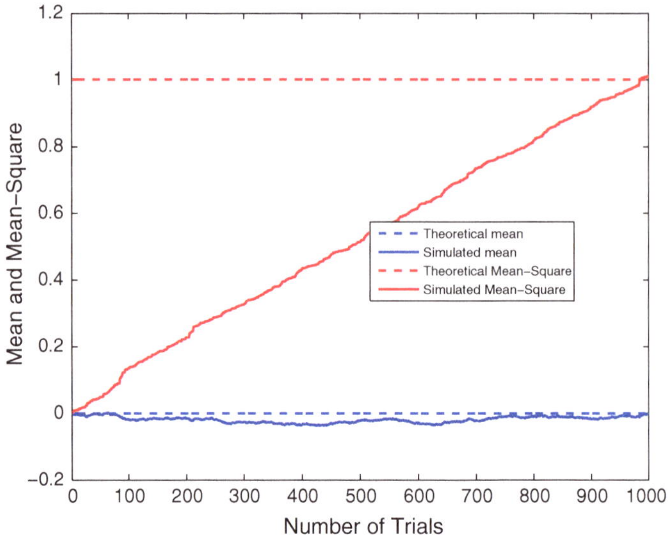

Fig. 1.26 The estimated mean and mean-square values as a function of the number of trials

An examination of the above equations reveals an important, yet overlooked point; that is, the statistical characteristics of a random signal must be available a priori in order to compute accurately the energy or power of the signal. For a large class of problems such statistical models are available. However, for a set of problems where one has to relay entirely on the experimental data (such may be the case for the biomedical signals), the underlying statistical models may be unknown. In that event, one has to rely heavily on data that is collected over extended periods of time. This approach, as seen in Fig. 1.26, assumes that the ensemble averages of a signal can be obtained from observing samples of the signal over long periods of time. Although this approach seems viable for all cases (and hence used routinely in practice), as we will see shortly, it may not yield a meaningful answer in some cases. This point is illustrated in the example below. Note that this constitutes the second scenario.

Example 11 For the signal $x(t) = A\cos(\omega_c t) + B\sin(\omega_c t)$, where A and B are a pair of i.i.d, zero-mean, Gaussian random variables with variance σ^2, find the power of this signal.

<u>Solution</u>: *Since the signal is a periodic signal, one may conclude that $x(t)$ is a power signal. Hence,*

$$\text{Average Power} = \lim_{T\to\infty} \frac{1}{T} E\left\{\int_{-\frac{T}{2}}^{\frac{T}{2}} |x(t)|^2\, dt\right\}$$

$$= E\left\{\lim_{T\to\infty}\frac{1}{T}\int_{-\frac{T}{2}}^{\frac{T}{2}}|x(t)|^2\,dt\right\}$$

$$= E\left\{\frac{A^2}{2}+\frac{B^2}{2}+\lim_{T\to\infty}\frac{A^2}{2T}\int_{-\frac{T}{2}}^{\frac{T}{2}}\cos(2\omega_c)\,dt\right.$$

$$\left.-\lim_{T\to\infty}\frac{B^2}{2T}\int_{-\frac{T}{2}}^{\frac{T}{2}}\cos(2\omega_c)\,dt\right\} \quad (1.100)$$

$$= E\left\{\frac{A^2}{2}+\frac{B^2}{2}\right\}$$

$$= \sigma^2 \quad (1.101)$$

However, if we were to conduct an experiment, the only tool at our disposal for assessing power would be to use the following expression:

$$\text{Average Power} = \lim_{T\to\infty}\frac{1}{T}\int_{-\frac{T}{2}}^{\frac{T}{2}}|x(t)|^2\,dt$$

$$= \frac{A^2}{2}+\frac{B^2}{2}, \quad (1.102)$$

which clearly yields an incorrect answer.

Although this is a unique example, it shatters a popular belief among practitioners in the field, which suggests that experimental observations are sufficient to assess essential parameters of a random signal, such as power, correlation, mean, or energy. Obviously, such an understanding is promoted by a lack of understanding of the underlying processes, which lead one to relay heavily on experimental observations rather than discovering the true statistical characteristics of the observed signal. To explain this discrepancy further, we need to consider the concept of 'ergodicity'. As will be seen shortly, ergodicity principle provides us with a set of conditions which must be satisfied in order for us to replace the ensemble averages of a signal with its time-average counterparts.

1.3.4 Ergodicity

As seen in the preceding example, the time-average process does not always lead to the ensemble average of a random signal. To examine this problem further, it is important to identify various ergodicity principles, which in simple words allow replacing ensemble averages with their time-average counterparts. In principle, a signal may be ergodic in mean, power, correlation, or distribution. This implies that one may obtain mean, power, correlation, or distribution of a random signal by merely observing sample paths of the random signal. More specifically, a signal is said to be ergodic in mean (or mean ergodic) if

$$E\{x[n]\} = m = \lim_{N\to\infty} \frac{1}{2N+1} \sum_{n=-N}^{N} x[n] \tag{1.103}$$

It becomes immediately obvious that for a mean ergodic process the ensemble average must be constant. Furthermore, if we consider the sum $y_N = \frac{1}{N}\sum_{n=1}^{N} x[n]$ as a random variable, then the above implies that, for mean ergodic signals, y_N must converge to m. Since y_N is a random variable with mean m (note that $E\{y_N\} = m$), the convergence implies that y_N must have a variance that goes to zero as $N \to \infty$. This form of convergence is known as "mean-square" convergence, where mean-square of error is expected to go to zero. That is,

$$\lim_{N\to\infty} E\left\{|y_N - m|^2\right\} \to 0 \tag{1.104}$$

which also implies that y_N must have a vanishing variance. Substituting for y_N leads to

$$\lim_{N\to\infty} E\left\{|y_N - m|^2\right\} = \lim_{N\to\infty} \frac{1}{(2N+1)^2} \sum_{n_2=-N}^{N} \sum_{n_1=-N}^{N} E\left\{x^*[n_1]\, x[n_2]\right\} - |m|^2 \tag{1.105}$$

For $\lim_{N\to\infty} E\left\{|y_N - m|^2\right\} \to 0,$ we must have

$$\lim_{N\to\infty} \frac{1}{(2N+1)^2} \sum_{n_2=-N}^{N} \sum_{n_1=-N}^{N} R_x[n_1, n_2] = |m|^2 \tag{1.106}$$

Therefore, one has to verify the above relationship in order to establish mean ergodicity. A *sufficient, but not necessary* condition for mean ergodicity, which leads one to an intuitive conclusion, is then

$$\lim_{|n_1-n_2|\to\infty} R_x[n_1, n_2] = |m|^2 = E^*\{x[n_1]\}\, E\{x[n_2]\}, \tag{1.107}$$

which simply implies that any two samples of a random signal should become uncorrelated as the separation in time of the two samples goes to infinity. This is a rather intuitive conclusion; that is, if we are to rely on time average to obtain statistical mean, then the signal must generate all possible sample paths of the signal over the course of large observation intervals. Hence, a "time" average will lead to an "ensemble" average if the observation interval is long enough to allow the observation of all the states of the process.

For the case where the mean is time-varying, but remains constant for long periods of time (one has to define "long" in this context) and the above condition is satisfied for each of those time intervals, one can obtain statistical mean for $N < \infty$. This process can be repeated for all intervals of time where the mean remains constant. The immediate implication of this observation is that one can obtain the statistical mean using time-domain analysis of a signal with time-varying mean if one can establish that the mean, although time-varying, is constant over long periods of time and the mean ergodicity condition is satisfied for all those time intervals.

For a WSS signal then (note that the first condition has already been satisfied, i.e., $E\{X_n\}$ is constant, so WSS implies that $R_x[n_1, n_2] = R_x[n_1 - n_2]$), the condition for ergodicity in mean reduces to

$$\lim_{N\to\infty} \frac{1}{2N+1} \sum_{n=-N}^{N} \left(1 - \frac{|n|}{2N+1}\right) R_x[n] = |m|^2 \tag{1.108}$$

Similarly, a sufficient condition is that $\lim_{n\to\infty} R_x[n] = |m|^2$. We can also go one step further and define a zero-mean process $x^{(e)}[n] = x[n] - m$. In order for $x[n]$ to be ergodic in mean,

$$\lim_{n\to\infty} R_{x^{(e)}}[n] = 0. \tag{1.109}$$

This means that a WSS is mean ergodic if samples of the signal separated by large time separations are uncorrelated.

Example 12 Let $x[n] = A\cos(2\pi nT) + z[n]$, where $z[n]$ is a zero mean, mean-ergodic random sequence. Determine whether $x[n]$ is mean-ergodic.

Solution: *We note that $E\{x[n]\} = A\cos(2\pi nT)$, which is a function of n (time). Hence, this process cannot be mean-ergodic. This can be verified further by considering*

$$\begin{aligned}
\lim_{N\to\infty} \frac{1}{2N+1} \sum_{n=-N}^{N} x[n] &= \lim_{N\to\infty} \frac{1}{2N+1} \sum_{n=-N}^{N} z[n] \\
&\quad + \lim_{N\to\infty} \frac{1}{2N+1} \sum_{n=-N}^{N} A\cos(2\pi nT) \\
&= \lim_{N\to\infty} \frac{1}{2N+1} \sum_{n=-N}^{N} z[n] \\
&= 0
\end{aligned} \tag{1.110}$$

The last step in the above equation is due to the fact that $z[n]$ is assumed to be a mean-ergodic process with zero mean.

Hence,

$$\lim_{N\to\infty} \frac{1}{2N+1} \sum_{n=-N}^{N} x[n] \neq E\{x[n]\} = A\cos(2\pi nT). \tag{1.111}$$

The above example is noteworthy in that it demonstrates that one of the most common signals used in many disciplines does not satisfy the condition for mean ergodicity. Fortunately, addition of a random element, which is routinely observed in pratice, to the above signal results in ergodicity not only in mean, but also in higher statistics. We consider this example next.

Example 13 Let $x[n] = A\cos(2\pi nT + \theta) + z[n]$, where $z[n]$ is a zero mean, mean-ergodic random sequence and θ is a random variabel uniformly distrubuted on $[0, 2\pi]$. Determine whether $x[n]$ is mean-ergodic.

Solution: *Note that* $E\{x[n]\} = AE\{\cos(2\pi nT + \theta)\} = 0$, *which is not a function of n (time). Furthermore*

$$\begin{aligned}
\lim_{N\to\infty} \frac{1}{2N+1} \sum_{n=-N}^{N} x[n] &= \lim_{N\to\infty} \frac{1}{2N+1} \sum_{n=-N}^{N} z[n] \\
&\quad + \lim_{N\to\infty} \frac{1}{2N+1} \sum_{n=-N}^{N} A\cos(2\pi nT + \theta) \\
&= \lim_{N\to\infty} \frac{1}{2N+1} \sum_{n=-N}^{N} z[n] \\
&= 0
\end{aligned} \tag{1.112}$$

Hence,

$$\lim_{N\to\infty} \frac{1}{2N+1} \sum_{n=-N}^{N} x[n] = E\{x[n]\} = 0 \tag{1.113}$$

which implies that the signal is ergodic in mean.

In the above example, the addition of θ made the signal ergodic in mean. This is not an artificial manipulation. In practice, sinusoidal signals generated by oscillators are subject to oscillator instabilities, which manifest itself in the form of a random phase (in fact, a time-varying phase). Hence, we are safe to assume that the above model is accurate, which make the signals of this type ergodic in mean.

Next, we need to examine ergodicity in other senses.

To establish ergodicity in any other sense, we can use the above principle. For instance, if one is interested in computing correlation, then we need to consider a new process

$$z_q[n] = x[n]\, x^*[n+q]. \tag{1.114}$$

The question now is whether the time average of $z_q[n]$ converges to the mean of $z_q[n]$ (the correlation function). That is,

$$E\left\{z_q[n]\right\} \stackrel{?}{=} \lim_{N\to\infty} \frac{1}{2N+1} \sum_{n=-N}^{N} z_q[n]. \tag{1.115}$$

Note that the special case of $q = 0$ corresponds to the computation of power from the observed samples. That is, if a process is WSS, then the average power is

$$\begin{aligned}
\text{Average Power} &= \lim_{N\to\infty} \frac{1}{2N+1} E\left\{\sum_{n=-N}^{N} |x[n]|^2\right\} \\
&= \lim_{N\to\infty} \frac{1}{2N+1} E\left\{\sum_{n=-N}^{N} R_x[0]\right\} && (1.116) \\
&= R_x[0] && (1.117)
\end{aligned}$$

Hence, the questions is

$$\lim_{N\to\infty} \frac{1}{2N+1} \sum_{n=-N}^{N} |x[n]|^2 \stackrel{?}{=} R_c[0] \tag{1.118}$$

Since ergodicity in power is a special case of the ergodicity in correlation, we limit ourselves to establishing ergodicity in correlation. We have then reduced the problem of establishing ergodicity in correlation for $x[n]$ to establishing ergodicity in mean for $z_q[n]$ for all q. As noted above, the question we are interested in answering is

$$R_x[q] = E\left\{z_q[n]\right\} \stackrel{?}{=} \lim_{N\to\infty} \frac{1}{2N+1} \sum_{n=-N}^{N} z_q[n] \tag{1.119}$$

Similar to the previous case, $E\left\{z_q[n]\right\}$ must be a constant (not a function of n) for the above to hold true. This immediately implies that $x[n]$ must be WSS if $z_q[n]$ has a constant mean. In that case, we have

$$\lim_{N\to\infty} \frac{1}{2N+1} E\left\{\sum_{n=-N}^{N} z_q[n]\right\} = \lim_{N\to\infty} \frac{1}{2N+1} \sum_{n=-N}^{N} E\left\{z_q[n]\right\}$$

$$= \lim_{N\to\infty} \frac{1}{2N+1} \sum_{n=-N}^{N} R_x[q]$$
$$= R_x[q]. \tag{1.120}$$

That is, $\lim_{N\to\infty} \frac{1}{2N+1} \sum_{n=-N}^{N} z_q[n]$ is a random variable whose mean is $R_x[q]$. Using the previous argument for mean ergodicity, for ergodicity in correlation (note that we are considering $z_q[n]$ as our process), we must have

$$\lim_{N\to\infty} \frac{1}{(2N+1)^2} \sum_{n_2=-N}^{N} \sum_{n_1=-N}^{N} R_{z_q^{(e)}}[n_1, n_2] = 0 \tag{1.121}$$

where

$$z_q^{(e)}[n] = z_q[n] - R_x[q] \tag{1.122}$$

with

$$R_{z_q^{(e)}}[n_1, n_2] = E\left\{z_q^{(e)}[n_1]\, z_q^{(e)*}[n_2]\right\}. \tag{1.123}$$

Unfortunately, one requires the 4th order statistics of $x[n]$ to establish ergodicity in correlation for $x[n]$. This is a rather disappointing result since such statistical knowledge may not be available in practice, and hence, one may not be able to assess the validity of the measured power and correlation which have been obtained using a time-domain analysis. As was argued earlier, a sufficient, but not necessary condition for ergodicity in correlation reduces to

$$\lim_{|n_1-n_2|\to\infty} R_{z_q^{(e)}}[n_1, n_2] = 0 \text{ for all } q \tag{1.124}$$

which implies that the samples of $z_q[n]$ taken at large separations in time must be uncorrelated. An interesting observation which can made is that, if $x[n]$ has a constant mean, then the condition $E\{z_q[n]\} = R_x[n]$ implies that $x[n]$ must be WSS for us to proceed. That is, for a constant mean process, one has to check to make sure that the process is WSS before attempting to establish the condition for ergodicity in correlation. By the same token, if a process is not WSS, then it cannot be ergodic in correlation.

Earlier, we alluded to the fact that there exists a class of processes where ergodicity can readily be established in not only mean, but also higher statistics. Gaussian processes are such processes. For Gaussian processes, the knowledge of the second order statistics is sufficient to establish higher statistical moments of the signal. To elaborate, if $x[n]$ is a zero-mean, complex Gaussian process, then

$$R_{z_q^{(e)}}[n_1, n_2] = E\left\{x[n_1]\, x^*[n_1+q]\, x^*[n_2]\, x[n_2+q]\right\} + |R_x[q]|^2$$
$$- R_x^*[q]\, E\left\{x[n_1]\, x^*[n_1+q]\right\}$$

$$
\begin{aligned}
&- R_x[q]\, E\left\{x^*[n_2]\, x[n_2+q]\right\} \\
&= |R_x[n_1-n_2]|^2 + |R_x[q]|^2 + |R_x[q]|^2 - 2\,|R_x[q]|^2 \\
&= |R_x[n_1-n_2]|^2 \qquad (1.125)
\end{aligned}
$$

where the zero-mean and $E\left\{z_q[n]\right\} = R_x[q]$ assumptions imply that $x[n]$ must be WSS We have also assumed that the random signal $x[n]$ is "proper," which implies that

$$E\{x[n_1]\, x[n_2]\} = E\left\{x^*[n_1]\, x^*[n_2]\right\} = 0$$

for all n_1 and n_2. Hence, one can claim ergodicity in correlation for a WSS Gaussian process if the following holds true:

$$\lim_{|n_1-n_2|\to\infty} R_x[n_1-n_2] = 0$$

or

$$\lim_{n\to\infty} R_x[n] = 0.$$

Note that this is the condition for mean-ergodicity (since $x[n]$ is zero-mean $\lim_{n\to\infty} R_x[n] = |m|^2 = 0$). Hence, a WSS, Gaussian process is ergodic in mean and correlation if one can establish that it is mean ergodic. Since the additive noise in electronic systems is typically a zero-mean, WSS Gaussian process, the above conclusion is quite encouraging. Note that the Gaussian assumption offered some significant simplification. In the event that the signal is not Gaussian, we need to establish whether $x[n_1]\, x^*[n_1+q]$ and $x^*[n_2]\, x[n_2+q]$ for $|n_1-n_2| \to \infty$ and all q are uncorrelated.

Example 14 Let $x[n]$ denote a zero-mean Gaussian sequence with correlation function $R_x[n] = \begin{cases} \frac{\sigma^2}{n^2} & n > 0 \\ 1+\sigma^2 & n = 0 \end{cases}$. Show that this process is ergodic in mean and correlation.

Solution: *Clearly,* $\lim_{n\to\infty} R_x[n] = m^2 = 0$, *which implies that samples separated by large separations are uncorrelated (and in fact independent for Gaussian signals). Since mean ergodic, WSS Gaussian processes are also ergodic in correlation,* $x[n]$ *is ergodic in mean, correlation, and, consequently, power.*

Provided that the above condition can easily be satisfied in many practical problems, as electronic noise is present in most systems (including biomedical signal processing devices), there is a natural tendency among practitioners in this field to approximate noise processes by a Gaussian process. As we will show in later chapters, unfortunately, for a large class of biomedical signals this assumption is not justifiable. Next, we tackle an important problem. That is, in the field of biomedical signal processing, we often collect biomedical signals with unknown a prior pdf. Since statistical decision making is a key part of signal processing, often we are required to obtain such pdfs using observation of the signal over a limited period of time. Then, a histogram is generated to estimate the desired pdf. At this point, we do not have any reason to believe whether the pdf formed in this manner converges to the true pdf in any sense as we increase the observation interval. To address this problem, we consider the concept of ergodicity in distribution.

An approach similar to the above approach may be used to insure ergodicity in distribution. In that case, a new random sequence is defined as follows:

$$I[n] = \begin{cases} 1 & x[n] < x \\ 0 & \text{otherwise} \end{cases} \tag{1.126}$$

Then, the question we are concerned about is whether $E\{I_n\}$ can be obtained using

$$\lim_{N\to\infty} \frac{1}{2N+1} \sum_{n_2=-N}^{N} I[n]. \tag{1.127}$$

Note that

$$E\{I[n]\} = F_x(x). \tag{1.128}$$

In practice, $I[n]$ may be computed by observing the outcome of an experiment or a biomedical signal for all x. Then, the question is whether

$$\frac{1}{2N+1} \sum_{n_2=-N}^{N} I_n \overset{?}{\rightarrow} F_x(x) \tag{1.129}$$

for all x as $N \rightarrow \infty$. As you can tell, this is the aim of anyone who is interested in estimating the pdf of an unknown process using sample path observations. Using the mean ergodicity condition and applying it to the problem at hand, a process is ergodic in distribution if we have

$$\lim_{N\to\infty} \frac{1}{(2N+1)^2} \sum_{n_2=-N}^{N} \sum_{n_1=-N}^{N} R_I[n_1, n_2] = F_x^2(x). \tag{1.130}$$

Again, a sufficient but not necessary condition for the above to hold is that

$$\lim_{|n_1-n_2|\to\infty} R_I[n_1, n_2] = F_x^2(x),$$

which implies that $I[n_1]$ and $I[n_2]$ must become " independent" when $|n_1 - n_2|$ is large. This is a far more difficult condition to satisfy. That is, we need to know a priori that observations separated by large distances of time are independent before attempting to obtain the pdf of an unknown process using the commonly use histogram method. In practice, we insure randomization by randomizing parts of the experiment that is under our control. This is a key aspect of biomedical signal processing when observation is a signal with unknown statistical characteristics.

An interesting observation which can be made here is that if samples of the signal taken at large time separations are independent (and therefore uncorrelated), then we can establish ergodicity in mean, correlation, and distribution. This is an important result which happens to agree with an intuitive assessment of the problem.

We also note that for WSS Gaussian processes, for which $\lim_{n\to\infty} R_x[n] \to |m|^2$, *one can establish that samples of the processes separated by large time separations are independent (note that uncorrelated samples in a Gaussian processes are also independent). Hence, a WSS, Gaussian random process where* $\lim_{n\to\infty} R_x[n] \to |m|^2$ *is ergodic in mean, correlation, and distribution.*

In conclusion, a sufficient condition for using experimental results to assess statistical measures of a signal is that any two samples of the signal must become independent as the separation in time of the two samples goes to infinity. This observation also underscores the importance of extending the observation interval (simulation length or the length of the experimental observation) well beyond the interval of time over which the random signal exhibits a non-negligible correlation.

So far, we have addressed the issue of power and energy using time-domain analysis. A question that can be raised here is whether we can consider a function which describes the distribution of signal power in the Fourier domain. Such a function is the power spectrum density (PSD), which will be discussed next.

1.3.5 Power Spectrum Density (PSD)

A useful tool in analyzing random signals is PSD. The obvious usefulness of this function resides in the fact that this function depicts the distribution of a random signal's power over a desired frequency spectrum. This is of particular importance to a system designer whose constraint is often bandwidth. In the biomedical field, the concentration of power in a given frequency band may have clinical significance. Hence, it is important to consider this function carefully.

For a WSS signal, the power spectrum density is defined as

$$S_x[\omega] - DTFT\{R_x[n]\} = \sum_{n=-\infty}^{\infty} R[n]\, e^{-jn\omega} \tag{1.131}$$

with its inverse

$$R[n] = \frac{1}{2\pi}\int_{-\pi}^{\pi} S_x[\omega]\, e^{j\omega n} d\omega. \tag{1.132}$$

where $\omega = \Omega T_s$ (T_s denotes sampling time). Similarly, in continuous-time domain,

$$S_x(\Omega) = FT\{R_x(\tau)\} = \int_{-\infty}^{\infty} R_x(\tau)\, e^{-j\Omega\tau} d\tau \tag{1.133}$$

where $FT\{\}$ is the Fourier transform of the enclosed. Given the Fourier transform relationship,

$$R_x(\tau) = \frac{1}{2\pi}\int_{-\infty}^{\infty} S_x(\Omega)\, e^{j\Omega\tau} d\Omega \tag{1.134}$$

Note that power P can be computed as $P = R_x(0) = \frac{1}{2\pi}\int_{-\infty}^{\infty} S_x(\Omega)\, d\Omega$, or $P = R[0] = \frac{1}{2\pi}\int_{-\pi}^{\pi} S_x[\omega]\, d\omega$. That is, the area underneath of the $S_x[\omega]$ or $S_x(\Omega)$ curves is the power of the signal, hence the name power spectrum density.

At this point, it seems that we are only concerned with WSS signals. However, it will be shown shortly that PSD may also be defined for cyclostationary or other signals as well using a slightly modified definition. Before discussing this topic further, let us consider some examples of PSD.

Example 15 Let $x[n] = A\sin[n\Omega_0 T_s + \theta] + z[n]$ where $z[n]$ is a zero-mean, WSS random sequence with auto-correlation $R_z[n] = \exp[-\alpha|n|]$. θ is uniformly distributed over $[-\pi, \pi]$. Find the PSD of $x[n]$.

Solution: *We can easily show that*

$$\begin{aligned} R_x[n_1, n_2] = {} & E\{A\sin[n_1\Omega_0 T_s + \theta]\, A\sin[n_2\Omega_0 T_s + \theta]\} \\ & + R_z[n_1, n_2]. \end{aligned} \tag{1.135}$$

Note that we have arrived at this result since it is assumed that $z[n]$ *is independent of* $A\sin[n\Omega_0 T_s + \theta]$ *and* $z[n]$ *is zero mean. The above leads to*

$$\begin{aligned} R_x[n_1, n_2] &= \frac{A^2}{2}\cos((n_1 - n_2)\,\Omega_0 T_s) + R_z[n_1 - n_2] \\ &= R_x[n_1 - n_2] \end{aligned} \tag{1.136}$$

So,

$$\begin{aligned} R_x[n] &= \frac{A^2}{2}\cos(n\Omega_0 T_s) + R_z[n] \\ &= \frac{A^2}{2}\cos(n\Omega_0 T_s) + \exp[-\alpha|n|]. \end{aligned} \tag{1.137}$$

We then have

$$\begin{aligned} S_x[\omega] &= DTFT\{R_x[n]\} \\ &= \frac{A^2\pi}{2}\sum_{m=-\infty}^{\infty}\{\delta(\omega - \Omega_0 T_s - 2m\pi) + \delta(\omega + \Omega_0 T_s - 2m\pi)\} \\ &\quad + \frac{1 - e^{-2\alpha}}{1 + e^{-2\alpha} - 2e^{-\alpha}\cos(\omega)}, \end{aligned} \tag{1.138}$$

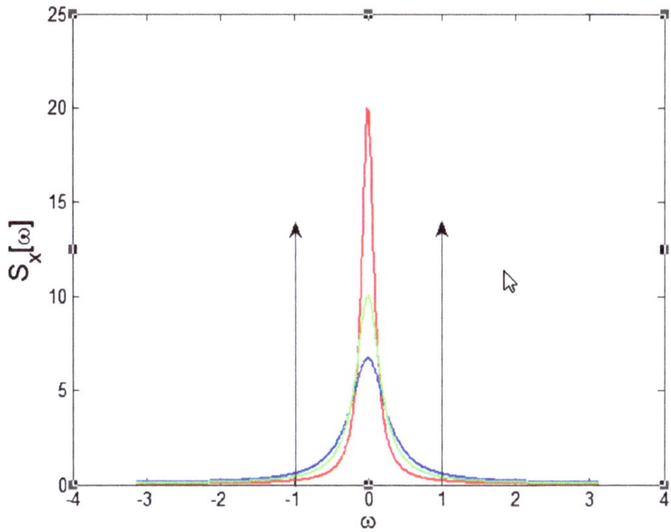

Fig. 1.27 PSD of $A \sin [n\Omega_0 T_s + \theta] + z[n]$ with $\alpha = 0.3$ (*blue*), 0.2 (*green*), and 0.1 (*red*). $\Omega T_s = 1$

where $\delta(\omega)$ is the dirac delta function and we have used the identity $\sum_{n=-\infty}^{\infty} e^{jn\omega} = 2\pi \sum_{m=-\infty}^{\infty} \delta(\omega - 2\pi m)$. Finally, the last term in the above is the DTFT of $\exp[-\alpha |n|]$ *(show this). Clearly, one sees the periodicity with respect to ω (as expected) with a period of 2π. This PSD is shown in Fig. 1.27, which is limited to the range$[-\pi, \pi]$. This figure repeats for all other intervals of the same duration in ω. It is assumed that $\Omega_0 T_s = 1$. Note that, as $\alpha \to 0$, the PSD of additive noise $z[n]$ approaches the characteristics of a highly correlated noise (such as a slow-varying dc level), which results in a dirac-delta-like PSD centered at 0. Note that the PSD reveals the spectral purity of the "signal" part of the above waveform as two dirac-delta functions centered at ± 1 rad.*

A key assumption of the above formulation is that the signal of interest is WSS. As we observed in previous sections, a large class of signals in practice are classified as cyclostationary. Furthermore, biomedical signals are often exhibit a time-varying characteristics, making them non-WSS. In that event, a more general definition of PSD must be used. In general, PSD is given by

$$S_x[\omega] = \lim_{N\to\infty} \frac{1}{2N+1} E\left\{ |x_N[\omega]|^2 \right\} \tag{1.139}$$

where

$$x_N[\omega] = \sum_{n=-N}^{N} x[n] e^{-j\omega n}. \tag{1.140}$$

Equivalently,

$$S_x(\Omega) = \lim_{T_{sa}\to\infty} \frac{1}{T_{sa}} E\left\{\left|x_{T_{sa}}(\Omega)\right|^2\right\} \tag{1.141}$$

where

$$x_{T_{sa}}(\Omega) = \int_{-\frac{T_{sa}}{2}}^{\frac{T_{sa}}{2}} x(t)\, e^{-j\Omega t} dt. \tag{1.142}$$

For cyclostationary signals, the above reduces to (discrete case)

$$S_x[\omega] = \lim_{N\to\infty} \frac{1}{2N+1} \sum_{n_1=-N}^{N} \sum_{n_2=-N}^{N} E\left\{x^*[n_1]\, x[n_1]\right\} e^{-j\omega(n_2-n_1)} \tag{1.143}$$

which is

$$S_x[\omega] = \lim_{N\to\infty} \frac{1}{2N+1} \sum_{n_1=-N}^{N} \sum_{n_2=-N}^{N} R_x[n_1, n_2]\, e^{-j\omega(n_2-n_1)}. \tag{1.144}$$

Now, without loss of generality, let $N = MN_p$ for some M. Then,

$$S_x[\omega] = \lim_{N\to\infty} \frac{1}{2N+1} \sum_{n_1=-MN_p}^{MN_p} \sum_{n_2=-MN_p}^{MN_p} R_x[n_1, n_2]\, e^{-j\omega(n_2-n_1)}. \tag{1.145}$$

Let $n_2 - n_1 = n$ and $n_2 + n_1 = m$, then $n_1 = \frac{m-n}{2}$ and $n_2 = \frac{m+n}{2}$. We then have

$$S_x[\omega] = \lim_{N\to\infty} \frac{1}{2N+1} \sum_{n=-2MN_p}^{2MN_p} \sum_{m=-2MN_p+|n|}^{2MN_p-|n|} R_x\left[\frac{m-n}{2}, \frac{m+n}{2}\right] e^{-j\omega n}. \tag{1.146}$$

Using the identity $\frac{m+n}{2} = \frac{m-n}{2} + n$ (note that both $\frac{m+n}{2}$ and $\frac{m+n}{2}$ are integers by definition)

$$S_x[\omega] = \lim_{N\to\infty} \frac{1}{2N+1} \sum_{n=-MN_p}^{MN_p} \sum_{m=-2MN_p+|n|}^{2MN_p-|n|} R_x\left[\frac{m-n}{2}, \frac{m-n}{2} + n\right] e^{-j\omega n}. \tag{1.147}$$

Let us make one more change of variable. Namely, let $\frac{m-n}{2} = m'$ (once again, by definition, $\frac{m-n}{2}$ is an integer), which results in

$$S_x[\omega] = \lim_{N\to\infty} \frac{1}{2N+1} \sum_{n=-2MN_p}^{2MN_p} \sum_{m'=-MN_p-\frac{n-|n|}{2}}^{MN_p-\frac{n+|n|}{2}} R_x\left[m', m'+n\right] e^{-j\omega n}. \tag{1.148}$$

The above can be expanded as

$$S_x[\omega] = \lim_{N\to\infty} \frac{1}{2N+1}\{ \sum_{n=-2MN_p}^{0} \sum_{m'=-MN_p-n}^{MN_p} R_x\left[m', m'+n\right] e^{-j\omega n} + \sum_{n=1}^{2MN_p} \sum_{m'=-MN_p}^{MN_p-n} R\left[m', m'+n\right] e^{-j\omega n}\}. \quad (1.149)$$

Assuming that the signal is cyclostationary, $R\left[m', m'+n\right]$ is a periodic function of m' with a period of N_p for cyclostationary signals. Realizing that $N \to \infty$ is equivalent to $M \to \infty$ and that

$$\sum_{m'=-MN_p-n}^{MN_p} R_x\left[m', m'+n\right] \to 2MN_p\langle R_x\left[m', m'+n\right]\rangle_{m'} \text{ as } M \to \infty$$

and

$$\sum_{m'=-MN_p}^{MN_p-n} R_x\left[m', m'+n\right] \to 2MN_p\langle R_x\left[m', m'+n\right]\rangle_{m'} \text{as } M \to \infty,$$

where $\langle R_x\left[m', m'+n\right]\rangle_{m'} = \frac{1}{N_p}\sum_{m'=0}^{N_p-1} R_x\left[m', m' + n\right]$, we have

$$S_x[\omega] = \lim_{M\to\infty} \sum_{n=-2MN_p}^{2MN_p} \frac{2MN_p}{2MN_p+1}\langle R_x\left[m', m'+n\right]\rangle_{m'} e^{-j\omega n}, \quad (1.150)$$

which for large M reduces to

$$S_x[\omega] = \lim_{M\to\infty} \sum_{n=-2MN_p}^{2MN_p} \langle R_x[m, m+n]\rangle_m e^{-j\omega n} \quad (1.151)$$

or

$$S_x[\omega] = \sum_{n=-\infty}^{\infty} \langle R_x[m, m+n]\rangle_m e^{-j\omega n}. \quad (1.152)$$

Hence,

$$S_x[\omega] = DTFT\{\langle R_x[m, m+n]\rangle_m\} = \sum_{n=-\infty}^{\infty} \langle R_x[m, m+n]\rangle_m e^{-j\omega n} \quad (1.153)$$

where $\langle R_x[m, m+n]\rangle_m = \frac{1}{N_p}\sum_{m=0}^{N_p-1} R_x[m, m+n]$. This is a key result which provides the PSD of cyclostationary random sequences. Simply stated, the PSD of a cyclostationary signal is the Fourier transform of the average of the correlation function (which is a periodic function itself).

Unfortunately, if the signal does not exhibit any form of stationarity, as is the case for many biomedical signals, we need to go back to

$$\boxed{S_x[\omega] = \lim_{N\to\infty} \tfrac{1}{2N+1} \textstyle\sum_{n_1=-N}^{N} \sum_{n_2=-N}^{N} R_x[n_1, n_2]\, e^{-j\omega(n_2-n_1)}} \quad (1.154)$$

in order to obtain the PSD of signals with finite power.

For the continuous time case,

$$S_x(\Omega) = \lim_{M\to\infty} \frac{1}{MT_p} \int_{-\frac{T_{sa}}{2}}^{\frac{T_{sa}}{2}} \int_{-\frac{T_{sa}}{2}}^{\frac{T_{sa}}{2}} E\left\{x^*(t_1)\, x(t_2)\right\} e^{-j\Omega(t_2-t_1)} dt_1 dt_2 \quad (1.155)$$

which leads to this general definition (can be used for non-stationary signals with finite power)

$$\boxed{S_x(\Omega) = \lim_{M\to\infty} \tfrac{1}{MT_p} \textstyle\int_{-\frac{T_{sa}}{2}}^{\frac{T_{sa}}{2}} \int_{-\frac{T_{sa}}{2}}^{\frac{T_{sa}}{2}} R_x(t_1, t_2)\, e^{-j\Omega(t_2-t_1)} dt_1 dt_2.} \quad (1.156)$$

Assuming that $T_{sa} = MT_p$,

$$S_x(\Omega) = \lim_{M\to\infty} \frac{1}{MT_p} \int_{-\frac{MT_p}{2}}^{\frac{MT_p}{2}} \int_{-\frac{MT_p}{2}}^{\frac{MT_p}{2}} R_x(t_1, t_2)\, e^{-j\Omega(t_2-t_1)} dt_1 dt_2. \quad (1.157)$$

Letting $t_2 - t_1 = \tau$ and $t_2 + t_1 = s$ (Jacobian for this transformation is 2),

$$S_x(\Omega) = \lim_{M\to\infty} \frac{1}{2MT_p} \int_{-MT_p}^{MT_p} \int_{-MT_p+|\tau|}^{MT_p-|\tau|} R_x\left(\frac{s-\tau}{2}, \frac{s-\tau}{2} + \tau\right) e^{-j\Omega\tau} ds\, d\tau \quad (1.158)$$

Let $\frac{s-\tau}{2} = s'$, then

$$S_x(\Omega) = \lim_{M\to\infty} \frac{1}{2MT_p} \int_{-MT_p}^{MT_p} \int_{\frac{-MT_p+|\tau|-\tau}{2}}^{\frac{MT_p-|\tau|-\tau}{2}} R_x\left(s', s'+\tau\right) e^{-j\Omega\tau} ds'\, d\tau \quad (1.159)$$

$$= \lim_{M\to\infty} \frac{1}{2MT_p} \int_0^{MT_p} \int_{\frac{-MT_p}{2}}^{\frac{MT_p}{2}-\tau} R_x\left(s', s'+\tau\right) e^{-j\Omega\tau} ds' d\tau$$

$$+ \int_{-MT_p}^{0} \int_{\frac{-MT_p}{2}-\tau}^{\frac{MT_p}{2}} R_X\left(s', s'+\tau\right) e^{-j\Omega\tau} ds' d\tau$$

$$= \lim_{M\to\infty} \frac{1}{2MT_p} \int_0^{MT_p} \left[\int_{\frac{-MT_p}{2}}^{\frac{MT_p}{2}-\tau} R_x\left(s', s'+\tau\right) ds' \right] e^{-j\Omega\tau} d\tau$$

$$+ \int_{-MT_p}^{0} \left[\int_{\frac{-MT_p}{2}-\tau}^{\frac{MT_p}{2}} R_x\left(s', s'+\tau\right) ds' \right] e^{-j\Omega\tau} d\tau. \tag{1.160}$$

If $x(t)$ is cyclostationary, $R_x\left(s', s'+\tau\right)$ is a periodic function of s' with period T_p. Hence, the terms inside the brackets are given by

$$\left[\int_{\frac{-MT_p}{2}}^{\frac{MT_p}{2}-\tau} R_x\left(s', s'+\tau\right) ds \right] \to MT_p \langle R_x\left(s', s'+\tau\right)\rangle_{s'} \text{as } M \to \infty$$

$$\left[\int_{\frac{-MT_p}{2}-\tau}^{\frac{MT_p}{2}} R_x\left(s', s'+\tau\right) ds \right] \to MT_p \langle R_x\left(s', s'+\tau\right)\rangle_{s'} \text{ as } M \to \infty$$

where $\langle R_x(t, t+\tau)\rangle_t = \frac{1}{T_p} \int_{-\frac{T_p}{2}}^{\frac{T_p}{2}} R_x(t, t+\tau)\, dt$. Hence,

$$S_x(\Omega) = \lim_{M\to\infty} \int_0^{MT_p} \langle R_x\left(s', s'+\tau\right)\rangle_{s'} e^{-j\Omega\tau} d\tau$$

$$+ \int_{-MT_p}^{0} \langle R_x\left(s', s'+\tau\right)\rangle_{s'} e^{-j\Omega\tau} d\tau, \tag{1.161}$$

which reduces to

$$\boxed{S_x(\Omega) = \int_{-\infty}^{\infty} \langle R_x(t, t+\tau)\rangle_t e^{-j\Omega\tau} d\tau = FT\left\{\langle R_x(t, t+\tau)\rangle_t\right\}.} \tag{1.162}$$

This is a key result that provides the PSD of a cyclostationary random process.

Example 16 A popular class of signals in signal processing world is the linearly modulated signal, commonly expressed as

$$x(t) = \sum_{q=-\infty}^{\infty} I_q p(t - qT), \tag{1.163}$$

where I_q is an information-bearing WSS random sequence with known statistics and $p(t)$ denotes the signaling "pulse" shape, which is defined over $[0, T]$ (full-response signaling) or longer (partial-response signaling). Let T denote the symbol duration. Find the PSD of this signaling scheme.

Solution: *In addition to signal processing field, this signal can be used in tissue optics to measure the response of tissue to laser-generated optical pulses in order to study the behavior of the dispersive tissue medium. Hence, an understanding of this signal's spectral behavior is of some interest. To arrive at the desired result, let us start with the autocorrelation of the signal, which is given by* (1.162). *According to this equation,*

$$\begin{aligned} R_x(t, t+\tau) &= E\left\{x^*(t)\, x(t+\tau)\right\} \\ &= E\left\{\sum_{q_1=-\infty}^{\infty} \sum_{q_2=-\infty}^{\infty} I_{q_1}^* I_{q_2} p^*(t - q_1 T)\, p(t+\tau - q_2 T)\right\} \\ &= \sum_{q_1=-\infty}^{\infty} \sum_{q_2=-\infty}^{\infty} R_I[q_1, q_2]\, p^*(t - q_1 T) \\ &\quad \times p(t+\tau - q_2 T) \end{aligned} \tag{1.164}$$

Now, realizing that the sequence I_n is WSS,

$$R_x(t, t+\tau) = \sum_{q_1=-\infty}^{\infty} \sum_{q_2=-\infty}^{\infty} R_I[q_2 - q_1]\, p^*(t - q_1 T)\, p(t+\tau - q_2 T) \tag{1.165}$$

This assumption, regardless of the pulse shape, leads to cyclostationarity of $x(t)$. Note that

$$\begin{aligned} R_x(t+mT, t+\tau+mT) &= \sum_{q_1=-\infty}^{\infty} \sum_{q_2=-\infty}^{\infty} R_I[q_2 - q_1] \\ &\quad \times p^*(t + mT_s - q_1 T)\, p(t + mT_s + \tau - q_2 T) \end{aligned} \tag{1.166}$$

Making the change of variables $q_1' = q_1 - mT$ and $q_2' = q_2 - mT$ lead to

$$\begin{aligned} R_x(t+mT, t+\tau+mT) &= \sum_{q_1'=-\infty}^{\infty} \sum_{q_2'=-\infty}^{\infty} R_I\left[q_2' - q_1'\right] \\ &\quad \times p^*\left(t - q_1' T\right) p\left(t+\tau - q_2' T\right) = R_x(t, t+\tau). \end{aligned} \tag{1.167}$$

Furthermore,

$$E\{x(t)\} = m_x(t) = \bar{I} \sum_{q=-\infty}^{\infty} p(t - qT) \tag{1.168}$$

where $E\left\{I_q\right\} = \bar{I}$ *(*I_q *is WSS, and hence, its mean must be constant). The reader can easily verify that*

$$m_x(t) = m_x(t + nT)$$

for all integer n*. Since* $m_x(t)$ *and* $R_x(t, t+\tau)$ *are periodic in* t *with a period of* T*,* $x(t)$ *is cyclostationary. Therefore, to compute its PSD, we need to obtain*

$$\langle R_x(t, t+\tau)\rangle_t = \sum_{q_1=-\infty}^{\infty} \sum_{q_2=-\infty}^{\infty} R_I[q_2 - q_1] \times \langle p^*(t - q_1 T)\, p(t + \tau - q_2 T)\rangle_t \qquad (1.169)$$

$$\langle R_x(t, t+\tau)\rangle_t = \lim_{T_{sa}\to\infty} \frac{1}{T_{sa}} \sum_{q_1=-\infty}^{\infty} \sum_{q_2=-\infty}^{\infty} R_I[q_2 - q_1] \times \int_{-\frac{T_{sa}}{2}}^{\frac{T_{sa}}{2}} \int_{-\infty}^{\infty} \int_{-\infty}^{\infty} P^*(\Omega_1)\, P(\Omega_2)\, e^{-j\Omega_1(t - q_1 T)} e^{j\Omega_2(t+\tau-q_2 T)} \times d\Omega_1 d\Omega_2 dt \qquad (1.170)$$

Taking the time average,

$$\langle R_x(t, t+\tau)\rangle_t = \lim_{T_{sa}\to\infty} \frac{1}{T_{sa}} \sum_{q_1=-\infty}^{\infty} \sum_{q_2=-\infty}^{\infty} R_I[q_2 - q_1] \times \int_{-\infty}^{\infty} \int_{-\infty}^{\infty} P^*(\Omega_1)\, P(\Omega_2) \int_{-\frac{T_{sa}}{2}}^{\frac{T_{sa}}{2}} e^{-j(\Omega_1 - \Omega_2)t} dt \times e^{j\Omega_1 q_1 T} e^{j\Omega_2(\tau - q_2 T)} d\Omega_1 d\Omega_2. \qquad (1.171)$$

Realizing that $\lim_{T_{sa}\to\infty} \int_{-\frac{T_{sa}}{2}}^{\frac{T_{sa}}{2}} e^{-j(\omega_1 - \omega_2)t} dt = \delta(\omega_1 - \omega_2)$*, the above reduces to*

$$\langle R_x(t, t+\tau)\rangle_t = \lim_{T_{sa}\to\infty} \frac{1}{T_{su}} \sum_{q_1=-\infty}^{\infty} \sum_{q_2=-\infty}^{\infty} R_I[q_2 - q_1] \times \int_{-\infty}^{\infty} |P(\Omega)|^2\, e^{j\Omega(q_1 - q_2)T} e^{j\Omega\tau} d\Omega \qquad (1.172)$$

Assuming that $T_{sa} = 2LT$ *and making the change of variable* $q = q_2 - q_1$ *(note that the summation is over* $2L$ *symbols, and one of the summations after the change of variable results in* $2L$ *identical terms),*

$$\langle R_x(t, t+\tau)\rangle_t = \lim_{T_{sa}\to\infty} \frac{1}{T_{sa}} 2L \sum_{q=-\infty}^{\infty} R_I[q] \\ \times \int_{-\infty}^{\infty} |P(\Omega)|^2 e^{j\Omega(q_1-q_2)T} e^{j\Omega\tau} d\Omega, \quad (1.173)$$

which reduces to

$$\langle R_x(t, t+\tau)\rangle_t = \frac{1}{T}\int_{-\infty}^{\infty} |P(\Omega)|^2 \sum_{q=-\infty}^{\infty} R_I[q]\, e^{j\Omega(q_1-q_2)T} e^{j\Omega\tau} d\Omega \quad (1.174)$$

Since $\sum_{q=-\infty}^{\infty} R_I[q]\, e^{j\Omega(q_1-q_2)T} = S_I[\omega]\,|_{\omega=\Omega T}$, *we arrive at*

$$\langle R_x(t, t+\tau)\rangle_t = \int_{-\infty}^{\infty} \frac{1}{T} |P(\Omega)|^2 S_I[\Omega T]\, e^{j\Omega\tau} d\Omega \\ = FT^{-1}\left\{\frac{1}{T}|P(\Omega)|^2 S_I[\Omega T]\right\} \quad (1.175)$$

Hence, the Fourier transform of $\langle R_x(t, t+\tau)\rangle_t$, *which is the PSD of* $x(t)$, *is given by*

$$\boxed{S_x(\Omega) = FT\left\{FT^{-1}\left\{\tfrac{1}{T}|P(\Omega)|^2 S_I[\Omega T]\right\}\right\} = \tfrac{1}{T} S_I[\Omega T]\,|P(\Omega)|^2}. \quad (1.176)$$

Example 17 In the previous example, find the PSD of $x(t) = \sum_{q=-\infty}^{\infty} I_q\, p(t-qT)$ when it is samples at the rate of $\frac{1}{T_s} = \frac{\mu}{T}$.

Solution: *In the previous example, it was shown that* $x(t)$ *is a cyclostationary signal. Now, consider that* $x(t)$ *is sampled* μ *times per* T *seconds. Hence, sampling interval* T_s *is given by* $T_s = \frac{T}{\mu}$. *We now have*

$$x[n] = x(nT_s) = \sum_{q=-\infty}^{\infty} I_q\, p(nT_s - q\mu T_s) \\ = \sum_{q=-\infty}^{\infty} I_q\, p[n-\mu q] \quad (1.177)$$

We first find the correlation function of $x[n]$, *which is given by*

$$R_x[m, m+n] = E\left\{x^*[m]\, x[m+n]\right\} \\ = \lim_{N\to\infty} \sum_{q_2=-N}^{N} \sum_{q_1=-N}^{N} E\left\{I_{q_1}^* I_{q_2}\right\} p^*[m-\mu q_1] \\ \times p[m+n-\mu q_2] \quad (1.178)$$

where $P[n] = P(nT_s)$. *Assuming that* $E\left\{I_{q_1}^* I_{q_2}\right\} = R_I[q_2 - q_1]$ *(a WSS sequence), (let* $q = q_2 - q_1$ and $q_s = q_2 + q_1$)

$$R_x[m, m+n] = \lim_{N\to\infty} \sum_{q=-N}^{N} \sum_{q_s=-2N+|q|}^{2N-|q|} \times R_I[q]\, p^*\left[m - \mu\frac{q_s - q}{2}\right] p\left[m + n - \mu\frac{q_s + q}{2}\right] \tag{1.179}$$

Averaging the correlation function over m, expressing $p[n]$ *in terms of the its inverse DFT (IDFT), and replacing* $R_I[q]$ *with* $\frac{1}{2\pi}\int_{-\pi}^{\pi} S_x[\omega_1]\, e^{jq\omega_1} d\omega_1$*, where* $S_x[\omega_1]$ *denotes the PSD of the WSS sequence* I_n*, we arrive at*

$$\begin{aligned}
&\langle R_x[m, m+n]\rangle_m \\
&= \frac{1}{(2\pi)^3} \lim_{N\to\infty} \sum_{q=-N}^{N} \sum_{q_s=-2N+|q|}^{2N-|q|} \int_{-\pi}^{\pi}\int_{-\pi}^{\pi}\int_{-\pi}^{\pi} S_I[\omega_1]\, P^*[\omega_2] \\
&\times P[\omega_3]\, \langle e^{jq\omega_1 - j\left(m - \mu\frac{q_s-q}{2}\right)\omega_2 + j\left(m+n-\mu\frac{q_s+q}{2}\right)\omega_3}\rangle_m \\
&\times d\omega_1 d\omega_2 d\omega_3,
\end{aligned} \tag{1.180}$$

which can be re-written as

$$\begin{aligned}
&\langle R_x[m, m+n]\rangle_m \\
&= \frac{1}{(2\pi)^3} \lim_{N\to\infty} \sum_{q=-N}^{N} \int_{-\pi}^{\pi}\int_{-\pi}^{\pi}\int_{-\pi}^{\pi} S_I[\omega_1]\, P^*[\omega_2]\, P[\omega_3] \\
&\times \langle e^{-jm(\omega_2-\omega_3)}\rangle_m e^{jq\omega_1 + jn\omega_3} e^{-j\mu\frac{q}{2}(\omega_2+\omega_3)} \\
&\times \sum_{q_s=-2N+|q|}^{2N-|q|} e^{j\mu\frac{q_s}{2}(\omega_2-\omega_3)} d\omega_1 d\omega_2 d\omega_3.
\end{aligned} \tag{1.181}$$

Realizing that $\lim_{N\to\infty}\sum_{q_s'=-N+\frac{|q|}{2}}^{N-\frac{|q|}{2}} e^{j\mu q_s'(\omega_2-\omega_3)} = 2\pi\sum_{l=-\infty}^{\infty}\delta(\mu(\omega_2 - \omega_3)$ $- 2\pi l)$ *(*$\delta(\omega)$ *is the dirac-delta function), which is* $\frac{2\pi}{\mu}\delta(\omega_2 - \omega_3)$ *for the range* $|\omega_2 - \omega_3| \le \pi$*), the above reduces to*

$$\begin{aligned}
&\langle R_x[m, m+n]\rangle_m \\
&= \frac{1}{(2\pi)^2\mu} \lim_{N\to\infty} \sum_{q=-N}^{N} \int_{-\pi}^{\pi}\int_{-\pi}^{\pi}\int_{-\pi}^{\pi} S_I[\omega_1]\, P^*[\omega_2]\, P[\omega_3]
\end{aligned}$$

$$\times \left\langle e^{-jm(\omega_2-\omega_3)} \right\rangle_m e^{jq\omega_1+jn\omega_3} e^{-j\mu\frac{q}{2}(\omega_2+\omega_3)} \delta\left(\omega_2-\omega_3\right) d\omega_1 d\omega_2 d\omega_3$$
$$= \frac{1}{(2\pi)^2\mu} \sum_{q=-\infty}^{\infty} \int_{-\pi}^{\pi}\int_{-\pi}^{\pi} S_I\left[\omega_1\right] \left|P\left[\omega_2\right]\right|^2$$
$$\times e^{jq\omega_1+jn\omega_2-j\mu q\omega_2} d\omega_1 d\omega_2. \quad (1.182)$$

Summing over q,

$$\left\langle R_x\left[m, m+n\right]\right\rangle_m$$
$$= \frac{1}{(2\pi)^2\mu} \int_{-\pi}^{\pi}\int_{-\pi}^{\pi} S_I\left[\omega_1\right] \left|P\left[\omega_2\right]\right|^2 e^{jn\omega_2}$$
$$\times \sum_{q=-\infty}^{\infty} e^{jq(\omega_1-\mu\omega_2)} d\omega_1 d\omega_2$$

Once again, realizing that $\sum_{q=-\infty}^{\infty} e^{jq(\omega_1-\mu\omega_2)} = 2\pi \sum_{l=-\infty}^{\infty} \delta\left(\omega_1-\mu\omega_2-2\pi l\right)$, we have

$$\left\langle R_x\left[m, m+n\right]\right\rangle_m$$
$$= \frac{1}{2\pi\mu} \sum_{l=-\infty}^{\infty} \int_{-\pi}^{\pi}\int_{-\pi}^{\pi} S_I\left[\omega_1\right] \left|P\left[\omega_2\right]\right|^2 e^{jn\omega_2}$$
$$\times \delta\left(\omega_1-\mu\omega_2-2\pi l\right) d\omega_1 d\omega_2. \quad (1.183)$$

Finally, we have

$$\left\langle R_x\left[m, m+n\right]\right\rangle_m$$
$$= \frac{1}{2\pi\mu} \int_{-\pi}^{\pi} \sum_{l=-\infty}^{\infty} S_I\left[\mu\omega_2-2\pi l\right] \left|P\left[\omega_2\right]\right|^2 e^{jn\omega_2} d\omega_2;$$
$$= \frac{1}{2\pi} \int_{-\pi}^{\pi} S_x\left[\omega\right] e^{jn\omega} d\omega; \quad (1.184)$$

The above implies that the PSD for the signaling examined in this example is given by

$$S_x\left[\omega\right] = \frac{1}{\mu} S_I\left[\mu\omega\right] \left|P\left[\omega\right]\right|^2; \ |\omega| \le \pi \quad (1.185)$$

which states that the PSD of a linearly modulated signal is the product of the PSD of random sequence I_n *and the absolute value squared DTFT of the pulse shape. Obviously, if one does not consider oversampling (i.e., when one obtains only one sample per symbol duration or* $\mu = 1$*), the above reduces to the more familiar expression*

$$\boxed{S_x\left[\omega\right] = S_I\left[\omega\right] \left|P\left[\omega\right]\right|^2; \ |\omega| \le \pi} \quad (1.186)$$

Note that, in the continuous domain, the PSD of (1.163) *is given by (see* (1.176)*)*

$$S_x(\Omega) = \frac{1}{T} S_I[\Omega T]\,|P(\Omega)|^2 \tag{1.187}$$

Now, realizing that $P[\omega] = \frac{1}{T_s} P\left(\frac{\omega}{T_s}\right)$ *(verify this) and noting that the PSD of the sampled version of a signal is given in terms of the PSD of its continuous-time domain counterpart as* $S_x[\omega] = \frac{1}{T_s} S_x\left(\frac{\omega}{T_s}\right)$*, we arrive at*

$$\begin{aligned} S_x[\omega] &= \frac{1}{T_s} S_x\left(\frac{\omega}{T_s}\right) = \frac{1}{T_s T} S_I\left[\frac{\omega}{T_s} T\right] \left|P\left(\frac{\omega}{T_s}\right)\right|^2 \\ &= \frac{1}{\mu T_s^2} S_I[\mu\omega]\,|P[\omega]|^2\, T_s^2 \\ &= \frac{1}{\mu} S_I[\mu\omega]\,|P[\omega]|^2, \end{aligned} \tag{1.188}$$

which is identical to (1.185).

The consequence the above examples is profound. To elaborate, if the intent is to study the behavior of a dispersive medium (such as tissue) in response to electrical or optical pulses, the above results suggest that the PSD of the signal is impacted by the pulse shape (subject to dispersion) and the modulating sequence in an independent manner. Hence, by manipulating the spectral characteristics of the sequence, one can assess the impact of environment on the pulse shape. This may also be of value from another perspective; that is, if the intend is to send information to an implanted device within a biological entity, then the dispersive nature of the biological environment must be compensated for in order to successfully communicate with the device. The compensation can be achieved by designing the PSD of the modulating sequence to achieve the desired spectral shaping, and hence, equalize the impact of environment.

1.3.6 Signal Space Representation

1.3.6.1 Karhunen-Loeve (K-L) Representation

The definition of signals presented in this chapter encompasses a broad range of signals. However, for digital signal processing applications focusing on biomedical signals, the signals of interest are power limited. That is, we need to concern ourselves with signals which yield a finite peak as well as average power levels. Given this restriction, the signal of interest lend themselves to orthogonal function expansion. More precisely, we assume that $x(t)$ can be expressed in terms of a number of known basis functions $\phi^{(m)}(t)$; $m = 0, 1, \ldots, \infty$. Since we treat all signals as random with deterministic signals as a special case of random signals, we assume that $x(t)$ has the correlation function $R_x(t_1, t_2)$. Note that WSS condition is not assumed here. Without proof, we state *Mercer's theorem* (see the reference chapter for further information), which states that one can express $R_x(t_1, t_2)$ as

$$R_x(t_1, t_2) = E\left\{x(t_1)\, x^*(t_2)\right\} = \sum_{m=0}^{\infty} \lambda^{(m)} \phi^{(m)}(t_1)\, \phi^{(m)*}(t_2) \tag{1.189}$$

where $\lambda^{(m)}$ and $\phi^{(m)}(t)$ denote the *eigenvalue* and *eigenfunctions*, respectively, of the random signal $x(t)$. Furthermore, one can express $x(t)$ in terms of the basis functions $\phi^{(m)}(t)$; this expansion is known as *Karhunen-Loeve* (K-L) expansion and is given by

$$x(t) =_{ms} \sum_{m=0}^{\infty} x^{(m)} \phi^{(m)}(t) \tag{1.190}$$

where

$$x^{(m)} = \int_{-\infty}^{\infty} x(t)\, \phi^{(m)*}(t)\, dt \tag{1.191}$$

with

$$\int_{-\infty}^{\infty} \phi^{(m)}(t)\, \phi^{(n)*}(t)\, dt = \begin{cases} 0 & n \neq m \\ 1 & n = m \end{cases}. \tag{1.192}$$

Mercer's theorem also provides a means of computing the eigenfunctions and eigenvalues using integral equations. That is,

$$\int_{-\infty}^{\infty} R_x(t_1, t_2)\, \phi^{(m)}(t_2)\, dt_2 = \lambda^{(m)} \phi^{(m)}(t_1)\,;\ m = 0, 1, \ldots \tag{1.193}$$

Needless to point out that the ease of computing the eigenvalues and eigenfunctions depend entirely on the complexity of the correlation function. In most signal processing applications, the signal is filtered before any signal processing is performed. This action results in a correlation function that is shaped by the impulse response of the front-end filter.

Furthermore, the most important aspect of this expansion is the correlation function of the expansion $X^{(m)}$, which is given by

$$E\left\{x^{(n)} x^{(m)*}\right\} = \begin{cases} \lambda^{(m)} & n = m \\ 0 & n \neq m \end{cases}. \tag{1.194}$$

This implies that the expansion coefficients become uncorrelated when K-L expansion is used to express a random signal. We also note that the equality in (1.190) is in mean-square (ms) sense. That is, for the K-L expansion,

$$\lim_{M\to\infty} E\left\{\left|\sum_{m=0}^{M} x^{(m)}\phi^{(m)}(t) - x(t)\right|^2\right\} \to 0 \tag{1.195}$$

This is rather disappointing for those of us who are used to having an equality in place of ms equality (for instance, Fourier series expansion allows an equality instead of ms equality). However, provided that most signals encountered in the signal processing world (including biomedical signals) are random in nature, we restrict ourselves to this notation. Furthermore, in discrete-time domain

$$x[n] =_{ms} \sum_{m=-\infty}^{\infty} x^{(m)}\phi^{(m)}[n] \tag{1.196}$$

where $\phi^{(m)}[n] = \phi^{(m)}(nT_s)$, and

$$\mathbf{x}[n] =_{ms} \sum_{m=-\infty}^{\infty} x^{(m)}\phi^{(m)}[n] \tag{1.197}$$

with $\phi^{(m)}[n] = \left[\phi^{(m)}(nT_s), \phi^{(m)}((n+1)T_s), \ldots, \phi^{(m)}\left(\left(n+l_p-1\right)T_s\right)\right]$. We caution the reader about the above notation, which distinguishes between $x[m]$ and $x^{(m)}$; namely, the former is the sample of $x(t)$ at mT_s while the latter refers to the projection of signal $x(t)$ onto the mth basis function.

It is important to note that even though biomedical signals are in general non-Gaussian, the measured biomedical signals using electronic means is plagued by this noise. Hence, a study of Gaussian random processes is important in the overall study of biomedical signals. Furthermore, we will introduce the concept of principle component analysis in a later chapter, where biomedical signals are expressed in terms of orthogonal functions. For these reasons, an understanding of K-L expansion is of paramount importance.

Example 18 For the signal $x(t) = \sum_{n=0}^{\infty} A_n \cos(n\omega_0 t + \theta_n)$, where $\theta_n's$ are known while $A_n's$ are zero mean, i.i.d random variables with $R_A[n,m] = \sigma_n^2\delta[m-n]$, find the K-L expansion of $x(t)$.

Solution: *We first find the correlation function of* $x(t)$*, which is given by*

$$\begin{aligned} R_x(t_1,t_2) &= \sum_{m=0}^{\infty}\sum_{n=0}^{\infty} E\left\{A_n A_m^*\right\}\cos\left(n\omega_0 t_1 + \theta^{(n)}\right)\cos\left(m\omega_0 t_2 + \theta^{(m)}\right) \\ &= \sum_{m=0}^{\infty}\sum_{n=0}^{\infty} R_A[n-m]\cos\left(n\omega_0 t_1 + \theta^{(n)}\right)\cos\left(m\omega_0 t_2 + \theta^{(m)}\right) \end{aligned}$$

$$= \sum_{n=0}^{\infty} \frac{\sigma_n^2}{2} \left\{ \sqrt{2} \cos\left(n\omega_0 t_1 + \theta^{(n)} \right) \right\} \left\{ \sqrt{2} \cos\left(n\omega_0 t_2 + \theta^{(n)} \right) \right\} \tag{1.198}$$

which by direct examination leads to the following eigenvalues and eigenfunctions: $\lambda_n = \frac{\sigma_n^2}{2}$ *and* $\Phi^{(n)}(t) = \sqrt{2}\cos(n\omega_0 t + \theta_n)$; $n = 0, 1, \ldots,$ *respectively. Hence,* $x(t)$*'s K-L expansion is* $x(t) = \sum_{n=0}^{\infty} \frac{A_n}{\sqrt{2}} \sqrt{2}\cos(n\omega_0 t + \theta_n)$. *This can be further verified by realizing that*

$$\begin{aligned} &\int_{-\infty}^{\infty} R_x(t_1, t_2)\, \phi^{(n)}(t_2)\, dt_2 \\ &= \int_{-\infty}^{\infty} \sum_{m=0}^{\infty} \frac{\sigma_m^2}{2} \sqrt{2}\cos\left(m\omega_0 t_1 + \theta^{(m)} \right) \\ &\quad \times \sqrt{2}\cos\left(m\omega_0 t_2 + \theta^{(m)} \right) \sqrt{2}\cos\left(n\omega_0 t_2 + \theta^{(n)} \right) dt_2 \\ &= \frac{\sigma_n^2}{2} \sqrt{2}\cos\left(n\omega_0 t_1 + \theta^{(m)} \right) = \lambda_n \phi^{(n)}(t) \end{aligned} \tag{1.199}$$

This approach gives us the same eigenvalues and eigenfunctions as those obtained by direct examination of the correlation function.

Example 19 Find the eigenfunction and eigenvalues of a white Gaussian random signal with two-sided power spectrum density of $\frac{N_0}{2}$ W/Hz.

Solution: *In this case, using inverse Fourier transform,*

$$R_x(t_1, t_2) = \frac{N_0}{2} \delta(t_1 - t_2). \tag{1.200}$$

A signal with this property is known as "white" noise since its PSD is constant over all frequencies and two samples separated by 'any' distance in time are uncorrelated. This signal is often observed in electronic devices subject to thermal noise or signals collected by a communications antenna in the absence of any signal. For this correlation function, the integral equation degrades. That is,

$$\int_{-\infty}^{\infty} R_x(t_1, t_2)\, \phi^{(n)}(t_2)\, dt_2 = \frac{N_0}{2} \phi^{(n)}(t_1) \tag{1.201}$$

which implies that any function can act as an eigenfunction for this signal with all eigenvalues are fixed at $\frac{N_0}{2}$. *That is,*

$$E\left\{ \left| x^{(m)} \right|^2 \right\} = \frac{N_0}{2} \text{ for all } m \tag{1.202}$$

Furthermore, from K-L expansion,

$$E\left\{x^{(n)}x^{(m)*}\right\} = 0 \text{ for } n \neq m \tag{1.203}$$

The key result here is that $x^{(m)\prime}s$ are Gaussian (since a linear operation is performed to obtain them from $x(t)$), and since $x^{(m)\prime}s$ are uncorrelated due to K-L expansion, they also become independent. The significance of this result is that any white Gaussian signal can be expressed in terms of any set of orthogonal functions and the projection of the signal onto those orthogonal functions results in a white Gaussian random sequence. Furthermore, the variances of the expansion terms are identical, leading to a condition known as "spherical symmetry". This definition stems from the fact that the joint Gaussian pdf of $x^{(m)\prime}s$ in an infinite-dimensional space will remain constant over the surface of an infinite-dimensional hypersphere.

Example 20 Find the K-L expansion of

$$r(t) = A\cos(\omega t + \theta) + n(t) \tag{1.204}$$

where $n(t)$ is an additive, zero-mean, white Gaussian noise (AWGN) with two sided PSD of $\frac{N_0}{2}$.Assume that A and θ are know.

Solution: *Since $A\cos(\omega t + \theta)$ is a known signal and any signal can be used as eigenfunction of a white noise, we choose*

$$\phi^{(m)}(t) = \sqrt{2}\cos(m\omega t + \theta) \text{ for all } m \tag{1.205}$$

Note that this set of orthogonal functions can fully describe $A\cos(\omega t + \theta)$ and $n(t)$. With these eigenfunctions, we see that

$$\begin{aligned} r^{(m)} &= \int_{-\infty}^{\infty} r(t)\,\phi^{(m)}(t)\,dt \\ &= \begin{cases} \frac{\sqrt{2}}{2}A + n^{(m)} & m = 1 \\ n^{(m)} & m \neq 1 \end{cases} \end{aligned} \tag{1.206}$$

where $n^{(m)}$ is a zero-mean Gaussian random variable with variance $\frac{N_0}{2}$. Hence, realizing that $\lambda^{(m)} = E\left\{\left|r^{(m)}\right|^2\right\}$, we have

$$\lambda^{(m)} = \begin{cases} \frac{A^2}{2} + \frac{N_0}{2} & m = 1 \\ \frac{N_0}{2} & m \neq 1 \end{cases}. \tag{1.207}$$

Note that the power associated with the "signal" part of $r(t)$ is $\frac{A^2}{2}$, and this is reflected in the expression for the eigenvalue for $m = 1$. This implies that projection of $r(t)$ onto this 'eigenspace' (which is defined as a space with its axes defined *by the eigenfunctions) resulted in signal being present in only one dimension of the space and absent in the rest, while noise, as expected, is present in all dimensions. Furthermore, the projections are all uncorrelated independent Gaussian random variables.*

Now we are in a position to define the concept of signal space. Based on what has been presented in this section, one can describe a random signal in terms of orthogonal basis functions. Since these functions are known (it is assume that we have been successful in solving the necessary integral equations suggested by *Mercer's theorem* to obtain $\phi^{(m)}(t)$ for all m), we can replace the random signal with its vector representation. To that end, let us first introduce a vector representation of the K-L projections, given by

$$\mathbf{x}^{(N)} = \left[x^{(0)}, x^{(1)}, \ldots, x^{(N)}\right] \tag{1.208}$$

Given this definition, then

$$\lim_{N\to\infty} \mathbf{x}^{(N)} \to_{ms} x(t) \tag{1.209}$$

This transformation provides one with a vector space representation of the signal, which is also referred to as the signal space representation of the signal. Note that this representation is exact for non-random signals while it is only valid in ms sense for random signals.

It is important to realize that we now have two representation of a signal. That is,

$$\lim_{l_v\to\infty} \mathbf{x}[n] \Longleftrightarrow x(t) \tag{1.210}$$

(sampled in time domain) and

$$\lim_{N\to\infty} \mathbf{x}^{(N)} \Longleftrightarrow {}_{ms}x(t) \tag{1.211}$$

(projection onto the basis functions of a signal space), with two entirely distinct definitions. The former representation may not offer any significant statistical information, whereas the latter is a mean-square-sense-accurate representation of the random signal as $N \to \infty$ (albeit, it may not provide a complete statistical knowledge of the signal in non-Gaussian cases). Hence, caution must be exercised in using the time-domain vector representation to analyze a random signal. We will elaborate on this representation in the context of Sampling Theorem next. Before doing so, however,we need to elaborate on a subtle, yet important issue. To elaborate, in the above formulation we have assumed that the signals extend over an infinite time interval. As will be seen in the ensuing chapters, in all fields, the signal observation interval (also known as the symbol time in the communications field) is finite in duration. Such a scenario will result in an K-L expansion over a finite time interval, leading to the following formulation:

$$R_x(t_1, t_2) = \sum_{m=0}^{\infty} \lambda^{(m)} \phi^{(m)}(t_1) \phi^{(m)*}(t_2)$$
$$\text{for } t_1, t_2 \in [0, T], \tag{1.212}$$

$$x^{(m)} = \int_0^T x(t) \phi^{(m)*}(t) \, dt, \tag{1.213}$$

and

$$\int_0^T R_x(t_1, t_2) \phi^{(m)}(t_2) \, dt_2 = \lambda^{(m)} \phi^{(m)}(t_1); \quad m = 0, 1, \ldots, \; t_1 \in [0, T] \tag{1.214}$$

where in this case T is the observation interval.

1.3.6.2 Mean-Square Sense Interpolation

In the previous section, we discussed interpolation for deterministic signals. This operation is of particular importance when one is faced with the problem of interpolating a signal using samples obtained at a rate lower than what is desired. The problem takes an interesting turn when we encounter random signals, where a simple upsampling and filtering may not lead to a meaningful result. This problem is encountered when one is asked to interpolate samples of an observed signal in biomedical or other signal processing scenarios where the observed signal takes on a random nature. Then, how does one approach this problem? Based upon the above discussion, needless to say that we need to take a probabilistic approach involving statistics of the random signal to answer this questions. In order to correctly address the problem, we need to re-examine the intent of interpolation. That is, we intend to interpolate when the following information is at our disposal:

$$x[-\infty], \ldots, x[n-1], x[n], x[n+1], \ldots x[\infty] \tag{1.215}$$

where $x[n] = x(nT_s)$. Now, we are interested in another random signal

$$x_{us}[p] = x(pT_{us}) \tag{1.216}$$

where $T_{us} = \frac{2\pi}{\Omega_{us}}$ with $\Omega_{us} > \Omega_s = \frac{2\pi}{T_s}$. As noted in the previous sections, for most practical problems, $M = \frac{T_s}{T_{us}}$, leading to $x[n] = x_{us}[nM]$. We need to emphasize that the equality in the preceding equation is in the mean-square sense (as one would expect from the treatment of random signals). So far, this representation is quite similar to that of the deterministic signals. The next step, however, requires a different approach. To elaborate, using the FT, we argued that a perfect interpolation can be performed by means of low pass filtering. In the case before us, this approach cannot be used without establishing a criteria for the effectiveness of the interpolation process. A metric that has been used in the previous sections can be used here as well; that is, we can establish mean-square error as the criterion for convergence. To that end, let the "estimate" of $x_{us}[p]$ be $\hat{x}_{us}[p]$. Then, we have

$$\hat{x}_{us}[p] = \sum_{n=-N}^{N} a_p[n]\, x[n]\,; \; -NM \leq p \leq NM \tag{1.217}$$

where $a_p[n]$'s are selected such that

$$\lim_{N\to\infty} E\left\{\left|x_{us}[p] - \hat{x}_{us}[p]\right|^2\right\} \to 0 \tag{1.218}$$

That is,

$$\lim_{N\to\infty} \hat{x}_{us}[p]_N \to_{ms} x_{us}[p]$$

Since $N = \infty$ does not lead to a tractable result, we try to minimize $E\left\{\left|x_{us}[p] - \hat{x}_{us}[p]\right|^2\right\}$ for a finite N (the limit on N may be due to hardware, delay, or other system engineering restrictions). That is, we are provided with

$$\mathbf{x}[-N] = [x[-N], x[-N+1], \ldots, x[N]]^{\mathsf{T}}$$

(in this case, $l_v = 2N + 1$). Let us define $\mathbf{a}_p = \left[a_p[-N],\; a_p[-N+1], \ldots, a_p[N]\right]^{\mathsf{T}}$. Then,

$$\boxed{\hat{x}_{us}[p] = \mathbf{x}^{\mathsf{T}}[-N]\,\mathbf{a}_p} \tag{1.219}$$

where $\mathbf{a}_p^{\mathsf{T}}$ denotes the transpose of $\mathbf{a}_p$. This implies that, given an observation interval $(-NT_s, NT_s)$ and observed samples of the signal $x(t)$ obtained at nT_s with $n \in [-N, N]$, can we "estimate" the samples of $x(t)$ at $t = \frac{m}{M}T_s$ with $m \in [-NM, NM]$? With little effort (namely, setting the derivatives of $E\left\{\left|x_{us}[p] - \hat{x}_N[p]\right|^2\right\}$ with respect to $a_p[n]'s$ to zero), we arrive at the following set of equations:

$$E\left\{x^*[n]\left(x_{us}[p] - \hat{x}_{us}[p]\right)\right\} = 0 \text{ for all } n \text{ in } [-N, N],$$

which leads to

$$E\left\{x^*[n]\left(x_{us}[p] - \sum_{q=-N}^{N} a_p[q]\, x[q]\right)\right\} = 0;\, n \in [-N, N] \text{ and } p \in [-NM, NM]. \tag{1.220}$$

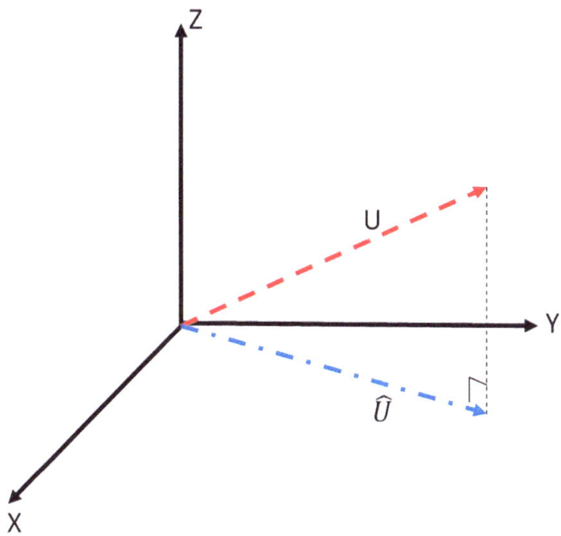

Fig. 1.28 A graphical representation of the linear minimum-mean-square estimation process

The above result is of significant importance as it underscores an important principle; that is, for the *linear minimum mean-square estimation (LMSE)* (note that we have used a linear operation to estimate $x_{us}[p]$), the error

$$\left(x_{us}[p] - \sum_{q=-N}^{N} a_p[q]\, x[q] \right) \tag{1.221}$$

must be '*orthogonal*' to the observations, *i.e.*, $x^*[n]'s$. A graphical representation of this process is depicted in Fig. 1.28. In this case, we are interested in estimating the vector U when only data available to us are confined to the two-dimensional (x, y) plane. The reader can verify that the best estimate for U is $\hat{U}$ where the error vector $U - \hat{U}$ is orthogonal to the (x, y) plane. This constitutes the best estimate of U in terms of minimizing MSE (that is, minimizing $\left|U - \hat{U}\right|^2$). Note that any vector in the (x, y) plane will be a linear combination of the observation in the (x, y) plane. This further implies that error must remain orthogonal to the observation in order for us to achieve an estimate with minimizes MSE. Obviously, if the condition of 'linearity' is relaxed, better solutions can be obtained. In that event, the solution will no longer be limited to the (x, y) plane. In contrast, a linear solution will be limited to the (x, y) plane as any linear combination of vectors in (x, y) plane will be limited to the (x, y) plane.

In many problems the estimate that minimizes MSE is not a linear estimator (perhaps, not limited to the (x, y) plane). A linear estimator, however, enjoys ease of implementation, which may be of critical importance in a real-time system.

Getting back to the interpolation problem, we realize that

$$E\left\{x^{*}[n]\,x_{us}[p]\right\} = R_x\left(nT_s, \frac{p}{M}T_s\right) \tag{1.222}$$

and

$$E\left\{x^{*}[n]\,x[q]\right\} = R_x\left(nT_s, qT_s\right). \tag{1.223}$$

Hence,

$$E\left\{R_x\left(nT_s, \frac{p}{M}T_s\right) - \sum_{q=-N}^{N} a_p[q]\,R_x\left(nT_s, qT_s\right)\right\} = 0;\ n \in [-N, N] \quad \text{and all } p \in [-NM, NM] \tag{1.224}$$

In other words, we have the following:

$$\mathbf{r}_p - \mathbf{R}_x[-N]\,\mathbf{a}_p = \mathbf{0};\ p \in [-NM, NM] \tag{1.225}$$

where

$$\mathbf{r}_p = \left[R_x\left(-NT_s, \frac{p}{M}T_s\right), R_x\left((-N+1)T_s, \frac{p}{M}T_s\right), \ldots, R_x\left(NT_s, \frac{p}{M}T_s\right)\right]^{\mathsf{T}} \tag{1.226}$$

and

$$\mathbf{R}_x[-N] = E\left\{\mathbf{x}[-N]\,\mathbf{x}^{H}[-N]\right\} \tag{1.227}$$

is the correlation matrix for $\mathbf{x}[-N]$. Now, we can obtain $\mathbf{a}_p$ using

$$\boxed{\mathbf{a}_p = \mathbf{R}_x^{-1}[-N]\,\mathbf{r}_p;\ p \in [-NM, NM]}. \tag{1.228}$$

As can be seen, the above solution hinges upon the existence of $\mathbf{R}_x^{-1}[-N]$. Hence, the condition for the existence of a solution is the non-singularity of the matrix $\mathbf{R}_x[-N]$. This condition is satisfied for the covariance matrix of all positive definite random signals (note that $\mathbf{R}_x[-N]$ is the covariance matrix of $\mathbf{x}[-N]$, which is assumed to be zero mean). For a random signal that is WSS, we have a major simplification. That is, $\mathbf{R}_x[-N] = \mathbf{R}_x$ (with $l_v = 2N+1$), which leads to

$$\mathbf{a}_p = \mathbf{R}_x^{-1}\mathbf{r}_p;\ p \in [-NM, NM].$$

Since a large class of signals can be considered to be WSS, we consider WSS signals first. We, then, extend the results to the case of non-WSS signals. Hence, $\mathbf{R}_x[-N]$ is replaced with $\mathbf{R}_x$ first. It becomes immediately obvious that the major stumbling block in the above analysis is the computation of $\mathbf{R}_x^{-1}$ for large N. We note that since $\mathbf{R}_x$ is a Hermitian matrix, it can be described in terms of its eigenvectors and eigenvalues as

$$\mathbf{R}_x = \mathbf{U}\Lambda\mathbf{U}^{H} \tag{1.229}$$

where $\mathbf{U} = \left[\phi_1, \ldots, \phi_{l_v}\right]$ and $\Lambda = \begin{bmatrix} \lambda_1 & 0 & 0 & 0 \\ 0 & \lambda_2 & 0 & 0 \\ 0 & 0 & \ddots & 0 \\ 0 & 0 & 0 & \lambda_{l_v} \end{bmatrix}$ with ϕ_j and λ_j denoting the jth eigenvector and jth eigenvalues, respectively, of the matrix $\mathbf{R}_x$. Note that $\mathbf{R}_x$ is an $l_v \times l_v$ matrix, and hence it has l_v eigenvectors and l_v eigenvalues. From the definition of eigenvectors and eigenvalues, we have

$$\mathbf{R}_x \phi_j = \lambda_j \phi_j \tag{1.230}$$

for $j = 1, 2, \ldots, l_v$. For a Hermitian matrix, U is unitary. That is, $\mathbf{U}\mathbf{U}^H = I$ (identity matrix). Hence, $U^{-1} = \mathbf{U}^H$. This further implies that,

$$\mathbf{R}_x^{-1} = \mathbf{U}\Lambda^{-1}\mathbf{U}^H \tag{1.231}$$

where $\Lambda^{-1} = \begin{bmatrix} \lambda_1^{-1} & 0 & 0 & 0 \\ 0 & \lambda_2^{-1} & 0 & 0 \\ 0 & 0 & \ddots & 0 \\ 0 & 0 & 0 & \lambda_{l_v}^{-1} \end{bmatrix}$. Since $\mathbf{R}_x$ is positive definite, $\lambda_j > 0$ for all j, and hence Λ^{-1} exists. The problem that remains is to compute the eigenvalues and eigenvectors of $\mathbf{R}_x^{-1}$. We then have

$$\mathbf{a}_p = \mathbf{U}\Lambda^{-1}\mathbf{U}^H \mathbf{r}_p;\ p \in [-NM, NM]. \tag{1.232}$$

This leads to

$$\hat{x}_{us}[p] = \mathbf{x}^{\mathsf{T}}[-N]\,\mathbf{U}\Lambda^{-1}\mathbf{U}^H \mathbf{r}_p$$

Finally, we can compute the MSE (this is the smallest MSE achievable using the linear interpolation described above) is

$$\begin{aligned} \text{MSE} &= E\left\{\left|x_{us}[p] - \hat{x}_{us}[p]\right|^2\right\} = E\left\{x_{us}^*[p]\left(x_{us}[p] - \hat{x}_{us}[p]\right)\right\} \\ &= R_x\left(\frac{pT_s}{M}, \frac{pT_s}{M}\right) - \mathbf{a}_p \mathbf{R}_x^{-1}[-N]\,\mathbf{a}_p^H \end{aligned} \tag{1.233}$$

which for a WSS process reduces to

$$\text{MSE} = R_x(0) - \mathbf{a}_p \mathbf{R}_x^{-1} \mathbf{a}_p^H \tag{1.234}$$

For the case of non-WSS, the solutions that are obtained will not longer be unique. That is, for each block of observation $\mathbf{x}[-N]$, $\mathbf{r}_p$ and $\mathbf{a}_p$ are computed using $\mathbf{R}_x[-N]$. The inverse matrix $\mathbf{R}_x^{-1}[-N]$ can also be computed using the eigenvalue and eigenvector method. The issue, however, will be that the solutions $\mathbf{U}$, Λ, and $\mathbf{a}_p$

will have to be computed for each block of $2N+1$ observation, making this solution impractical for high-speed environments. However, for scenarios where data can be analyzed over long periods of time, such as biomedical applications, this restriction becomes insignificant.

Example 21 A zero-mean random signal $x(t)$ is sampled at a rate of 1 kHz between 0 and 2 ms. The correlation function of this signal is given by

$$R_x(t_1, t_2) = 0.1\exp(-500|t_1 - t_2|)$$

Using the above samples, we like to obtain the samples of this signal over the same interval of time when the sampling rate is 2 kHz. In particular, we are interested in reconstructing the sample of the signal at $t = 0.5$ ms. Find the interpolation coefficients and obtain the resulting MSE.

Solution: *Since $T_{us} = 0.5$ ms and $T_s = 1$ ms, $M = 2$. Furthermore, we can verify that the signal is WSS with*

$$\mathbf{R}_x = \begin{bmatrix} 0.1 & 0.1\exp(-0.5) & 0.1\exp(-1) \\ 0.1\exp(-0.5) & 0.1 & 0.1\exp(-0.5) \\ 0.1\exp(-1) & 0.1\exp(-0.5) & 0.1 \end{bmatrix} \tag{1.235}$$

Also,

$$\mathbf{r}_p = \left[0.1e^{-0.25p}, 0.1e^{-0.25(2-p)}, 0.1e^{-0.25(4-p)}\right] \tag{1.236}$$

Hence,

$$\begin{aligned} \mathbf{a}_p &= \mathbf{R}_x^{-1}\mathbf{r}_p \rightarrow \\ \mathbf{a}_p &= \begin{bmatrix} 15.8 & -9.5 & 0 \\ -9.5 & 21.6 & -9.5 \\ 0 & -9.5 & 15.8 \end{bmatrix} \\ &\times \left[0.1e^{-0.25p}, 0.1e^{-0.25(2-p)}, 0.1e^{-0.25(4-p)}\right]^{\mathsf{T}} \end{aligned}$$

In order for us to go any further, one needs the value for p (p is an integer $\in [0, 4]$). For the problem at hand, we are interested in $p = 1$ (note that $p = 0$ corresponds to $t = 0$ whereas $p = 4$ corresponds to $t = 2$ ms). Hence,

$$\begin{aligned} \mathbf{a}_1 &= \begin{bmatrix} 15.8 & -9.5 & 0 \\ -9.5 & 21.6 & -9.5 \\ 0 & -9.5 & 15.8 \end{bmatrix} \left[0.1e^{-0.25}, 0.1e^{-0.25}, 0.1e^{-0.75}\right]^{\mathsf{T}} \\ &= [0.48,\ 0.48,\ 0]^{\mathsf{T}} \end{aligned}$$

Therefore,

$$\hat{x}_{us}[1] = 0.48x[0] + 0.48x[1]$$

It would be important to consider the mean-square error for this interpolation, which is given by

$$\begin{aligned}
\text{MSE} &= E\left\{\left(\hat{x}_{us}[1] - x[1]\right)^2\right\} = E\left\{(0.48x[0] - 0.52x[1])^2\right\} \\
&= (0.48)^2\, 0.1 + (0.52)^2\, 0.1 \\
&\quad - 2 \times 0.52 \times 0.48 \times 0.1 \exp(-0.5) \\
&= 0.02.
\end{aligned}$$

Although not zero, the MSE is relatively small (as one would expect). Also, one can obtained MSE using

$$\begin{aligned}
\text{MSE} &= R_x(0) - \mathbf{a}_p \mathbf{R}_x^{-1} \mathbf{a}_p^H \\
&= 0.1 - 0.08 \\
&= 0.02
\end{aligned}$$

which agrees with the direct calculation of MSE.

It is interesting to note that the variance of each sample in the above example is 0.1. Hence, the interpolation process has resulted in a 20 % increase in the variance of the estimated/interpolated samples. Although this appears to be rather disappointing, in the context of random signal estimation, this estimate yields the lowest possible mean-square error among all linear estimator/interpolators. One question that one can consider at this point is whether this estimate is the best estimate among all possible interpolators/estimators. In order to answer this question, we need to re-examine the parameters of our analysis here. From a statistical point of view and considering the MSE as the metric of interest, one can show that the minimum mean-square estimate of a signal is in general a nonlinear estimator. In fact, if the pdf of $x_{us}[p]$ conditioned on $\mathbf{x}[-N]$ is known, then an estimate that minimizes MSE is given by

$$x_{us,MMSE}[p] = E\{x_{us}[p] \,|\mathbf{x}[-N]\}$$

One quickly realizes that this approach requires a statistical knowledge of not only the available samples, but also those which we intend to estimate! In the event that such a statistical knowledge is available, there is yet another problem which we need to address. That is, the above estimation process may not lead to a linear processing of the available samples, which makes this approach even more unattractive. For this very reason, this approach is not pursued in many practical scenarios.

1.3.7 Mean-Square Sense Sampling Theorem

At the beginning of this section, we discussed Sampling Theorem. ST was originally concerned with the deterministic signals. However, since majority of signals observed in biomedical field can be classified as random signals, we need to extend the concept of ST to non-deterministic signal case. In order to address this problem in a proper manner, one has to treat this problem in a "mean-square sense." Sampling of a continuous-time random signal allows one to convert a random process into a random sequence. This approach may be viewed as a means of expressing a continuous-time signal using only its sampled values. For random signals, however, one has to tackle the non-trivial task of reconstructing a random signal from its samples. Similar to the deterministic case, we are concerned with finding the values of the signal at points in time other than those for which the samples are available. Let

$$x[n] = x(t)\,|_{t=nT_s} \tag{1.237}$$

where now $x(t)$ is a random process. The main distinction here is that we assume that the random process of interest is (or can be approximated as) WSS with a PSD that is band-limited. That is, we assume that

$$S_x(\Omega) \leq S_0 \text{ for } |\Omega| > \Omega_c \tag{1.238}$$

for some Ω_c (in rad/s), known as the cut-off frequency for the signal. In this definition, S_0 denotes the maximum allowable PSD (in watts/rad/s) in the reject band. Consequently, one can view Ω_c as the "bandwidth" of $x(t)$. Since $R_x[n] = E\{x[m]x^*[n+m]\} = R_x(\tau)\,|_{\tau=nT_s}$ and realizing that the Fourier Transform of $R_x(\tau)$ is band-limited, one can apply the Sampling Theorem to the correlation function, which is a deterministic function. That is, using Shannon's Sampling Theorem, one can reconstruct $R_x(\tau)$ using

$$R_x(\tau) = \lim_{N\to\infty} \sum_{m=-N}^{N} R_x[m]\,\mathrm{sinc}\,(\Omega_c(\tau - mT_s)) \tag{1.239}$$

if the sampling rate in rad/s satisfies: $\Omega_s = \frac{2\pi}{T_s} \geq 2\Omega_c$. In that event, using the Sampling Theorem, we can state that the Fourier Transform of the discrete sequence $R_x[n]$ (namely, the PSD of $x[n]$) is related to the Fourier Transform of the function $R_x(\tau)$ (namely, the PSD of $x(t)$) as follows:

$$S_x[\omega] = \frac{1}{T_s} S_x\left(\frac{\omega}{T_s}\right);\ |\omega| \leq \pi$$

Note that since $\Omega_s \geq 2\Omega_c$ (i.e., the Nyquist rate), the above sampling does not distort the PSD of the sampled signal.

Now, an interpolation of $x(t)$ at any t can be obtained using

$$\hat{x}_N(t) = \sum_{m=-N}^{N} x[m]\, b_\tau[m] \quad (1.240)$$

where

$$b_\tau[m] = \mathrm{sin\,c}\left(\Omega_c(\tau - mT_s)\right) \quad (1.241)$$

In fact, it can be shown that

$$\lim_{N\to\infty} \hat{x}_N(t) \to_{m.s} x(t).$$

The proof is rather straightforward. To elaborate, one can easily show that an estimate that minimizes MSE for linear interpolation must satisfy (set the derivatives of $E\left\{\left|\hat{x}_N(t) - x(t)\right|^2\right\}$ with respect to $b_t[n]'s$ to zero)

$$E\left\{x[m]\left(\hat{x}_N(t) - x(t)\right)\right\} = 0 \text{ for all } m, \quad (1.242)$$

or equivalently,

$$\lim_{N\to\infty} E\left\{\hat{x}_N(t)\left(\hat{x}_N(t) - x(t)\right)\right\} = 0 \quad (1.243)$$

That is, the error becomes orthogonal to the observation as N gets large. Note that

$$\begin{aligned} &E\left\{x[m]\left(\hat{x}_N(t) - x(t)\right)\right\} \\ &= \sum_{m'=-N}^{N} R_x\left((m-m')T_s\right) b_t\left[m'\right] - R_x(t - mT_s) \\ &= \sum_{m'=-N}^{N} R_x\left((m-m')T_s\right) \mathrm{sin\,c}\left(\Omega_c\left(t - m'T_s\right)\right) - R_x(t - mT_s) \quad (1.244) \end{aligned}$$

From (1.239), we have

$$\begin{aligned} R_x(t - mT) &= \lim_{N\to\infty} \sum_{m'=-N}^{N} R_x[m]\, \mathrm{sin\,c}\left(\Omega_c\left(t - mT - m'T_s\right)\right) \\ &= \lim_{N\to\infty} \sum_{m''=-N}^{N} R_x\left[m - m''\right] \mathrm{sin\,c}\left(\Omega_c\left(t - m''T\right)\right) \quad (1.245) \end{aligned}$$

Hence,

$$\lim_{N\to\infty} E\left\{x[m]\left(\hat{x}_N(t) - x(t)\right)\right\} = 0 \text{ for all } m \quad (1.246)$$

Furthermore,

$$\lim_{N\to\infty} E\left\{x(t)\left(\hat{x}_N(t) - x(t)\right)\right\} = 0 \qquad (1.247)$$

Since from (1.239),

$$R_x(0) = \lim_{N\to\infty} \sum_{m=-N}^{N} R_x[m] \operatorname{sin c}\left(\Omega_c (mT_s)\right). \qquad (1.248)$$

Hence, the interpolated offers a 'zero' MSE. This further implies that $x(t)$ *for any* t *has been estimated up to an event with probability* 0. For most engineering problems, this level of accuracy is quite satisfactory. However, similar to case of the perfect interpolation for deterministic signals, this approach becomes unrealizable as it requires infinite number of samples to achieve its goal.

1.3.7.1 Practical Implementation of Mean-Square Sense Sampling

A remedy to the above problem is to resort to a linear MSE technique (see the previous section). Once again, let

$$x[n] = x(t)\,|_{t=nT_s}$$

where now $x(t)$ is a random process. Furthermore, assume that a reconstruction mechanism has been considered as follows

$$\hat{x}(t) = x_N(t) = \sum_{n=-N}^{N} x[m-n]\,h[n] \text{ for } (m-1)\,T_s < t < mT_s. \qquad (1.249)$$

This notation implies that $\hat{x}(t)$ is assumed to be constant over the interval $t \in [(m-1)\,T_s, mT_s]$. This is not an unusual assumption since methods such as zero-hold interpolation make a similar assumption. Since $x_N(t)$ is a random process as well, we have to consider the statistical characteristics of the error ($x_e(t) = x(t) - \hat{x}(t)$) rather than its absolute value. Since mean-square value establishes the second order statistics of the error, it is reasonable to consider a solution where

$$\lim_{N\to\infty} E\left\{|x_e(t)|^2\right\} = \lim_{N\to\infty} E\left\{|x(t) - x_N(t)|^2\right\} \to 0$$

Note that this does not insure a "point-wise" convergence, but rather a statistical convergence in the mean-square sense. Unlike the deterministic way, we have no perfect reconstruction mechanism. Indeed, the above convergence does not insure sure or almost sure convergence. However, for many engineering problems, this convergence criterion establishes a workable framework within which one can obtain practical solutions. Hence, our task has been reduced to estimating $h[n]$ for all n. However, a more practical problem is to reconstruct the signal for a pre-defined interval of time. That is, we limit the scope of the problem to estimating $\hat{x}(t)$ for $(m-1)T_s < t < mT_s$ using $x[n]$ for $|n| \le N$. Note that the interval is an open interval as we confine the values of $\hat{x}(t)$ at $t = mT$ and $(m-1)T_s$ to $x[m]$ and $x[m-1]$, respectively. Since the intent of the interpolation is to obtain values of the signal between two known points, this approach is consistent with interpolation techniques for deterministic signals. In that event,

$$\hat{x}(t) = x_N(t) = \sum_{n=-N}^{N} x[m-n]\,h[n] \text{ for } (m-1)T_s < t < mT_s. \tag{1.250}$$

The mean-square error for this estimation is

$$\text{MSE} = E\left\{|x(t) - x_N(t)|^2\right\} \text{ for } (m-1)T_s < t < mT_s. \tag{1.251}$$

As noted previously, $h[n]'s$ which will minimize MSE must satisfy the following set of equations:

$$E\left\{x^*[l](x(t) - x_N(t))\right\} = 0 \text{ for } l \in [m-N, m+N] \tag{1.252}$$

$$\text{and } (m-1)T_s < t < mT_s. \tag{1.253}$$

This is a system of $2N+1$ equations with $2N+1$ unknowns. This equation is motivated by the orthogonality principle, which was explained earlier (see Fig. 1.28).

For the problem at hand, we have

$$R_x(lT_s, t) - \sum_{n=-N}^{N} R_x(lT_s, (m-n)T_s)\,h[n] = 0 \text{ for all } l \in [m-N, m+N] \tag{1.254}$$

$$\mathbf{R}_{m,N}\mathbf{h} = \mathbf{r}_{m,t} \tag{1.255}$$

where

$$\mathbf{r}_{m,t} = \left[R_x((m-N)T_s, t), \ldots, R_x((m+N)T_s, t)\right]^{\mathsf{T}}, \tag{1.256}$$

$$\mathbf{h} = \left[h[-N], \ldots, h[N]\right]^{\mathsf{T}}, \tag{1.257}$$

$$\left[\mathbf{R}_{m,N}\right]_{i,j} = R_x(iT_s, (m-j)T_s); i \in [m-N, m+N],$$

$$\text{and } j \in [-N, N+1, \ldots, N]. \tag{1.258}$$

Provided that $\mathbf{R}_{m,l}^{-1}$ exists ($\mathbf{R}_{m,l}$ is a $(2N+1)\times(2N+1)$ correlation matrix, and if the process is positive definite, $\mathbf{R}_{m,l}^{-1}$ exists), we have

$$\mathbf{h} = \mathbf{R}_{m,l}^{-1}\mathbf{r}_{m,t} \tag{1.259}$$

Note that, similar to the previous discussion on the problem of, this solution requires only the second order statistics of $x(t)$ to arrive at a solution. If we obtain the eigenvector and eigenvalues of the matrix $\mathbf{R}_{m,l}$, the above reduces to

$$\mathbf{h} = \mathbf{U}_{m,l}\Lambda_{m,l}^{-1}\mathbf{U}_{m,l}^{H}\mathbf{r}_{m,t} \tag{1.260}$$

where

$$\mathbf{U}_{m,l} = \left[\phi_1, \ldots, \phi_{2N+1}\right] \tag{1.261}$$

and

$$\Lambda_{m,l} = \begin{bmatrix} \lambda_1 & 0 & 0 & 0 \\ 0 & \lambda_2 & 0 & 0 \\ & & & \\ 0 & 0 & \ddots & 0 \\ 0 & 0 & 0 & \lambda_{2N+1} \end{bmatrix} \tag{1.262}$$

with ϕ_j and λ_j denoting the jth eigenvector and eigenvalues, respectively, of the matrix $\mathbf{R}_{m,l}$. Since N is a finite number, this approach results in an implementable (provided that $\mathbf{R}_{m,l}^{-1}$ exists and is available to us) solution.

This chapter provided us with a view of random signals and tools to study such signals. Although not exhaustive, the approach taken her tackled key issues in studying random signals, including power, energy, PSD, correlation, and estimation of random signals. We are now in a position to address the concept of systems which will be used to shape and process random signals. In particular, we will consider how linear systems shape PSD of random signals, which are encountered in biomedical engineering field.

Chapter 2
Linear and Nonlinear Systems

2.1 Linear Systems Theory

A magnificent tool for studying the behavior of practical systems is the linear system theory. Many devices and components routinely used to implement signal processing tools may be modelled (or at least approximated) as linear systems, allowing one to design and analyze complicated systems using frequency-domain approach. However, caution must be exercised when dealing with some elements of the problem under investigation that do not fit the profile. Many devices in the biomedical field, which presumed to be linear, operate as a linear device over only a limited range of parameters, imposing a restrict limitations on the range of validity of analyses that are conducted. We postpone a discussion on the subject of nonlinear systems to the ensuing section and proceed with a discussion of linear systems in this section.

Although we have discussed the concept of filtering in the previous chapter, a formal discussion of the concept of systems has not been provided so far. To that end, we begin by considering the impulse-response of a linear system, which fully characterizes the response of a linear system to any input signal using an integral equation. As the name implies, IR describes the response of a system to an impulse function.

Let $h(t, \tau)$ denote the response of a linear system at time t to a dirac delta function (*i.e.*, $\delta(t-\tau) = \begin{cases} \infty & t = \tau \\ 0 & \text{otherwise} \end{cases}$) applied to the system at $t - \tau$. Equivalently, in the discrete-time domain, we have $h[n, m]$ as the system output at time n to a delta function (*i.e.*, $\delta[n-m] = \begin{matrix} 1 & n = m \\ 0 & \text{otherwise} \end{matrix}$) at time $n - m$. The output of the linear system may then be described in terms of the input $x(t)$ (or $x[n]$) and $h(t, \tau)$ (or $h[n, m]$) as follows:

$$y[n] = \sum_{m=-\infty}^{\infty} h[n, m]\, x[n-m]\,. \tag{2.1}$$

K. Kiasaleh, *Biological Signals Classification and Analysis*,
Lecture Notes in Bioengineering, DOI 10.1007/978-3-642-54879-6_2

and in the continuous-time domain

$$y(t) = \int_{-\infty}^{\infty} h(t, \tau)\, x(t - \tau)\, d\tau. \tag{2.2}$$

To arrive at the above results, the following identities are used:

$$x(t) = \int_{-\infty}^{\infty} \delta(t - \tau)\, x(\tau)\, d\tau \tag{2.3}$$

and

$$x[n] = \sum_{m=-\infty}^{\infty} \delta[n - m]\, x[m]. \tag{2.4}$$

The reader may recognize these types of systems as linear and time-variant as the impulse response seems to be a function of "when" the impulse has been applied to the system. Note that $h(t, \tau)$ is a function of both t and τ. Similarly, $h[n, m]$ is also a function of m and n.

A large class of systems may be modelled (or at least approximated) as time-invariant, resulting in a noticeable reduction in the complexity of the above computations. For linear time-invariant (LTI) systems, $h(t, \tau) = h(\tau)$ and $h[n, m] = h[m]$. This simplification simply implies that the response of the system is a function of time separation between when the impulse function has been applied to the system and when the observation has been made. Many systems in use today may fit this model, although all systems experience minor fluctuations in their behavior over time.

The reader may verify that this simplification reduces the computation of the output of a linear system to that of a convolutional integration (or summation). Namely,

$$y[n] = h[n] \copyright x[n] = \sum_{m=-\infty}^{\infty} h[m]\, x[n - m] \tag{2.5}$$

and

$$y(t) = h(t) \copyright x(t) = \int_{-\infty}^{\infty} h(\tau)\, x(t - \tau)\, d\tau \tag{2.6}$$

where © denotes convolution. A further simplification, which is of practical interest, can be made to the above by realizing that a realizable linear system is causal, and hence $h(\tau) = 0$ for $\tau < 0$ or $h[m] = 0$ for $m < 0$. Hence, for a linear, time-invariant, and causal system, we have

$$y[n] = \sum_{m=0}^{\infty} h[m]\, x[n - m] \tag{2.7}$$

and

$$y(t) = \int_0^{\infty} h(\tau) x(t-\tau) d\tau. \tag{2.8}$$

For time-varying systems that are causal, we have

$$y[n] = \sum_{m=0}^{\infty} h[n,m] x[n-m]. \tag{2.9}$$

and in the continuous-time domain

$$y(t) = \int_0^{\infty} h(t,\tau) x(t-\tau) d\tau. \tag{2.10}$$

Linear systems must also provide bounded outputs for bounded inputs (BIBO) in order to be useful devices in signal processing. BIBO condition is satisfied if

$$\sum_{m=-\infty}^{\infty} |h[m]|^2 < \infty$$

or

$$\int_{-\infty}^{\infty} |h(\tau)| d\tau < \infty$$

A question that perhaps lingers in the mind of reader is whether there is a simple relationship between $h(\tau)$ and $h[m]$, and if so, what is the relationship? One is tempted to use $h[m] \stackrel{?}{=} h(mT_s)$ as the relationship between continuous and discrete time systems. However, as we will see shortly, that may not be the case. In order to explore this further, we need to resort to Fourier analysis.

2.1.1 System Function

It is widely accepted that the Fourier transform provides the most effective tool in linear system study. Although Laplace transform offers the benefit of taking initial conditions into account, the steady-state nature of most systems in practice makes the use of this transform less useful. Hence, we concern ourselves with FT analysis here. To that end, we define system function as

$$H[\omega] = \sum_{m=-\infty}^{\infty} h[m] e^{-j\omega m} \tag{2.11}$$

or, in the continuous-time domain,

$$H(\Omega) = \int_{-\infty}^{\infty} h(t)\, e^{-j\Omega t} dt. \tag{2.12}$$

An immediate and profound consequence of this transformation is the well-known and well-used relationship between input and output of a linear system, which is given by

$$Y[\omega] = H[\omega]\, X[\omega] \tag{2.13}$$

or

$$Y(\Omega) = H(\Omega)\, X(\Omega) \tag{2.14}$$

where

$$X[\omega] = \sum_{m=-\infty}^{\infty} x[m]\, e^{-j\omega m} \tag{2.15}$$

and

$$X(\Omega) = \int_{-\infty}^{\infty} x(t)\, e^{-j\Omega t} dt. \tag{2.16}$$

Note that FT has reduced the convolution operation for finding the output of the system to that of multiplication in the frequency domain. Such a simplification is not without cost; one has to obtain the FT of signal and impulse response and then convert the FT of the output back into the time domain. However, such a cost becomes trivial when we are dealing with cascaded systems. That is, when a series of q linear systems with IR of $h_1, h_2, \ldots, h_q$ are cascaded, the output can be found using

$$y[n] = h_1[n] \copyright h_2[n] \copyright \cdots \copyright h_n[q] \copyright x[n] \tag{2.17}$$

or

$$y(t) = h_1(t) \copyright h_2(t) \copyright \cdots \copyright h_n(t) \copyright x(t). \tag{2.18}$$

This involves involving the cumbersome convolutional summation or integration q times! This is by any standard a tedious task. With the use of system function, the output can be obtained using

$$Y[\omega] = H_1[\omega]\, H_2[\omega] \ldots H_q[\omega]\, X[\omega] \tag{2.19}$$

or

$$Y(\Omega) = H_1(\Omega)\, H_2(\Omega) \ldots H_q(\Omega)\, X(\Omega), \tag{2.20}$$

which is simply the product of the individual system functions by the FT of the input. Now, the cost of taking FT of the input and computing the inverse FT of the output becomes less burdensome for a large q.

2.1.2 Response of Linear Systems to Random Signals

A pertinent question to ask at this stage is whether a simple Fourier transform of a signal is sufficient to characterize a random process or sequence in frequency domain. The answer is clearly "no" as $X[\omega]$ and $X(\Omega)$ are both random variables. As noted in earlier sections, PSD is a more appropriate tool in spectrum analysis of random signals. We, then, are concerned with the PSD of the output of a linear system as a function of input PSD. To address this question, one has to consider the correlation of the output of the linear system as a function of the input correlation. We first consider the case of WSS input and LTI system. That is, let us consider

$$R_y[n,m] = E\left\{Y^*[n]\,Y[m]\right\} = \sum_{l_1=-\infty}^{\infty}\sum_{l_2=-\infty}^{\infty} E\left\{x^*[n-l_1]\,x[m-l_2]\right\} h^*[l_1]\,h[l_2] \tag{2.21}$$

Realizing that $E\{x^*[n-l_1]\,x[m-l_2]\} = R_x[n-l_1, m-l_2] = R_x[m-n+l_1-l_2]$,

$$\begin{aligned} R_y[n,m] &= \sum_{l_1=-\infty}^{\infty}\sum_{l_2=-\infty}^{\infty} R_x[m-n+l_1-l_2]\,h^*[l_1]\,h[l_2] \\ &= R_y[m-n] \end{aligned} \tag{2.22}$$

Furthermore,

$$\begin{aligned} E\{y[n]\} = m_y[n] &= \sum_{m=-\infty}^{\infty} h[n-m]\,E\{x[m]\} \\ &= m_x \sum_{m=-\infty}^{\infty} h[n-m] \\ &= c m_x \end{aligned} \tag{2.23}$$

where $c = \sum_{m=-\infty}^{\infty} h[m]$ is a constant and the mean of $x[n]$ is assumed to be constant since $x[n]$ is WSS. Since $m_y[n] = m_y$ and $R_y[n,m] = R_y[m-n]$, *the output of a LTI linear system with WSS input is another WSS signal.* Further, the correlation of the output of a LTI system in terms of the correlation of the WSS input signal is give by

$$\boxed{R_y[m] = E\left\{Y_n^* Y_m\right\} = \sum_{l_1=-\infty}^{\infty}\sum_{l_2=-\infty}^{\infty} R_x[m+l_1-l_2]\,h^*[l_1]\,h[l_2].} \tag{2.24}$$

Taking the FT of $R_y[m]$,

$$S_y[\omega] = \sum_{l_1=-\infty}^{\infty} \sum_{l_2=-\infty}^{\infty} \sum_{m=-\infty}^{\infty} R_x[m + l_1 - l_2] h^*[l_1] h[l_2] e^{-j\omega m} \tag{2.25}$$

Letting $m' = m + l_1 - l_2$

$$\begin{aligned} S_y[\omega] = &\sum_{l_1=-\infty}^{\infty} \sum_{l_2=-\infty}^{\infty} \sum_{m'=-\infty}^{\infty} R_X\left[m'\right] h^*[l_1] h[l_2] \\ &\times e^{-j\omega m'} e^{j\omega l_1} e^{-j\omega l_2} \end{aligned} \tag{2.26}$$

$$\begin{aligned} S_y[\omega] &= \left(\sum_{l_1=-\infty}^{\infty} h^*[l_1] e^{j\omega l_1} \right) \left(\sum_{m'=-\infty}^{\infty} R_X\left[m'\right] e^{-j\omega m'} \right) \\ &\quad \times \left(\sum_{l_2=-\infty}^{\infty} h[l_2] e^{-j\omega l_2} \right) \\ &= S_x[\omega] H[\omega] H^*[\omega] \end{aligned} \tag{2.27}$$

So,

$$\boxed{S_y[\omega] = |H[\omega]|^2 S_x[\omega]} \tag{2.28}$$

Similarly,

$$\begin{aligned} R_y(t_1, t_2) &= E\left\{y^*(t_1) y(t_2)\right\} \\ &= \int_{-\infty}^{\infty} \int_{-\infty}^{\infty} E\left\{x^*(t_1 - \tau_1) x(t_2 - \tau_2)\right\} h^*(\tau_1) h(\tau_2) d\tau_1 d\tau_2 \end{aligned} \tag{2.29}$$

which when $x(t)$ is WSS, reduces to

$$\begin{aligned} R_y(t_1, t_2) &= \int_{-\infty}^{\infty} \int_{-\infty}^{\infty} R_x(t_2 - t_1 - \tau_2 + \tau_1) h^*(\tau_1) h(\tau_2) d\tau_1 d\tau_2 \\ &= R_y(t_2 - t_1) \end{aligned} \tag{2.30}$$

Also, for a WSS input,

$$\begin{aligned} E\{y(t)\} = m_y(t) &= \int_{-\infty}^{\infty} E\{x(t-\tau)\}\, h(\tau)\, d\tau \\ &= m_x \int_{-\infty}^{\infty} h(\tau)\, d\tau \\ &= c' m_x \end{aligned} \tag{2.31}$$

where $c' = \int_{-\infty}^{\infty} h(\tau)\, d\tau$.

Hence, $y(t)$ is also WSS, and the PSD of $y(t)$ can be found as

$$\begin{aligned} S_y(\Omega) &= \int_{-\infty}^{\infty} R_y(\tau)\, e^{-j\Omega\tau} d\tau \\ &= \int_{-\infty}^{\infty}\int_{-\infty}^{\infty}\int_{-\infty}^{\infty} R_x(\tau - \tau_2 + \tau_1)\, h^*(\tau_1)\, h(\tau_2) \\ &\quad \times e^{-j\Omega\tau} d\tau d\tau_1 d\tau_2. \end{aligned} \tag{2.32}$$

Letting $\tau - \tau_2 + \tau_1 = \tau'$

$$\begin{aligned} S_y(\Omega) = &\int_{-\infty}^{\infty}\int_{-\infty}^{\infty}\int_{-\infty}^{\infty} R_x\left(\tau'\right) h^*(\tau_1)\, h(\tau_2) \\ &\times e^{-j\Omega\tau'} e^{j\Omega\tau_1} e^{-j\Omega\tau_2} d\tau' d\tau_1 d\tau_2 \end{aligned} \tag{2.33}$$

$$\begin{aligned} S_y(\Omega) = &\left(\int_{-\infty}^{\infty} R_x\left(\tau'\right) e^{-j\Omega\tau'} d\tau'\right)\left(\int_{-\infty}^{\infty} h^*(\tau_1)\, e^{j\Omega\tau_1} d\tau_1\right) \\ &\times \left(\int_{-\infty}^{\infty} h(\tau_2)\, e^{-j\Omega\tau_2} d\tau_2\right) \end{aligned} \tag{2.34}$$

which reduces to

$$\boxed{S_y(\Omega) = |H(\Omega)|^2 S_x(\Omega)} \tag{2.35}$$

Equations (2.28) and (2.35) are critical to system study as they shed light on how power of a signal is distributed in the frequency domain as a result of filtering. This is a very powerful tool for system engineering, where power distribution of a signal in the frequency domain is required to establish spectrum utilization.

Furthermore, or perhaps more importantly, these equations allow one to filter out undesired portion of the spectrum of a signal which has been corrupted by additive or other undesirable effects. Since almost all biomedical signals are non-stationary, one can only use the above results when the input signal is confined to a duration in time where its characteristics parallel those of a WSS signal. This implies that, although one cannot in general assume WSS condition for the input signal, the above results can be applied to signals over limited periods of time over which the signal characteristics are not subject to change. For this to be useful, however, one has to have a priori knowledge of the signal characteristics.

A critical question to ask at this stage is whether the above equations hold true for cyclostationary processes. This is of particular interest as most linearly modulated signals can be classified as cyclostationary. We, then, proceed to reconsider this problem when the input signal is cyclostationary. First, we establish that the output of a LTI system is cyclostationary when the input is the same. To that end, we consider the correlation function of the output, given by

$$\begin{aligned} R_y[n,m] &= E\left\{y^*[n]\,y[m]\right\} \\ &= \sum_{l_1=-\infty}^{\infty}\sum_{l_2=-\infty}^{\infty} R_x[l_1,l_2]\,h^*[n-l_1]\,h[m-l_2] \end{aligned} \tag{2.36}$$

$$\begin{aligned} R_y\left[n+lN_p, m+lN_p\right] &= \sum_{l_1=-\infty}^{\infty}\sum_{l_2=-\infty}^{\infty} R_x[l_1,l_2]\,h^*\left[n+lN_p-l_1\right] \\ &\quad\times h\left[m+lN_p-l_2\right] \end{aligned} \tag{2.37}$$

$$\begin{aligned} &= \sum_{l_1'=-\infty}^{\infty}\sum_{l_2'=-\infty}^{\infty} R_x\left[n-l_1'+lN_p, m+lN_p-l_2'\right] \\ &\quad\times h^*\left[l_1'\right]h\left[l_2'\right] \\ &= R_y[n,m] \end{aligned} \tag{2.38}$$

Furthermore, since

$$E\{y[n]\} = m_y[n] = \sum_{m=-\infty}^{\infty} h[n-m]\,E\{x[m]\}, \tag{2.39}$$

we have

$$\begin{aligned}E\left\{y\left[n+lN_p\right]\right\} &= m_y\left[n+lN_p\right]\\ &= \sum_{m=-\infty}^{\infty} h\left[n-m\right]E\left\{x\left[m+lN_p\right]\right\}\\ &= \sum_{m=-\infty}^{\infty} h\left[n-m\right]m_x\left[m+lN_p\right]\\ &= \sum_{m=-\infty}^{\infty} h\left[n-m\right]m_x\left[m\right] = m_y\left[n\right]\end{aligned}\tag{2.40}$$

Hence, the output is cyclostationary. The PSD of the output, hence, will be the FT of the average autocorrelation function. That is,

$$\begin{aligned}S_y\left[\omega\right] &= DTFT\left\{\langle R_y\left[m,m+n\right]\rangle_m\right\}\\ &= \sum_{n=-\infty}^{\infty}\sum_{l_1=-\infty}^{\infty}\sum_{l_2=-\infty}^{\infty}\langle R_x\left[m-l_1,m+n-l_2\right]\rangle_m\\ &\quad\times h^*\left[l_1\right]h\left[l_2\right]e^{-j\omega n}\end{aligned}\tag{2.41}$$

$$\begin{aligned}S_y\left[\omega\right] = \frac{1}{N_p}&\sum_{n=-\infty}^{\infty}\sum_{l_1=-\infty}^{\infty}\sum_{l_2=-\infty}^{\infty}\sum_{m=0}^{N_p-1}R_x\left[m-l_1,m+n-l_2\right]\\ &\times h^*\left[l_1\right]h\left[l_2\right]e^{-j\omega n}\end{aligned}\tag{2.42}$$

Adding $e^{\pm j\omega l_2}$ and $e^{\pm j\omega l_1}$,

$$\begin{aligned}S_y\left[\omega\right] = \frac{1}{N_p}&\sum_{n=-\infty}^{\infty}\sum_{l_1=-\infty}^{\infty}\sum_{l_2=-\infty}^{\infty}\sum_{m=0}^{N_p-1}R_x\left[m-l_1,m+n-l_2\right]\\ &\times e^{-j\omega(n+l_1-l_2)}h^*\left[l_1\right]e^{j\omega l_1}h\left[l_2\right]e^{-j\omega l_2}.\end{aligned}\tag{2.43}$$

Now, making the change of variable $m-l_1=m'$

$$\begin{aligned}S_y\left[\omega\right] = \frac{1}{N_p}&\sum_{n=-\infty}^{\infty}\sum_{l_1=-\infty}^{\infty}\sum_{l_2=-\infty}^{\infty}\sum_{m'=-l_1}^{N_p-1-l_1}R_x\left[m',m'+n-l_2+l_1\right]\\ &\times e^{-j\omega(n+l_1-l_2)}h^*\left[l_1\right]e^{j\omega l_1}h\left[l_2\right]e^{-j\omega l_2}\end{aligned}\tag{2.44}$$

Finally, making the change of variable $n' = n - l_2 + l_1$ and summing over n' first,

$$\begin{aligned} S_y[\omega] &= \sum_{l_1=-\infty}^{\infty} \sum_{l_2=-\infty}^{\infty} \sum_{n'=-\infty}^{\infty} \frac{1}{N_p} \sum_{m'=-l_1}^{N_p-1-l_1} R_x\left[m', m'+n'\right] \\ &\quad \times e^{-j\omega n'} h^*[l_1] e^{j\omega l_1} h[l_2] e^{-j\omega l_2} \\ &= \sum_{l_1=-\infty}^{\infty} \sum_{l_2=-\infty}^{\infty} \sum_{n'=-\infty}^{\infty} \langle R_x\left[m', m'+n'\right]\rangle_{m'} \\ &\quad \times e^{-j\omega n'} h^*[l_1] e^{j\omega l_1} h[l_2] e^{-j\omega l_2} \end{aligned} \tag{2.45}$$

Realizing that

$$\sum_{n'=-\infty}^{\infty} \langle R_x\left[m', m'+n'\right]\rangle_{m'} e^{-j\omega n'} = S_x[\omega] \tag{2.46}$$

for cyclostationary signals,

$$\begin{aligned} S_y[\omega] &= \sum_{l_1=-\infty}^{\infty} \sum_{l_2=-\infty}^{\infty} S_x[\omega] h^*[l_1] e^{j\omega l_1} h[l_2] e^{-j\omega l_2} \\ &= S_x[\omega] H^*(\omega) H[\omega] \\ &= S_x[\omega] |H(\omega)|^2 \end{aligned} \tag{2.47}$$

which leads to the same result as that for WSS signals! This is a very encouraging result in that it allows one to shape/filter the spectrum of cyclostationary signals using filtering in a manner that is identical to that used for WSS signals.

Similarly, in the continuous domain, we have

$$R_y(t_1, t_2) = \int_{-\infty}^{\infty} \int_{-\infty}^{\infty} R_x(t_1 - \tau_1, t_2 - \tau_2) h^*(\tau_1) h(\tau_2) d\tau_1 d\tau_2 \tag{2.48}$$

$$\begin{aligned} R_y\left(t_1 + lT_p, t_2 + lT_p\right) &= \int_{-\infty}^{\infty} \int_{-\infty}^{\infty} R_x\left(t_1 + lT_p - \tau_1, t_2 + lT_p - \tau_2\right) \\ &\quad \times h^*(\tau_1) h(\tau_2) d\tau_1 d\tau_2 \end{aligned} \tag{2.49}$$

Since $x(t)$ is cyclostationary,

$$R_x\left(t_1 + lT_p - \tau_1, t_2 + lT_p - \tau_2\right) = R_x(t_1 - \tau_1, t_2 - \tau_2), \tag{2.50}$$

and hence $R_y\left(t_1 + lT_p, t_2 + lT_p\right) = R_y\left(t_1, t_2\right)$. Also,

$$
\begin{aligned}
E\{y(t)\} = m_y(t) &= \int_{-\infty}^{\infty} E\{x(t-\tau)\} h(\tau)\, d\tau \\
E\left\{y\left(t + lT_p\right)\right\} &= m_y\left(t + lT_p\right) \\
&= \int_{-\infty}^{\infty} E\left\{x\left(t - \tau + lT_p\right)\right\} h(\tau)\, d\tau \\
&= \int_{-\infty}^{\infty} m_x\left(t + lT_p\right) h(\tau)\, d\tau \\
&= \int_{-\infty}^{\infty} m_x(t)\, h(\tau)\, d\tau \\
&= m_y(t)
\end{aligned}
\tag{2.51}
$$

Therefore, $y(t)$ is also cyclostationary and $S_y(\Omega)$ is given by

$$
\begin{aligned}
S_y(\Omega) &= FT\left\{\langle R_y(t, t+\tau)\rangle_t\right\} \\
&= \int_{-\infty}^{\infty}\int_{-\infty}^{\infty}\int_{-\infty}^{\infty} \langle R_x(t-\tau_1, t+\tau-\tau_2)\rangle_t e^{-j\Omega\tau} \\
&\quad \times h^*(\tau_1)\, h(\tau_2)\, d\tau_1 d\tau_2 d\tau.
\end{aligned}
\tag{2.52}
$$

Making a change of variables $t - \tau_1 = t'$

$$
\begin{aligned}
S_y(\Omega) = &\int_{-\infty}^{\infty}\int_{-\infty}^{\infty}\int_{-\infty}^{\infty} \langle R_x\left(t', t' + \tau + \tau_1 - \tau_2\right)\rangle_{t'} e^{-j\Omega\tau} \\
&\times h^*(\tau_1)\, h(\tau_2)\, d\tau_1 d\tau_2 d\tau
\end{aligned}
\tag{2.53}
$$

Now, adding $e^{\pm j\omega\tau_1}$ and $e^{\pm j\omega\tau_2}$

$$
\begin{aligned}
S_y(\Omega) = &\int_{-\infty}^{\infty}\int_{-\infty}^{\infty}\int_{-\infty}^{\infty} \langle R_x\left(t', t' + \tau + \tau_1 - \tau_2\right)\rangle_{t'} e^{-j\Omega(\tau+\tau_1-\tau_2)} \\
&\times h^*(\tau_1)\, e^{j\Omega\tau_1} h(\tau_2)\, e^{-j\Omega\tau_2} d\tau_1 d\tau_2 d\tau
\end{aligned}
\tag{2.54}
$$

Finally, making the change of variable $\tau' = \tau + \tau_1 - \tau_2$,

$$S_y(\Omega) = \int_{-\infty}^{\infty}\int_{-\infty}^{\infty}\int_{-\infty}^{\infty} \langle R_x(t', t' + \tau')\rangle_{t'} e^{-j\Omega\tau'} h^*(\tau_1) e^{j\Omega\tau_1} \times h(\tau_2) e^{-j\Omega\tau_2} d\tau' d\tau_1 d\tau_2 \tag{2.55}$$

and realizing that

$$\int_{-\infty}^{\infty} \langle R_x(t', t' + \tau')\rangle_{t'} e^{-j\Omega\tau'} d\tau' = S_x(\Omega), \tag{2.56}$$

we have

$$\begin{aligned} S_y(\Omega) &= \int_{-\infty}^{\infty}\int_{-\infty}^{\infty} S_x(\Omega) h^*(\tau_1) e^{j\Omega\tau_1} h(\tau_2) e^{-j\Omega\tau_2} d\tau' d\tau_1 d\tau_2 \\ &= S_x(\Omega) H(\Omega) H^*(\Omega) \\ &= S_x(\Omega) |H(\Omega)|^2 \end{aligned} \tag{2.57}$$

With this useful tool at hand, we can examine the output of linear systems when the input is a random process. In particular, this tool allows a systematic means of shaping the spectrum of signals. To illustrate this point, let us consider the following scenario: Let us assume that we are interested in computing the PSD of the signal

$$x(t) = \sum_{n=-\infty}^{\infty} I_n p(t - nT). \tag{2.58}$$

Based upon the above discussions, we are better off computing the PSD of

$$w(t) = \sum_{n=-\infty}^{\infty} I_n \delta(t - nT) \tag{2.59}$$

and then filter the above signal with a filter with an impulse response of $p(t)$. Finally, applying $S_y(\Omega) = S_x(\Omega)|H(\Omega)|^2$, we can obtain the PSD of the desired signal. To elaborate, one can view the random signal as the input to a linear system with impulse response $h(t) = p(t)$. It is rather easy to verify that

$$x(t) = w(t) \copyright h(t) \tag{2.60}$$

where © denotes convolution. Now, in the previous examples, we have shown that $w(t)$ is cyclostationary with period T. That is, $R_w(t, t+\tau)$ is a periodic function of t with period T. Hence, PSD of $w(t)$ can be found as

$$\begin{aligned} FT\{\langle R_w(t, t+\tau)\rangle_t\} &= \frac{1}{T}\int_{-\frac{T}{2}}^{\frac{T}{2}}\int_{-\infty}^{\infty}\sum_{n_1=-\infty}^{\infty}\sum_{n_1=-\infty}^{\infty} R_I(n_2-n_1) \\ &\quad \times \delta(t-n_1T)\,\delta(t+\tau-n_2T)\,e^{-i\Omega\tau}d\tau dt \\ &= \frac{1}{T}\int_{-\frac{T}{2}}^{\frac{T}{2}}\sum_{n_1=-\infty}^{\infty}\sum_{n_1=-\infty}^{\infty} R_I(n_2-n_1) \\ &\quad \times \delta(t-n_1T)\,e^{j\Omega(t-n_2T)}dt \\ &= \frac{1}{T}\sum_{n_1=-\infty}^{\infty}\sum_{n_1=-\infty}^{\infty} R_I(n_2-n_1)\,e^{-i\Omega T(n_2-n_1)} \\ &= \frac{1}{T}S_I[\Omega T] \end{aligned} \tag{2.61}$$

Now, using (2.35), we arrive at (1.176).

Example 1 Signal

$$x(t) = \sum_{n=-\infty}^{\infty} I_n p(t-nT)$$

with

$$p(t) = \begin{cases} 0.1\left(1-\cos\left(\frac{2\pi t}{T}\right)\right) & 0 \le t \le T \\ 0 & \text{otherwise} \end{cases}$$

(shown below)
is sampled at the rate of $\frac{10}{T}$ and the resulting signal is passed through a filter with the system function

$$H[\omega] = \begin{cases} 1; & |\omega| \le \frac{\pi}{8} \\ \frac{40}{3\pi}\left(-|\omega|+\frac{\pi}{5}\right); & \frac{\pi}{8} \le |\omega| \le \frac{\pi}{5} \\ 0; & \text{otherwise} \end{cases}. \tag{2.62}$$

$H[\omega] = \begin{cases} 1; & \text{if } |\omega| \leq \frac{\pi}{8} \\ \frac{40}{3\pi}\left(-|\omega| + \frac{\pi}{5}\right); & \text{if } \frac{\pi}{8} \leq |\omega| \leq \frac{\pi}{5} \\ 0; & \text{if} \end{cases}$. Furthermore, I_n is a zero-mean random sequence with $R_I[q] = \delta[q] + 0.5\delta[q-1] + 0.5\delta[q+1]$. Plot the PSD of the input and output of this filter.

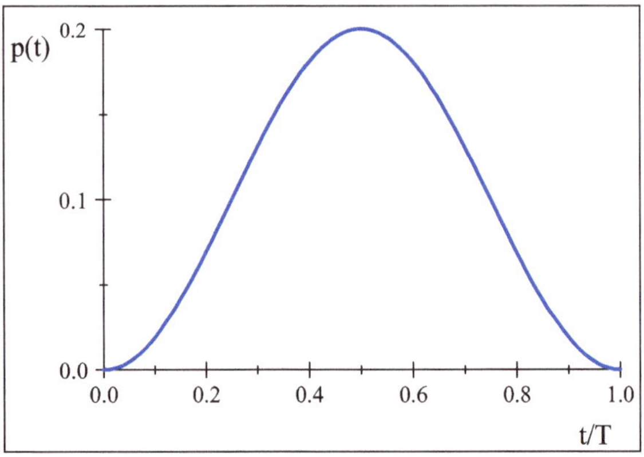

A plot of p(t) for Example 1

Solution: *Using* (1.185),

$$S_x[\omega] = \frac{1}{10} S_I[10\omega] |P[\omega]|^2; \; |\omega| \leq \pi$$

where

$$\begin{aligned} P[\omega] &= \sum_{n=0}^{10} P\left(\frac{nT}{10}\right) \exp(-j\omega n) \\ &= 0.1 \sum_{n=0}^{10} \left(1 - \frac{1}{2} e^{j\frac{\pi n}{5}} - \frac{1}{2}^{-j\frac{\pi n}{5}}\right) e^{-j\omega n} \\ &= 0.1 \frac{e^{-j11\omega} - 1}{e^{-j\omega} - 1} - \frac{0.1}{2} \frac{e^{-j\left(11\omega - \frac{11\pi}{5}\right)} - 1}{e^{-j\left(\omega - \frac{\pi}{5}\right)} - 1} - \frac{0.1}{2} \frac{e^{-j\left(11\omega + \frac{11\pi}{5}\right)} - 1}{e^{-j\left(\omega + \frac{\pi}{5}\right)} - 1}. \end{aligned} \tag{2.63}$$

Next, we examine the PSD of I_n, *which is given by*

$$S_I[\omega] = 1 + 0.5e^{-j\omega} + 0.5e^{+j\omega} = 1 + \cos(\omega).$$

The PSD of input is plotted below.

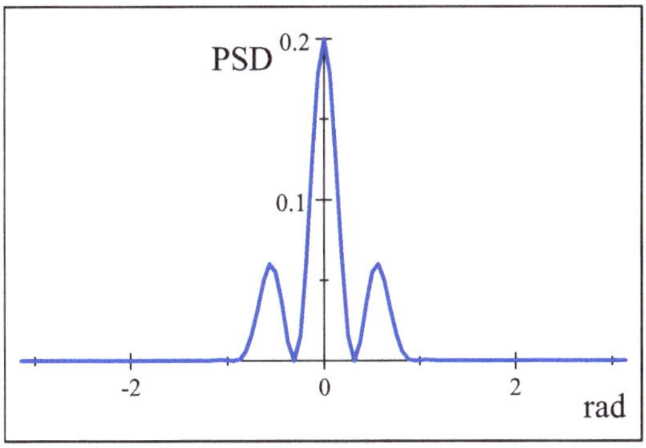

Input power spectrum as a function of ω (in rad).

The filter system function is depicted in the figure below.

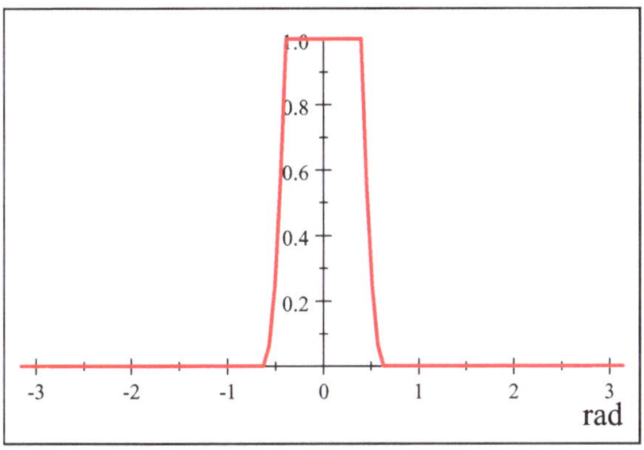

$H[\omega]$ as a function of $|\omega| \leq \pi$.

The output spectrum is given by $S_y[\omega] = S_x[\omega]\,|H(\omega)|^2$ *and is shown in the following figure.*

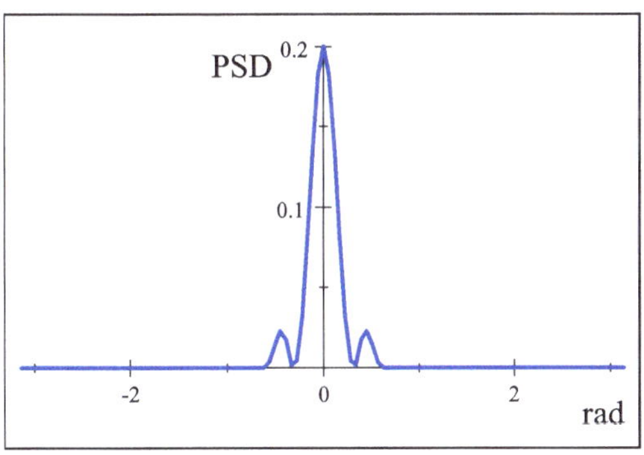

Output power spectrum as a function of ω (in rad).

Note that filtering has a profound impact on the shape of the signal spectrum, and hence it is a powerful tool in reshaping the frequency content of a random signal.

Example 2 Repeat the previous example when the pulse shape is change to a raised-cosine (RC) pulse, given by $p(t) = \sin\text{c}\left(\frac{t}{T}\right)\frac{\cos\left(\frac{\pi\beta t}{T}\right)}{1-\frac{4\beta^2 t^2}{T^2}}$ where $\sin\text{c}(x) = \frac{\sin(\pi x)}{\pi x}$ with β denoting the 'roll-off' factor of the pulse. Plot the output spectrum for $\beta = 0.25$ *and* $\beta = 0.5$.

Solution: *This pulse is used very frequently in digital communications (as we observed in the preceding sections) due to its desirable properties. A plot of the pulse shape for different values of* β *(known as roll-off factor) is depicted in Fig. 2.1. Note that this pulse assumes a value of zero for integer multiples of a symbol period. That is, for the signal* $x(t) = \sum_{n=-\infty}^{\infty} I_n p(t - nT)$*, we have* $x[n] = I_n$ *for all n. Hence, a sample at the correct "time" will yield a desired result. Also, for* $\beta = 0$*,* $p(t)$ *reduces to a* $\sin\text{c}(x)$ *function, which implies that the RC filter with* $p(t)$ *impulse response is an ideal lowpass filter.*

Given this pulse shape, $P[\omega]$ *is given by*

$$P[\omega] = \sum_{n-\infty}^{\infty} p(nT_s)\exp(-j\omega n) = \sum_{n-\infty}^{\infty} \sin\text{c}\left(\frac{nT_s}{T}\right)\frac{\cos\left(\frac{\pi n\beta T_s}{T}\right)}{1-\frac{4\beta^2 n^2 T_s^2}{T^2}} \rightarrow$$

$$P[\omega] = \begin{cases} T/T_s & |\omega| \le \frac{2\pi(1-\beta)T_s}{2T} \\ \frac{T}{2T_s}\left[1+\cos\left(\frac{\pi T}{\beta}\left[\frac{|\omega|}{2\pi T_s}-\frac{1-\beta}{2T}\right]\right)\right] & \frac{2\pi(1-\beta)T_s}{2T} \le |\omega| \le \frac{2\pi(1+\beta)T_s}{2T} \\ 0 & |\omega| > \frac{2\pi(1+\beta)T_s}{2T} \end{cases} \tag{2.64}$$

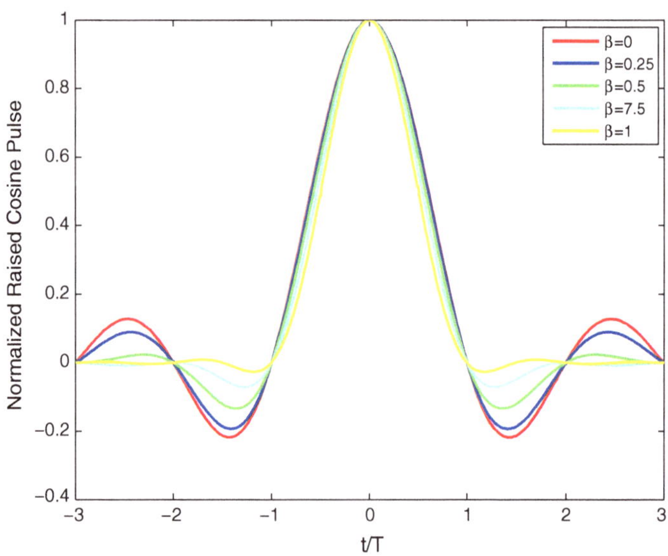

Fig. 2.1 Raised-cosine pulse for different roll-off factors

and is shown below for $T = 10T_s$ *and* $\beta = 0.5$ *and* 0.25. *Note that* β *has an impact on the 'bandwidth' of the pulse shape.*

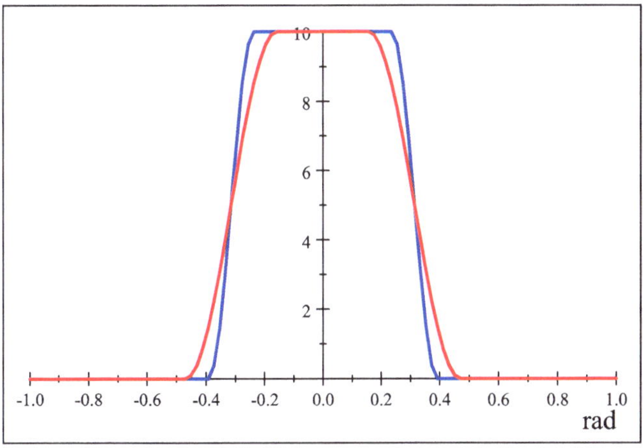

$P[\omega]$ *for RC pulse as a function of* ω *for* $\beta = 0.25$ *(blue) and* $\beta = 0.5$ *(red).*

The output PSD is shown below. The filter has almost entirely suppressed the impact of β *as the response of the filter is pronounced around* $\omega = 0$ *and the two PSDs are almost identical around* $\omega = 0$ *(see above).*

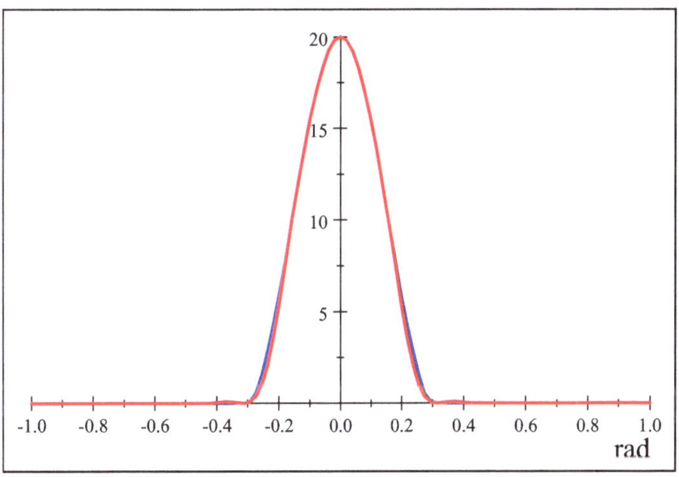

$S_y[\omega]$ *as a function of* ω *for RC pulse with* $\beta = 0.25$ *(blue) and* $\beta = 0.5$ *(red).*

Since linear system theory can be applied here, one can consider this problem as a problem of filtering a signal (such as $w(t)$ shown in this section), where the filtering consists of two stages. One is $H[\omega]$ and the other $P[\omega]$. From linear system theory, we realize that the ordering of the filtering operation is immaterial here. Note that the filtering operation in the above example behaves quite similar to the upsampling/interpolation scenario. To elaborate, let us consider the signal

$$\zeta_q = \sum_{n=-\infty}^{\infty} I_n U\left[q-n\right] \tag{2.65}$$

where $U\left[q\right]$ is the unit step response given by $U\left[q\right] = \begin{cases} 1 & q \geq 0 \\ 0 & \text{otherwise} \end{cases}$. If ζ_q is upsampled by a factor of μ (i.e., sampling interval is $T_s = \frac{T}{\mu}$), the resulting signal is given by

$$\eta_q = \zeta_{\frac{q}{\mu}} = \sum_{n=-\infty}^{\infty} I_n \delta\left[q - n\mu\right] \tag{2.66}$$

where $\delta\left[n\right] = \begin{cases} 1 & n = 0 \\ 0 & \text{otherwise} \end{cases}$. Obviously, filtering η_q with a filter with an impulse response $p\left[n\right]$ results in the sampled version of $x\left(t\right)$, given by (1.77). Now, if we choose to filter (2.66) using $H\left[\omega\right]$ first and then pass the resulting signal through a filter with the impulse response $p\left[n\right]$, an identical result to that of the above example is obtained. Hence, $x\left(t\right)$ may be viewed as the result of oversampling/interpolation/filtering of the signal ζ_q.

2.2 Response of Nonlinear Systems to Random Signals

In this section, we tackle a problem which does not allow the use of linear system theory, as discussed in previous section, to analyze the response of the system to random signals. For reasons which will become obvious shortly, these types of problems have received little attention in textbooks on the subject. Although initially one may conclude that nonlinear systems are of limited interest, a careful examination of some routine operations in biomedical or electronics applications in general contradicts this conclusion. To elaborate, virtually all linear devices are comprised of components which are designed to appear linear over a prescribed range of input parameters. For instance, power amplifiers, low-noise amplifiers, or mixers function as linear devices over a limited range of the input signal power level. Hence, one can be satisfied with the linear system theory approach toward characterizing the response of such systems to random inputs as long as one is confined to an appropriate range of system parameters. A more serious problem emerges when the intended operation is a nonlinear one. In biomedical signal processing, the signal of interest, which are used for diagnosis purposes, may experience a number of nonlinear effects before being detected by a receiving device. The nonlinearity may or may not be known to us. Either way, the recovered signal, which may be used to identify an underlying problem, has now gone through a nonlinear effect which will make the analysis of the signal rather difficult. It is also well known that biological systems behave in a 'chaotic' manner in a variety of scenarios. Epileptic brain response is a good example of such a scenario. The chaotic responses of systems may be attributed to their nonlinear characteristics. Hence, the response of nonlinear systems to random signals is of particular interest to those interested in studying chaotic systems. A complete study of nonlinear systems response to random signals is beyond the scope of this text. Therefore, we focus on some basic nonlinearity which can be used to model some common effects in electronic devices, which are often used to collect and process biomedical signals.

For instance, power-of-ν operation may be used to model a vast number of nonlinear devices. In this case, the input-output relationship of the system is given by

$$y[n] = |x[n]|^{\nu} \tag{2.67}$$

or

$$y(t) = |x(t)|^{\nu}. \tag{2.68}$$

The other model, which may allows delay propagation effect, is the delay-and-multiply (D&M) operation, given by

$$y[n] = x^{*}[n]\,x[n-m] \tag{2.69}$$

or

$$y(t) = x^{*}(t)\,x(t-T_d) \tag{2.70}$$

for some T_d and m. Note that $T_d = 0$ (or $m = 0$) leads to a special case of power-of-ν device, known as squaring device for $\nu = 2$. The results discussed in the previous sections will fail to model the responses of the system to random signals in these cases. One immediate observation which can be made is that the statistics of the output of such devices involve higher statistics of the input signal, which may or may not be readily available. In particular, when a non-Gaussian input signal is considered, these operations often lead to intractable results. A large class of signals in the engineering field can be modeled as Gaussian. In that case, the statistics of nonlinear signals, as noted earlier, can be obtained from its second order statistics. We, therefore, consider Gaussian signals in the remainder of this section. Although a full treatment of general signals is not presented here, the ensuing analysis presents a roadmap for further analysis of nonlinear devices with non-Gaussian inputs.

2.2.1 Nonlinear Processing of Gaussian Signals

We assume that $x[n]$ (or $x(t)$) is a "proper" zero-mean Gaussian random sequence (or process). That is,

$$E\{x[n]\,x[n-m]\} = 0 \tag{2.71}$$

for all n and m. Furthermore, the correlation function is defined as

$$R_x[n, n-m] = E\left\{x^*[n]\,x[n-m]\right\}. \tag{2.72}$$

The above implies that if

$$x[n] = x_r[n] + jx_i[n],$$

then

$$R_{x_i x_r}[n, n-m] = -R_{x_r x_i}[n, n-m] \tag{2.73}$$

and

$$R_{x_i}[n, n-m] = R_{x_r}[n, n-m]. \tag{2.74}$$

Similarly,

$$E\{x(t)\,x(t-\tau)\} = 0 \tag{2.75}$$

where

$$R_x(t, t-\tau) = E\left\{x^*(t)\,x(t-\tau)\right\} \tag{2.76}$$

This implies that if

$$x(t) = x_r(t) + jx_i(t), \tag{2.77}$$

then

$$R_{x_i x_r}(t, t-\tau) = -R_{x_r x_i}(t, t-\tau) \tag{2.78}$$

and

$$R_{x_i}(t, t-\tau) = R_{x_r}(t, t-\tau) \tag{2.79}$$

for all t and τ.

Since power is related to the second order statistics of a signal, we first consider the first two moments of a signal which has gone through a simple squaring or delay-and-multiply operation. Note that squaring operation is a special case of delay-and-multiply, and hence, we begin with this operation. To that end,

$$\begin{aligned} E\{y[n]\} &= E\left\{x^*[n]\,x[n-m]\right\} \\ &= R_x[n, n-m] \end{aligned} \tag{2.80}$$

or

$$E\{y(t)\} = R_x(t, t - T_d), \tag{2.81}$$

which implies that the first order statistics of the signal depends on the second order statistics of the input. To address the second order statistics of the output, we first re-write $y[n]$ (or $y(t)$) in terms of its the real and imaginary parts. That is,

$$\begin{aligned} y[n] &= y_r[n] + jy_i[n] \\ &= E\left\{x^*[n]x[n-m]\right\} \\ &= x_r[n]x_r[n-m] + x_i[n]x_i[n-m] \\ &\quad + j(x_r[n]x_i[n-m] - x_i[n]x_r[n-m]) \end{aligned} \tag{2.82}$$

or

$$\begin{aligned} y(t) &= y_r(t) + jy_i(t) \\ &= E\left\{y^*(t)y(t-T_d)\right\} \\ &= x_r(t)x_r(t-T_d) + x_i(t)x_i(t-T_d) \\ &\quad + j(x_r(t)x_i(t-T_d) - x_i(t)x_r(t-T_d)). \end{aligned} \tag{2.83}$$

Hence,

$$y_r[n] = x_r[n]x_r[n-m] + x_i[n]x_i[n-m] \tag{2.84}$$

$$y_i[n] = x_r[n]x_i[n-m] - x_i[n]x_r[n-m] \tag{2.85}$$

Similarly,

$$y_r(t) = x_r(t)x_r(t-T_d) + x_i(t)x_i(t-T_d) \tag{2.86}$$

$$y_i(t) = x_r(t)x_i(t-T_d) - x_i(t)x_r(t-T_d). \tag{2.87}$$

Hence, in general,

$$\begin{aligned} E\left\{y_r^2[n]\right\} &= E\left\{x_r^2[n]x_r^2[n-m]\right\} + E\left\{x_i^2[n]x_i^2[n-m]\right\} \\ &\quad + 2E\left\{x_r[n]x_r[n-m]x_i[n]x_i[n-m]\right\}, \end{aligned} \tag{2.88}$$

$$\begin{aligned} E\left\{y_i^2[n]\right\} &= E\left\{x_r^2[n]x_i^2[n-m]\right\} + E\left\{x_i^2[n]x_r^2[n-m]\right\} \\ &\quad - 2E\left\{x_r[n]x_i[n-m]x_i[n]x_r[n-m]\right\}, \end{aligned} \tag{2.89}$$

and

$$\begin{aligned} E\{y_i[n]\,y_r[n]\} &= E\left\{x_r^2[n]\,x_i[n-m]\,x_r[n-m]\right\} \\ &\quad - E\left\{x_i^2[n]\,x_i[n-m]\,x_r[n-m]\right\} \\ &\quad + E\left\{x_r[n]\,x_i[n]\,x_i^2[n-m]\right\} \\ &\quad - E\left\{x_r[n]\,x_i[n]\,x_r^2[n-m]\right\}. \end{aligned} \tag{2.90}$$

Similarly, for continuous time,

$$\begin{aligned} E\left\{y_r^2(t)\right\} &= E\left\{x_r^2(t)\,x_r^2(t-T_d)\right\} + E\left\{x_i^2(t)\,x_i^2(t-T_d)\right\} \\ &\quad + 2E\{x_r(t)\,x_r(t-T_d)\,x_i(t)\,x_i(t-T_d)\}, \end{aligned} \tag{2.91}$$

$$\begin{aligned} E\left\{y_i^2(t)\right\} &= E\left\{x_r^2(t)\,x_i^2(t-T_d)\right\} + E\left\{x_i^2(t)\,x_r^2(t-T_d)\right\} \\ &\quad - 2E\{x_r(t)\,x_i(t-T_d)\,x_i(t)\,x_r(t-T)\}, \end{aligned} \tag{2.92}$$

and

$$\begin{aligned} E\{y_r(t)\,y_i(t)\} &= E\left\{x_r^2(t)\,x_i(t-T_d)\,x_r(t-T_d)\right\} \\ &\quad - E\left\{x_i^2(t)\,x_i(t-T_d)\,x_r(t-T_d)\right\} \\ &\quad + E\left\{x_r(t)\,x_i(t)\,x_i^2(t-T_d)\right\} \\ &\quad - E\left\{x_r(t)\,x_i(t)\,x_r^2(t-T_d)\right\}. \end{aligned} \tag{2.93}$$

In general, the above equations can be applied to all signals. One glaring conclusion of the above is that the 4th order statistics of the signals are typically needed in order to proceed. Such statistical information may not be available to us. For Gaussian signals, however, significant simplification is possible. That is, for zero-mean Gaussian random processes, we have (use the characteristic function of a random vector to show this)

$$\begin{aligned} E\{x(t_1)\,x(t_2)\,x(t_3)\,x(t_4)\} &= E\{x(t_1)\,x(t_2)\}\,E\{x(t_3)\,x(t_4)\} \\ &\quad + E\{x(t_1)\,x(t_3)\}\,E\{x(t_2)\,x(t_4)\} \\ &\quad + E\{x(t_1)\,x(t_4)\}\,E\{x(t_2)\,x(t_3)\}. \end{aligned} \tag{2.94}$$

For the case the signal is non-zero mean, we can define

$$x(t) = m(t) + x'(t) \tag{2.95}$$

where $m(t) = E\{x(t)\}$ and $x'(t)$ is a zero-mean Gaussian random process which is statistically identical to $x(t)$. In that event,

$$\begin{aligned} E\{x(t_1)x(t_2)x(t_3)x(t_4)\} = & E\left\{x'(t_1)x'(t_2)\right\}E\left\{x'(t_3)x'(t_4)\right\} \\ & + m(t_1)m(t_2)m(t_3)m(t_4) \qquad (2.96) \\ & + m(t_1)m(t_2)R_{x'}(t_3,t_4) \\ & + m(t_1)m(t_3)R_{x'}(t_2,t_4) \\ & + m(t_1)m(t_4)R_{x'}(t_2,t_3) \\ & + m(t_2)m(t_3)R_{x'}(t_1,t_4) \\ & + m(t_2)m(t_4)R_{x'}(t_1,t_3) \\ & + m(t_3)m(t_4)R_{x'}(t_1,t_2)\,. \end{aligned} \tag{2.97}$$

where we have used the following property of zero-mean Gaussian random processes:

$$E\left\{x'(t_1)x'(t_2)x'(t_3)\right\} = 0 \tag{2.98}$$

for all t_1, t_2, and t_3. Hence, having access to the correlation of the process brings us back to computing $E\left\{x'(t_1)x'(t_2)\right\}E\left\{x'(t_3)x'(t_4)\right\}$ where $x'(t)$ is a zero-mean process.

For the following discussion, however, we assume that the Gaussian process is zero-mean (as would be the case for any electronic noise or additive noise observed in biomedical signals). Hence, for Gaussian signals, we have

$$\begin{aligned} E\left\{y_r^2[n]\right\} = & R_{x_r}[n,n]R_{x_r}[n-m,n-m] + 2R_{x_r}^2[n,n-m] \\ & + R_{x_i}[n,n]R_{x_i}[n-m,n-m] + 2R_{x_i}^2[n,n-m] \\ & + 2R_{x_r}[n,n-m]R_{x_i}[n,n-m] + 2R_{x_rx_i}[n,n-m]R_{x_rx_i}[n-m,n] \\ & + 2R_{x_rx_i}[n,n]R_{x_rx_i}[n-m,n-m]\,, \end{aligned} \tag{2.99}$$

$$\begin{aligned} E\left\{y_i^2[n]\right\} = & 2R_{x_rx_i}^2[n,n-m] + R_{x_r}[n,n]R_{x_i}[n-m,n-m] \\ & + 2R_{x_ix_r}^2[n,n-m] + R_{x_i}[n,n]R_{x_r}[n-m,n-m] \\ & - 2R_{x_r}[n,n-m]R_{x_i}[n,n-m] - 2R_{x_rx_i}[n,n]R_{x_rx_i}[n-m,n-m] \\ & - 2R_{x_rx_i}[n,n-m]R_{x_rx_i}[n-m,n]\,, \end{aligned} \tag{2.100}$$

and

$$\begin{aligned} & E\{y_r[n]y_i[n]\} = R_{x_r}[n,n-m]R_{x_rx_i}[n,n-m] + R_{x_r}[n,n]R_{x_rx_i}[n-m,n-m] \\ & + R_{x_r}[n-m,n]R_{x_rx_i}[n,n-m] - R_{x_i}[n,n-m]R_{x_ix_r}[n,n-m] \\ & - R_{x_i}[n,n]R_{x_ix_r}[n-m,n-m] - R_{x_i}[n-m,n]R_{x_ix_r}[n,n-m] \\ & + R_{x_i}[n,n-m]R_{x_rx_i}[n,n-m] + R_{x_i}[n-m,n-m]R_{x_rx_i}[n,n] \\ & + R_{x_i}[n,n-m]R_{x_ix_r}[n-m,n] - R_{x_r}[n,n-m]R_{x_ix_r}[n,n-m] \\ & - R_{x_r}[n-m,n-m]R_{x_rx_i}[n,n] - R_{x_r}[n,n-m]R_{x_rx_i}[n-m,n] \end{aligned} \tag{2.101}$$

Realizing (2.73) and (2.74), these equations reduce to

$$E\left\{y_r^2[n]\right\} = 6R_{x_r}^2[n, n-m] + 2R_{x_r}[n, n]\, R_{x_r}[n-m, n-m] - 2R_{x_r x_i}^2[n, n-m] + 2R_{x_r x_i}[n, n]\, R_{x_r x_i}[n-m, n-m], \quad (2.102)$$

$$E\left\{y_i^2[n]\right\} = 6R_{x_r x_i}^2[n, n-m] + 2R_{x_r}[n, n]\, R_{x_r}[n-m, n-m] - 2R_{x_r}^2[n, n-m] - 2R_{x_r x_i}[n, n]\, R_{x_r x_i}[n-m, n-m] \quad (2.103)$$

and

$$E\left\{y_r[n]\, y_i[n]\right\} = 8R_{x_r}[n, n-m]\, R_{x_r x_i}[n, n-m] \quad (2.104)$$

Similarly,

$$\begin{aligned} E\left\{y_r^2(t)\right\} = &\; R_{x_r}(t, t)\, R_{x_r}(t-T_d, t-T_d) + 2R_{x_r}^2(t, t-T_d) \\ &+ R_{x_i}(t, t)\, R_{x_i}(t-T_d, t-T_d) + 2R_{x_i}^2(t, t-T_d) \\ &+ 2R_{x_r}(t, t-T_d)\, R_{x_i}(t, t-T_d) + 2R_{x_r x_i}(t, t)\, R_{x_r x_i}(t-T_d, t-T_d) \\ &+ 2R_{x_r x_i}(t, t-T_d)\, R_{x_i x_r}(t, t-T_d), \end{aligned} \quad (2.105)$$

$$\begin{aligned} E\left\{y_i^2(t)\right\} = &\; R_{x_r}(t, t)\, R_{x_i}(t-T_d, t-T_d) + 2R_{x_r x_i}^2(t, t-T_d) \\ &+ R_{x_i}(t, t)\, R_{x_r}(t-T_d, t-T_d) + 2R_{x_i x_r}^2(t, t-T_d) \\ &- 2R_{x_r}(t, t-T_d)\, R_{x_i}(t, t-T_d) - 2R_{x_r x_i}(t, t)\, R_{x_i x_r}(t-T_d, t-T_d) \\ &- 2R_{x_r x_i}(t, t-T_d)\, R_{x_i x_r}(t, t-T_d), \end{aligned} \quad (2.106)$$

and

$$\begin{aligned} E\left\{y_r(t)\, y_i(t)\right\} = &\; E\left\{x_r^2(t)\, x_r(t-T_d)\, x_i(t-T_d)\right\} \\ &- E\left\{x_i^2(t)\, x_r(t-T_d)\, x_i(t-T_d)\right\} \\ &+ E\left\{x_r(t)\, x_i(t)\, x_i^2(t-T_d)\right\} \\ &- E\left\{x_r(t)\, x_i(t)\, x_r^2(t-T_d)\right\} \end{aligned} \quad (2.107)$$

Substituting for the expectations in terms of the correlation function of $x_r(t)$ and $x_i(t)$,

$$\begin{aligned} E\left\{y_r(t)\, y_i(t)\right\} = &\; R_{x_r}(t, t)\, R_{x_r x_i}(t-T_d, t-T_d) \\ &+ 2R_{x_r}(t, t-T_d)\, R_{x_r x_i}(t, t-T_d) \\ &- R_{x_i}(t, t)\, R_{x_r x_i}(t-T_d, t-T_d) \end{aligned}$$

$$\begin{aligned}
& -2R_{x_i}(t,t-T_d)\,R_{x_i x_r}(t,t-T_d) \\
& +R_{x_i}(t-T_d,t-T_d)\,R_{x_r x_i}(t,t) \\
& +2R_{x_i}(t,t-T_d)\,R_{x_r x_i}(t,t-T_d) \\
& -R_{x_r}(t-T_d,t-T_d)\,R_{x_r x_i}(t,t) \\
& -2R_{x_r}(t,t-T_d)\,R_{x_i x_r}(t,t-T_d)\,, \qquad (2.108)
\end{aligned}$$

which reduce to

$$\begin{aligned}
E\left\{y_r^2(t)\right\} = {} & R_{x_r}(t,t)\,R_{x_r}(t-T_d,t-T_d) + 2R_{x_r}^2(t,t-T_d) \\
& + R_{x_i}(t,t)\,R_{x_i}(t-T_d,t-T_d) + 2R_{x_i}^2(t,t-T_d) \\
& + 2R_{x_r}(t,t-T_d)\,R_{x_i}(t,t-T_d) + 2R_{x_r x_i}(t,t)\,R_{x_r x_i}(t-T_d,t-T_d) \\
& + 2R_{x_r x_i}(t,t-T_d)\,R_{x_i x_r}(t,t-T_d)\,, \qquad (2.109)
\end{aligned}$$

and

$$\begin{aligned}
& E\left\{y_r^2(t)\right\} = 6R_{x_r}^2(t,t-T_d) + 2R_{x_r}(t,t)\,R_{x_r}(t-T_d,t-T_d) \\
& -2R_{x_r x_i}^2(t,t-T_d) + 2R_{x_r x_i}(t,t)\,R_{x_r x_i}(t-T_d,t-T_d)\,, \qquad (2.110)
\end{aligned}$$

$$\begin{aligned}
& E\left\{y_i^2(t)\right\} = 6R_{x_r x_i}^2(t,t-T_d) + 2R_{x_r}(t,t)\,R_{x_r}(t-T_d,t-T_d) \\
& -2R_{x_r}^2(t,t-T_d) - 2R_{x_r x_i}(t,t)\,R_{x_r x_i}(t-T_d,t-T_d)\,, \qquad (2.111)
\end{aligned}$$

and

$$E\left\{y_r(t)\,y_i(t)\right\} = 8R_{x_r}(t,t-T_d)\,R_{x_r x_i}(t,t-T_d)\,.$$

It becomes immediately obvious that the output of this device is a complex Gaussian random sequence with correlated real and imaginary parts. In most practical problems, the real and imaginary parts of the input complex Gaussian random sequence (or process) are uncorrelated. Hence, $R_{X_i X_r}[m,n] = 0$ for all n and m. We then have

$$E\left\{y_r^2[n]\right\} = 6R_{x_r}^2[n,n-m] + 2R_{x_r}[n,n]\,R_{x_r}[n-m,n-m]\,, \qquad (2.112)$$

$$E\left\{y_i^2[n]\right\} = 2R_{x_r}[n,n]\,R_{x_r}[n-m,n-m] - 2R_{x_r}^2[n,n-m] \qquad (2.113)$$

and

$$E\left\{y_r[n]\,y_i[n]\right\} = 0 \qquad (2.114)$$

which, interestingly enough, implies that the real and imaginary parts of $y[n]$ also become uncorrelated. In the continuous time domain,

$$E\left\{y_r^2(t)\right\} = 6R_{x_r}^2(t, t - T_d) + 2R_{x_r}(t, t)\, R_{x_r}(t - T_d, t - T_d) \tag{2.115}$$

$$E\left\{y_i^2(t)\right\} = 2R_{x_r}(t, t)\, R_{x_r}(t - T_d, t, t - T_d) - 2R_{x_r}^2(t, t - T_d) \tag{2.116}$$

$$E\left\{y_r(t)\, y_i(t)\right\} = 0. \tag{2.117}$$

2.2.1.1 WSS Gaussian Input Signal

In this case, which is often encountered when one uses electronic circuits and devices to observe a biomedical signal, the additive signal once processed by the nonlinearity will have the following properties:

$$\begin{aligned} E\left\{y_r^2[n]\right\} &= 6R_{x_r}^2[m] + 2R_{x_r}^2[0] \\ &\quad - 2R_{x_r x_i}^2[m] + 2R_{x_r x_i}^2[0], \end{aligned} \tag{2.118}$$

$$\begin{aligned} E\left\{y_i^2[n]\right\} &= 6R_{x_r x_i}^2[m] + 2R_{x_r}^2[0] \\ &\quad -2R_{x_r}^2[m] - 2R_{x_r x_i}^2[0] \end{aligned} \tag{2.119}$$

and

$$E\left\{y_r[n]\, y_i[n]\right\} = 8R_{x_r}[m]\, R_{x_r x_i}[m] \tag{2.120}$$

Furthermore, for the continuous-time domain,

$$\begin{aligned} E\left\{y_r^2(t)\right\} &= 6R_{x_r}^2(T_d) + 2R_{x_r}^2(0) \\ &\quad - 2R_{x_r x_i}^2(T_d) + 2R_{x_r x_i}^2(0), \end{aligned} \tag{2.121}$$

$$\begin{aligned} E\left\{y_i^2(t)\right\} &= 6R_{x_r x_i}^2(T_d) + 2R_{x_r}^2(0) \\ &\quad - 2R_{x_r}^2(T_d) - 2R_{x_r x_i}^2(0), \end{aligned} \tag{2.122}$$

and

$$E\{y_r(t)\, y_i(t)\} = 8R_{x_r}(T_d)\, R_{x_r x_i}(T_d)\,.$$

A further simplification occurs when the input signal is a white random process. In that case, the real and imaginary part of $x[n]$ (or $x(t)$) are uncorrelated and $R_x[n] = 0$ for $n \neq 0$ (or $R_x(\tau) = 0$ for $\tau \neq 0$). Hence,

$$E\left\{y_r^2[n]\right\} = E\left\{y_i^2[n]\right\} = 2R_{x_r}^2[0] \tag{2.123}$$

and

$$E\{y_r[n]\, y_i[n]\} = 0. \tag{2.124}$$

In the continuous domain, we arrive at similar results. That is,

$$E\left\{y_i^2(t)\right\} = E\left\{y_r^2(t)\right\} = 2R_{x_r}^2(0) \tag{2.125}$$

$$E\{y_r(t)\, y_i(t)\} = 0. \tag{2.126}$$

We will examine this case further in the next section.

2.2.2 Nonlinear Processing of WSS Gaussian Processes

As noted earlier, the Gaussian signal case is of particular interest as this noise is often present in all electronic measurements. We briefly tackled this problem earlier. In this section, we consider this case in some detail. Let us consider $y[n]$, given by

$$\begin{aligned} y[n] = {} & \{x_r[n]\, x_r[n-m] + x_i[n]\, x_i[n-m]\} \\ & + j\{x_r[n]\, x_i[n-m] - x_i[n]\, x_r[n-m]\} \end{aligned} \tag{2.127}$$

If one assumes that the input sequence is a zero mean, WSS (a condition which is often satisfied in practice) Gaussian process with the real and imaginary parts of the process being uncorrelated with identical correlation functions (i.e., $R_{x_r}[m] = R_{x_i}[m]$), we have the following simplifications:

$$E\left\{y_r^2[n]\right\} = 6R_{x_r}^2[m] + 2R_{x_r}^2[0] \tag{2.128}$$

$$E\left\{y_i^2[n]\right\} = 2R_{x_r}^2[0] - 2R_{x_r}^2[m] \tag{2.129}$$

and

$$E\left\{y_r[n]\,y_i[n]\right\} = 0. \tag{2.130}$$

Similarly, for the continuous-time case,

$$E\left\{y_r^2(t)\right\} = 6R_{x_r}^2(T_d) + 2R_{x_r}^2(0) \tag{2.131}$$

$$E\left\{y_i^2(t)\right\} = 2R_{x_r}^2(0) - 2R_{x_r}^2(T_d)$$

and

$$E\left\{y_r(t)\,y_i(t)\right\} = 0.$$

This situation is perhaps the most commonly encountered scenario observed in practice, making this analysis of significant importance. As we observed in the previous chapter, a time-domain random process is often projected onto two or more orthogonal eigenfunctions of the process, resulting in uncorrelated random variables (or sequence). For a Gaussian signal, this further implies that the projection results in an independent, identically-distributed (i.i.d) Gaussian random sequence. We then view the discrete case as not only the discrete "time" case, but also the case where the random sequence is the by-product of the projection of the process onto its eigenfunctions, resulting in an uncorrelated random sequence.

Before we discuss the continuous-time case, we need to obtain the correlation of $y[n]$ in order to find the power spectrum density of the output. For this case, we have

$$R_y[n_2, n_1] = E\left\{y^*[n_2]\,y[n_1]\right\} \tag{2.132}$$

$$\begin{aligned} R_y[n_2, n_1] = E\{&(x_r[n_1]\,x_r[n_1 - m] + x_i[n_1]\,x_i[n_1 - m] \\ &+ j\,(x_r[n_1]\,x_i[n_1 - m] - x_i[n_1]\,x_r[n_1 - m])) \\ &\times (x_r[n_2]\,x_r[n_2 - m] + x_i[n_2]\,x_i[n_2 - m] \\ &- j\,(x_r[n_2]\,x_i[n_2 - m] - x_i[n_2]\,x_r[n_2 - m]))\} \end{aligned} \tag{2.133}$$

This expression is rather involved. Although significant simplification will result if one assumes a WSS signal, for the sake of completeness, we will carry out the analysis for non-WSS case and then provide an expression for the case of WSS signal as a special case. We note that although WSS case is quite common, in some cases, the interest may be to obtain long term correlation function of the signal. In such cases, the underlying process that dictates the behavior of the random process may not lend itself to stationarity. In such cases, one has to assumed that the process is no longer WSS. In that event,

$$
\begin{aligned}
R_y[n_2, n_1] = {} & E\{x_r[n_1]x_r[n_1 - m,]x_r[n_2]x_r[n_2 - m]\} \\
& + E\{x_i[n_1]x_i[n_1 - m]x_i[n_2]x_i[n_2 - m]\} \\
& + E\{x_r[n_1]x_r[n_1 - m]x_i[n_2]x_i[n_2 - m]\} \\
& + E\{x_r[n_2]x_r[n_2 - m]x_i[n_1]x_i[n_1 - m]\} \\
& + E\{x_r[n_1]x_i[n_1 - m]x_r[n_2]x_i[n_2 - m]\} \\
& + E\{x_i[n_1]x_r[n_1 - m]x_i[n_2]x_r[n_2 - m]\} \\
& - E\{x_r[n_1]x_i[n_1 - m]x_i[n_2]x_r[n_2 - m]\} \\
& - E\{x_i[n_1]x_r[n_1 - m]x_r[n_2]x_i[n_2 - m]\} \\
& + jE\{x_r[n_1]x_i[n_1 - m]x_r[n_2]x_r[n_2 - m]\} \\
& - jE\{x_i[n_1]x_r[n_1 - m]x_r[n_2]x_r[n_2 - m]\} \\
& + jE\{x_r[n_1]x_i[n_1 - m]x_i[n_2]x_i[n_2 - m]\} \\
& - jE\{x_i[n_1]x_r[n_1 - m]x_i[n_2]x_i[n_2 - m]\} \\
& - jE\{x_r[n_2]x_i[n_2 - m]x_r[n_1]x_r[n_1 - m]\} \\
& + jE\{x_i[n_2]x_r[n_2 - m]x_i[n_1]x_i[n_1 - m]\} \\
& - jE\{x_r[n_2]x_i[n_2 - m]x_i[n_1]x_i[n_1 - m]\} \\
& + jE\{x_i[n_2]x_r[n_2 - m]x_i[n_1]x_i[n_1 - m]\}
\end{aligned} \tag{2.134}
$$

Noting that $x_r[n]$ and $x_i[n]$ are independent (since uncorrelated), zero-mean Gaussian random sequences and that

$$
E\{x_r[n_1]x_r[n_2]x_r[n_3]\} = E\{x_i[n_1]x_i[n_2]x_i[n_3]\} = 0 \tag{2.135}
$$

for all n_1, n_2, and n_3, we have

$$
\begin{aligned}
R_y[n_2, n_1] = {} & 2E\{x_r[n_1]x_r[n_1 - m,]x_r[n_2]x_r[n_2 - m]\} \\
& + 2E\{x_r[n_1]x_r[n_1 - m]\}E\{x_i[n_2]x_i[n_2 - m]\} \\
& + 2E\{x_i[n_1 - m]x_i[n_2 - m]\}E\{x_r[n_2]x_r[n_1]\} \\
& - 2E\{x_r[n_1]x_r[n_2 - m]\}E\{x_i[n_2]x_i[n_1 - m]\}
\end{aligned} \tag{2.136}
$$

For WSS Gaussian signal case (note that $R_{x_r}[n]$ is an even function since $x_r[n]$ is real. Note that, by definition, $R_{x_r}[n] = R^*_{x_r}[-n]$, and since $x_r[n]$ is real, $R_{x_r}[n] = R^*_{x_r}[n]$. Hence, $R_{x_r}[n] = R_{x_r}[-n]$ for real random sequence).

$$
R_y[n_2, n_1] = 4R^2_{x_r}[m] + 4R^2_{x_r}[n_1 - n_2] \tag{2.137}
$$

Furthermore,

$$
\begin{aligned}
m_y[n] &= E\{y[n]\} = E\{x_r[n]x_r[n - m] + x_i[n]x_i[n - m]\} \\
&= 2R_{x_r}[m]
\end{aligned} \tag{2.138}
$$

which implies that the output sequence remains WSS! Interestingly enough, the output sequence now possesses a non-zero mean even though the input signal is a zero-mean signal. This is not a surprising result since the average of the output of a squaring device, for instance, is proportional to the power of the input signal, which is non-zero for all power signals. Hence, such a device may be viewed as an energy or power detector. There are other (and perhaps more interesting) aspects of the output of nonlinear devices which will be explored in the ensuing chapters.

We also note that

$$\begin{aligned} E\left\{|y[n]|^2\right\} &= E\left\{y_r^2[n]\right\} + E\left\{y_i^2[n]\right\} \\ &= 2R_{x_r}^2[0] + 6R_{x_r}^2[m] + 2R_{x_r}^2[0] - 2R_{x_r}^2[m] \\ &= 4R_{x_r}^2[m] + 4R_{x_r}^2[0], \end{aligned} \tag{2.139}$$

which is consistent with the result given by $R_Y[n_1, n_1]$ in (2.137). Now, if one is interested in the covariance function of the process

$$\begin{aligned} K_y[n_1, n_2] &= R_y[n_1, n_2] - m_y[n_1]\, m_y[n_2] \\ &= 4R_{x_r}^2[n_1 - n_2] \end{aligned} \tag{2.140}$$

For the continuous-time domain, using (2.115), (2.115), and (2.117), we have

$$E\left\{y_r^2(t)\right\} = 2R_{x_r}^2(0) + 6R_{x_r}^2(T_d) \tag{2.141}$$

$$E\left\{y_i^2(t)\right\} = 2R_{x_r}^2(0) - 2R_{x_r}^2(T_d) \tag{2.142}$$

$$R_{y_r y_i}(t_1, t_2) = 0. \tag{2.143}$$

Similarly, using the previous results shown above, we have the following autocorrelation and covariance functions for the output of the delay-and-multiply device for WSS Gaussian input, respectively:

$$R_y(\tau) = 4R_{x_r}^2(T_d) + 4R_{x_r}^2(\tau) \tag{2.144}$$

$$m_y = 2R_{x_r}(T_d) \tag{2.145}$$

$$K_y(\tau) = 4R_{x_r}^2(\tau) \tag{2.146}$$

2.2.3 Output PSD of DM Devices with WSS Gaussian Input

Since PSD is a key tool in studying the behavior of random signals in the frequency domain, we proceed to investigate the output of the nonlinear device discussed above from the PSD point of view. For the discrete domain, using (2.137), we have

$$\begin{aligned} S_y[\omega] &= \sum_{n=-\infty}^{\infty} R_y[n]\, e^{-j\omega n} \\ &= 8\pi R_{x_r}^2[m]\,\delta(\omega) + 4S_{x_r}[\omega] \otimes S_{x_r}[\omega] \end{aligned} \tag{2.147}$$

where $\otimes$ is convolution in the frequency domain. As noted earlier, the power spectrum will have to contain a dc term corresponding to the power of the input signal. Furthermore, due to the convolution operation, the spectrum in the output suffers from a "smearing" or dispersion effect. This is a common trademark of nonlinear devices where the extend of the spectrum content is expanded due to the nonlinear operation.

For continuous-time domain, we have a similar set of results. That is,

$$\begin{aligned} S_y(\Omega) &= \int_{-\infty}^{\infty} R_y(\tau)\, e^{-j\omega\tau} d\tau \\ &= 8\pi R_{x_r}^2(T_d)\,\delta(\Omega) + 4S_{x_r}(\Omega) \otimes S_{X_r}(\Omega) \end{aligned} \tag{2.148}$$

Example 3 Find the PSD of the output of a delay-and-multiple device for the signal

$$x(t) = \sum_{n=-\infty}^{\infty} I[n]\, p(t - nT) \tag{2.149}$$

where $p(t) = \frac{1}{\sqrt{2\pi\sigma^2}} \exp\left\{-\frac{(t-T/2)^2}{2\sigma^2}\right\}$ with $\sigma = \frac{T}{7}$ and I_n is a zero-mean, white sequence which takes on $\{-1, 1\}$ with equal probability. Plot the PSD for $T_d = \frac{T}{7}$, T.

Solution: *We recognize $x(t)$ as a linearly modulated signal, and hence of some considerable importance to us. For this signal, we have obtained the PSD in previous examples. That is, from* (1.176),

$$S_x(\Omega) = \frac{1}{T} S_I[\Omega T]\, |P(\Omega)|^2 \tag{2.150}$$

where, in this case,

$$S_x(\Omega) = \frac{1}{T} |P(\Omega)|^2 \tag{2.151}$$

since the sequence I_n is a white sequence. Furthermore, $|P(\Omega)|^2 = \exp\left(-\sigma^2\Omega^2\right)$. *Figure* 2.2 *depicts a sample of the signal for* $I[n] = [+1, +1, -1, -1, -1, -1, +1, +1, +1, -1]$. *The value of σ is small enough (as compared with T) so that the adjacent symbol pulses do not overlap significantly, resulting in negligible inter-symbol interference. In Fig.* 2.3*, the PSD of this signal is depicted. As can be seen, the spectrum stays significant over a bandwidth that stretches from* $\frac{-15}{T}$ *to* $\frac{15}{T}$*, resulting in a bandwidth that is almost 30 times the symbol rate. Hence, this pulse shape seems inefficient for a bandwidth-limited application. For biomedical signal processing, however, bandwidth is not a significant consideration. Hence, such pulses can be used to study the behavior of dispersive medium, such as tissue, where the medium will cause the broadening of the pulse. In tissue optics, the use of optical pulses with Gaussian temporal characteristics is common. It is noteworthy that this means of signaling (or signals similar to this) are used to generate ultra wide-band signals (UWB) in signal processing. Now, considering the DM device, the output for such a device is depicted in Fig.* 2.4 *for a delay of* $T_d = T$. *Note that the output in this case resembles the original signal, and hence no significant information can be extracted from the signal. Note that* $R_{x_r}(T_d = T) = 0$, *which implies that the output of DM is a zero-mean signal. A different result emerges when* $T_d = \frac{T}{5}$. *From Fig.* 2.5*, we can see that a delay substantially smaller than the symbol time result in a signal whose average value is not zero. In fact, for* $T_d = 0$, *the mean in the output is* $m_Y = 2R_{x_r}(0) = 2P_x$, *and hence the DM device may be used to measure the power of the input signal. More importantly, the signal Fig.* 2.5 *seems to be independent of the modulating sequence and exhibits a periodic behavior at the symbol rate.*

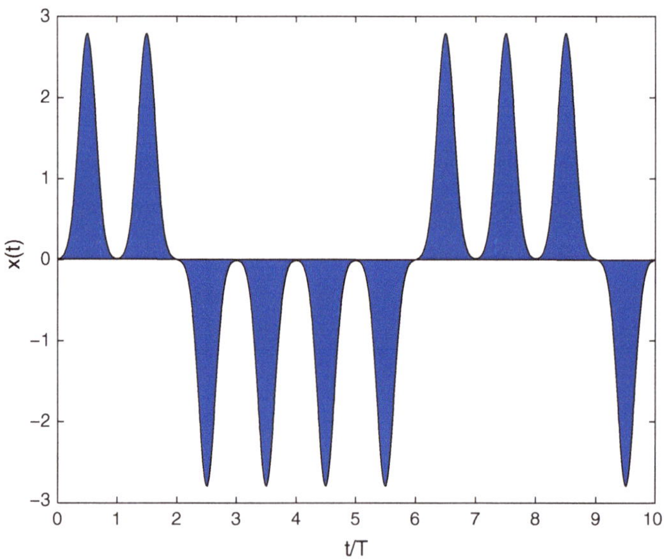

Fig. 2.2 Linearly modulate signal with Gaussian pulse shape for $I_n = [+1, +1, -1, +1, +1, -1, +1, -1, -1]$

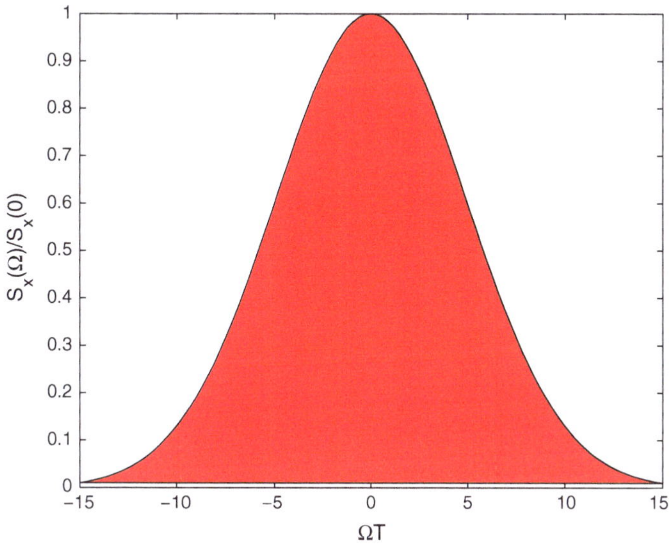

Fig. 2.3 PSD of the linearly modulated signal with Gaussian pulse

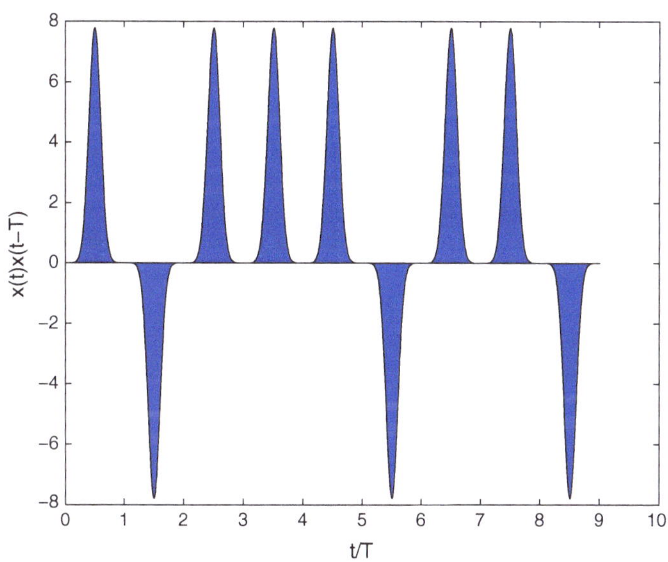

Fig. 2.4 Response of DM device to a linearly modulated Gaussian signal. Delay is assumed to be one symbol duration

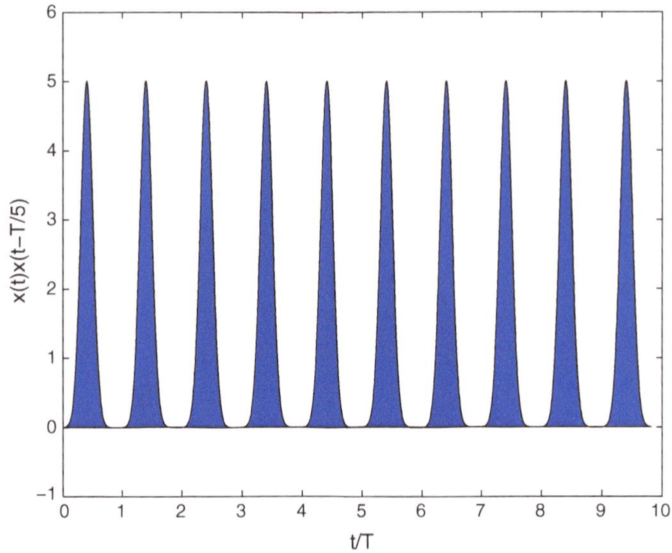

Fig. 2.5 Response of DM device to a linearly modulated Gaussian signal. Delay is assumed to be 20 % of the symbol duration

2.3 Systems with Signal + Noise

In the previous chapters, we have treated signals as random, with deterministic signals classified as a special class of random signals. In practice, electronic and other additive processes often present themselves at the output of an electronic measurement device as an "additive" noise. In biomedical signal processing domain, the additive noise is primarily due to electronic measurement processes, although some underlying biological processes may contribute to this noise. Additive noise is, in general, assumed to be "wideband" as compared with the signal of interest (or the signal that is being measured). That is, the PSD of the additive noise typically spreads over a bandwidth that is larger (if not substantially larger) than that of the "signal" portion of the observed signal.

For this reason, it is instructive to examine this scenario in some depth. Without loss of generality, we consider the discrete-time scenario as most modern systems dealing with biomedical signals take advantage of DSP systems, and hence the very first step of the signal processing chain is a sampling device. Hence, for most signals consider in this text, we arrive at the following signal at the output of the sampling device:

$$x[n] = y[n] + w[n] \tag{2.152}$$

where $y[n]$ denotes the "desired" signal to be measured while $w[n]$ denotes an additive, zero-mean noise (or unwanted) signal that is due to thermal noise in electronic devices or other unknown wideband noise processes in the observation chain. In some scenarios, where the measured signal is transmitted over a wireless channel (it is possible that the biomedical signal is sent over a wireless channel), $w[n]$ is due to background noise (or white noise) that is present in wireless communication links. It is important to note that $y[n]$ may very well be a random signal with known statistical characteristics, whereas $w[n]$ is typically assume to be an additive, zero-mean Gaussian noise signal with known (or measurable) statistics. In the remainder of this section, we assume that all measured signals adhere to this model and that the additive Gaussian noise is an ever present fixture of biomedical signal processing. In almost all cases of interests, it is assumed a statistical independent between $y[n]$ and $w[n]$. We follow the same approach hereafter. This does not limit us in any way as one can account for nonlinearity and other measurement impairments via modeling $y[n]$ in a proper fashion. Namely, $y[n]$ may be viewed as a nonlinear mapping of a desired signal.

It is often of interest to study means of reducing the impact of $w[n]$ on the observation signal. We will address this problem from a spectral point of view followed by a time-domain analysis. To that end, we need to study the PSD of $x[n]$. This is, in general, a difficult problem since one may not have any form of stationarity for $y[n]$. There is, however, one assumption that can be made; that is, $w[n]$ is an additive Gaussian noise, which is independent of $y[n]$. For the scenario that $y[n]$ and $w[n]$ are WSS,

$$R_x[n] = R_y[n] + R_w[n] \tag{2.153}$$

which leads to

$$S_x[\omega] = S_y[\omega] + S_w[\omega] \tag{2.154}$$

where $S_y[\omega]$ and $S_w[\omega]$ are the PSDs' of $y[n]$ and $w[n]$, respectively. This result confirms an intuitive assessment of this problem; that is, when two signals are independent, the power of the sum of the two signals is the sum of the individual power levels. This observation will certainly become invalid when there is nonzero correlation between the two signals. That is, when $y[n]$ and $w[n]$ are not only WSS, but also are jointly WSS

$$\begin{aligned} R_x[n] &= E\left\{x^*[m]\,x[m+n]\right\} \\ &= R_y[n] + R_w[n] + R_{yw}[n] + R_{wy}[n] \\ &= R_y[n] + R_w[n] + R_{yw}[n] + R_{yw}^*[-n] \end{aligned} \tag{2.155}$$

which leads to

$$S_x[\omega] = S_y[\omega] + S_w[\omega] + 2\,\mathrm{Re}\left\{S_{yw}[\omega]\right\} \tag{2.156}$$

with $S_{yw}[\omega]$ denoting the FT of $R_{yw}[n]$ (this is known as the cross PSD).

Now, if we relax the WSS condition imposed on $y[n]$, but keep $w[n]$ as an independent, zero-mean Gaussian noise that is WSS (this is always satisfied in practice), one can use the general definition of PSD to arrive at

$$S_x[\omega] = S_y[\omega] + S_w[\omega] \tag{2.157}$$

where now

$$S_y[\omega] = \lim_{N\to\infty} \frac{1}{2N+1} E\left\{|y_N[\omega]|^2\right\} \tag{2.158}$$

with $y_N[\omega] = \sum_{n=-N}^{N} y[n]\,e^{-j\omega n}$. Hence, the concept of "additive" noise in this case literally implies a scenario where PSD of noise is added to that of the signal.

Example 4 A signal $x[n]$ consists of the sum of a zero-mean, Gaussian noise and a sinusoidal signal with a random phase (uniformly distributed over the range $[-\pi, \pi]$), amplitude A, and frequency f_c Hz, which has been sampled at 4 times the Nyquist rate. The additive noise is independent of the sinusoidal signal and has the following property: $E\{w[n]\,w^*[m]\} = \sigma^2\delta[n-m]$. Find the PSD of $x[n]$.

Solution: *From the definition of the problem,*

$$S_x[\omega] = S_y[\omega] + S_w[\omega] \tag{2.159}$$

and

$$S_w[\omega] = \sigma^2 \text{ for all } \omega \tag{2.160}$$

and

$$\begin{aligned} R_y[n,m] &= E\left\{A^2 \sin(2\pi n f_c T_s + \theta)\sin(2\pi m f_c T_s + \theta)\right\} \\ &= \frac{A^2}{2}\cos(2\pi(n-m) f_c T_s) \end{aligned} \tag{2.161}$$

where $T_s = \frac{1}{8f_c}$. *In arriving at this, it is assumed that* $y[n] = A\sin(2\pi n f_c T_s + \theta)$. *Hence,*

$$\begin{aligned} S_y[\omega] &= \frac{A^2\pi}{2}\delta(\omega - 2\pi f_c T_s) + \frac{A^2\pi}{2}\delta(\omega + 2\pi f_c T_s) \\ &= \frac{A^2\pi}{2}\delta\left(\omega - \frac{2\pi}{8}\right) + \frac{A^2\pi}{2}\delta\left(\omega + \frac{2\pi}{8}\right) \end{aligned} \tag{2.162}$$

Furthermore,

$$S_x[\omega] = \sigma^2 + \frac{A^2\pi}{2}\delta\left(\omega - \frac{\pi}{4}\right) + \frac{A^2\pi}{2}\delta\left(\omega + \frac{\pi}{4}\right),$$

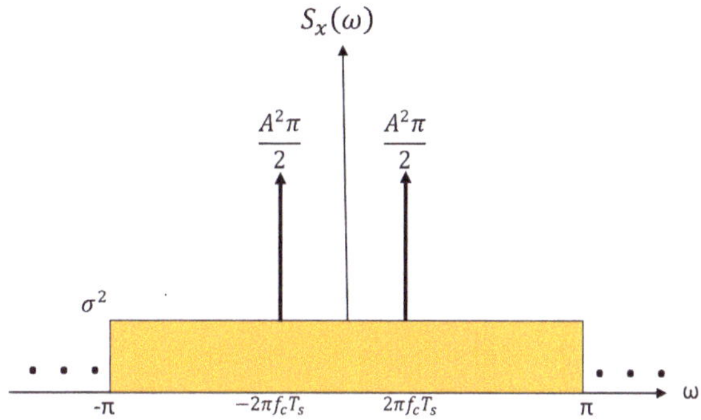

Fig. 2.6 The PSD of signal that has been corrupted by AWGN

which is depicted in Fig. 2.6. Note that the additive noise appears as a "white" noise, occupying the entire spectrum, while signal is narrow-band, confined to a very small (in this case, 0) portion of the spectrum. This figure offers a unique perspective as far as the concept of signal detection in noise is concerned; although in time domain one cannot distinguish between the signal and the additive noise, in the frequency domain, the presence of signal can be established in the fact of wideband noise.

Before we leave this section, it is helpful to consider the above example from the "detection of signal in noise" point of view. That is, how can we detect (or extract) signal from the signal+noise scenario described above? To that end, let us consider the above example in time domain. Figures 2.7 and 2.8 depict the signal+noise scenario in time domain where $A = 1$, $\sigma^2 = 0.25$, $\theta = 0$, and $f_c T_s = 1/8$ for 1000 and 300

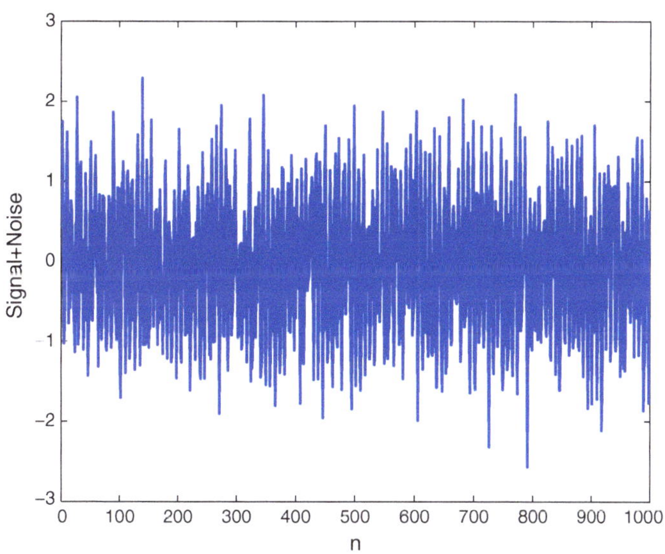

Fig. 2.7 1000-sample view of the signal plus noise waveform for Example 4

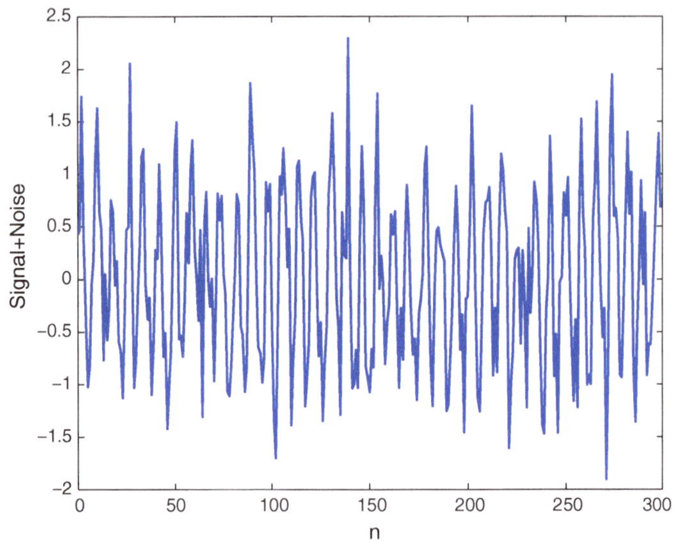

Fig. 2.8 300-sample view of the signal plus noise waveform for Example 4

samples, respectively. A quick glance at these figures reveal a sobering fact; even for a situation where the signal magnitude (A) is 2 times the standard deviation of noise (σ), the periodic signal is entirely buried in noise (which makes the detection of signal rather hard, if not impossible). By comparison, we observed in the previous example that the spectral analysis, such as PSD, allow one to discover a signal which may be hidden in a wideband noise with relative ease. That is, one can clearly identify the signal in Fig. 2.6 from noise, while this task will be cumbersome, if not impossible, using the time series of the signal. In fact, if one is asked the question "*How do you recover the signal here?*," the response of "*I will use a bandpass filter with a bandwidth that is as small as possible with its center frequency fixed at* $2\pi f_c T_s = \frac{\pi}{4}$ *rad*" will be the most likely answer one would receive in view of Fig. 2.6. As we will see in later sections that the above answer is indeed not too far from an "optimum" solution for the detection of signals buried in a wideband noise.

2.3.1 Signal-to-Noise Ratio (SNR)

One reasonable question to ask at this stage is whether one can devise a method to extract $y[n]$ as "accurately" as possible by observing $x[n]$? The term "accurately" is ambiguous at this point as we have not established a metric for accuracy. One well-known and widely used metric is the concept of signal-to-noise ratio (SNR), which as to do with the ration of the "power" of signal to that of the noise. In general, SNR may be defined as follows:

$$SNR_x = \frac{P_s}{P_n} \tag{2.163}$$

where P_s and P_n are the power levels of signal and "noise" portions of $x[n]$, respectively. For the additive scenario considered above (i.e., $x[n] = y[n] + w[n]$),

$$SNR_x = \frac{P_y}{P_w} = \frac{\int_{-\pi}^{\pi} S_y[\omega]\, d\omega}{\int_{-\pi}^{\pi} S_w[\omega]\, d\omega} \tag{2.164}$$

This formulation assumes that the PSD of $y[n]$ and $w[n]$ are available. In time domain, we have

$$SNR_x = \frac{\langle R_y[n,n]\rangle_n}{\langle R_w[n,n]\rangle_n} = \frac{\langle E\left\{|y[n]|^2\right\}\rangle_n}{\langle E\left\{|w[n]|^2\right\}\rangle_n} \tag{2.165}$$

where $\langle z[n]\rangle_n = \lim_{N\to\infty} \frac{1}{2N\_+1} \sum_{n=-N}^{N} z[n]$ may be viewed as the time-average of the enclosed. Note that, one can define an instantaneous SNR as

$$SNR_x[n] = \frac{R_y[n,n]}{R_w[n,n]}, \tag{2.166}$$

and when the additive noise is WSS (i.e., $R_w[n,n] = R_w[0]$), as is the case for most practical scenarios,

$$SNR_x = \langle SNR_x[n]\rangle_n. \tag{2.167}$$

That is,

$$\begin{aligned} SNR_x &= \langle SNR_x[n]\rangle_n \\ &= \langle \frac{R_y[n,n]}{R_w[0]}\rangle_n = \frac{\langle R_y[n,n]\rangle_n}{R_w[0]} \\ &= \frac{\int_{-\pi}^{\pi} S_y[\omega]\, d\omega}{\int_{-\pi}^{\pi} S_w[\omega]\, d\omega} \end{aligned} \tag{2.168}$$

and SNR_x may be viewed as the average SNR. Now, if we perform any filtering operation on $x[n]$, we have

$$SNR_{x_o} = \frac{P_{y_o}}{P_{w_o}} = \frac{\int_{-\pi}^{\pi} S_y[\omega]\, |H_o(\omega)|^2\, d\omega}{\int_{-\pi}^{\pi} S_w[\omega]\, |H_o(\omega)|^2\, d\omega}, \tag{2.169}$$

where we have assumed that ($\circ$ denotes convolution in time)

$$\begin{aligned} x_o[n] &= x[n] \circ h[n] \\ &= y[n] \circ h[n] + w[n] \circ h[n] \\ &= y_o[n] + w_o[n], \end{aligned} \tag{2.170}$$

2.3.2 Matched and Optimum Filtering

A question that comes up at this stage is whether there is an operation that can improve the SNR of $x[n]$. That is, is there an operation that can maximize SNR_x? In general, if such operation exits, then it may very well be a nonlinear operation. For the sake of implementability, we only consider linear operations instead. This is a significant assumption with non-trivial consequence of sacrificing performance for the sake of implementability.

It is noteworthy that *when $w[n]$ is a zero-mean additive Gaussian noise, the operation that will maximize the SNR is indeed a linear operation.* Before we address that problem, we will consider another approach for achieving the best "performance" in the face of additive noise system. As noted in the previous chapter, in addition to SNR, one can establish another criterion for extracting (or estimating) signal in the face of noise: that is, the concept of a system that minimizes mean-square error was shown to yield practical architectures. For the scenario at hand, this implies finding a filter that minimizes the mean-square error in estimating $y[n]$ by observing $x[n]$, given by

$$\begin{aligned} \text{MSE} &= E\left\{|e[n]|^2\right\} \\ &= E\left\{|y[n] - x[n] \circ h_o[n]|^2\right\}. \end{aligned} \tag{2.171}$$

Before we proceed any further, we need to address the stationarity of the processes involved. In particular, we assume that $y[n]$ and $w[n]$ are both WSS (or at least behave as WSS over the range of observation). This assumption is rather restrictive as signals observed in biomedical field do not tend to lend themselves to stationarity However, as was noted, for short observation intervals where the underlying elements that control the statistical behavior of the signal are somewhat stationary, the results presented in the following can be applied to such problems.

As shown previously, an estimate that minimizes the MSE is obtained by using the orthogonality principle, which states that such an estimate renders the error "orthogonal" to the observation. That is,

$$E\left\{e[n]\, x^*[n]\right\} = 0. \tag{2.172}$$

Carrying out the expectation, we arrive at $E\{w[n]\, y^*[m] = 0 \text{ for all } n, m\}$

$$E\left\{e[n]x^*[n]\right\} = E\left\{(y[n] - x[n] \circ h_o[n])\, x^*[n]\right\}$$
$$= R_y[0] - \sum_{m=-\infty}^{\infty} R_y[m]\, h_o[m]$$
$$- \sum_{m=-\infty}^{\infty} R_w[m]\, h_o[m] \tag{2.173}$$

Taking this into the FT domain,

$$E\left\{e[n]x^*[n]\right\} = \frac{1}{2\pi}\int_{-\pi}^{\pi} S_y[\omega] - H_o[\omega]\left\{S_y[\omega] + S_w[\omega]\right\} d\omega \tag{2.174}$$

In order for this to be zero,

$$H_o^{(\mathrm{WH})}[\omega] = \frac{S_y[\omega]}{S_y[\omega] + S_w[\omega]} \tag{2.175}$$

This filter is due to Wiener-Hof. The main issue here is whether this filter is causal (therefore, implementable). This all depends on the PSD of signal and noise. For this filter, the MSE (which is the minimum MSE) is given by

$$\begin{aligned}
\mathrm{MSE}_{\mathrm{min}} &= E\left\{\left|y[n] - x[n] \circ h_o^{(\mathrm{WH})}[n]\right|^2\right\} \\
&= E\left\{\left(y[n] - x[n] \circ h_o^{(\mathrm{WH})}[n]\right) y^*[n]\right\} \\
&= R_y[0] - \sum_{m=-\infty}^{\infty} R_y[m]\, h_o^{(\mathrm{WH})}[m] \\
&= \frac{1}{2\pi}\int_{-\pi}^{\pi} S_y[\omega]\, d\omega - \frac{1}{2\pi}\int_{-\pi}^{\pi} S_y[\omega]\, H_o^{(\mathrm{WH})}[\omega]\, d\omega \\
&= \frac{1}{2\pi}\int_{-\pi}^{\pi} S_y[\omega]\left[1 - H_o^{(\mathrm{WH})}[\omega]\right] d\omega \\
&= \frac{1}{2\pi}\int_{-\pi}^{\pi} \frac{S_y[\omega]\, S_w[\omega]}{S_y[\omega] + S_w[\omega]}\, d\omega
\end{aligned} \tag{2.176}$$

The above results allow one to find an optimum filtering operation. Furthermore, one can obtain the error in estimating the signal using the suggested filtering operation.

Example 5 For the previous example, give an expression for system function of the optimum filter (a filter that maximizes SNR) and comment on its implementability. If the filter is not implementable, suggest a realizable system and compare its output SNR to that of the optimum filter.

Solution: *Based on the above analysis (and Eq. (2.175)), we have*

$$H_o^{(WH)}[\omega] = \frac{A^2\pi\delta\left(\omega - \frac{2\pi}{8}\right) + A^2\pi\delta\left(\omega + \frac{2\pi}{8}\right)}{A^2\pi\delta\left(\omega - \frac{2\pi}{8}\right) + A^2\pi\delta\left(\omega + \frac{2\pi}{8}\right) + 2\sigma^2} \tag{2.177}$$

This is clearly not an easily implementable filter, but an examination of the filter response provides a valuable insight; given the behavior of a delta function, we arrive at the following realization of the filter:

$$H_o^{(WH)}[\omega] = \begin{cases} 1 & |\omega| = \frac{\pi}{4} \\ 0 & \text{otherwise} \end{cases} \tag{2.178}$$

This implies that the optimum is a "perfect," zero bandwidth bandpass filter centered at the input frequency! This makes an intuitive sense as the filter should not consider any portion of the spectrum over which the signal is not present. However, such a filter is not realizable. Instead, one can come up with a sub-optimum strategy. That is a BPF with finite, but small bandwidth. Furthermore, linear phase shift must be introduced to insure causality. Specifically, we suggest (the phase does not impact the SNR)

$$\left|H_o^{(AF)}(\omega)\right| = \begin{cases} 1 & \left||\omega| - \frac{\pi}{4}\right| \le \Delta\omega \\ 0 & \text{otherwise} \end{cases} \tag{2.179}$$

This results in the output SNR

$$\begin{aligned} SNR_o^{(AF)} &= \frac{\frac{A^2}{2}}{2\sigma^2\frac{\Delta\omega}{2\pi}} = \frac{A^2/2}{\sigma^2 2\Delta\omega/2\pi} \\ &= \frac{A^2/2}{N_0} = \frac{A^2/2}{\sigma^2/2\pi}\frac{1}{2\Delta\omega} \end{aligned} \tag{2.180}$$

where $N_0 = \sigma^2 2\Delta\omega/2\pi$ *may be viewed as the noise power. It is interesting to note that the noise power increases with* $\Delta\omega$ *(reducing SNR) without improving the signal power. Hence, smaller this parameter, better would be the output SNR. The SNR of the optimum filter is given by*

$$\begin{aligned} SNR_o^{(WH)} &= \frac{\int_{-\pi}^{\pi} S_y[\omega]\left|H_o^{(WH)}[\omega]\right|^2 d\omega}{\int_{-\pi}^{\pi} S_w[\omega]\left|H_o^{(WH)}[\omega]\right|^2 d\omega} \\ &= \frac{A^2/2}{0} = \infty \end{aligned}$$

This implies that the optimum filter in this case extracts the signal "perfectly" without any loss (captures all of the signal energy while rejecting all of the noise contribution). This is apparent from the description of the filter. The approximate filter still has a problem; namely, it has an infinite number of taps and is not realizable. A remedy is to introduce delay (linear phase shift) to achieve causality for a truncated response case. That is, in order for the filter to be causal, i.e., $h'[n] = 0$ *for* $n < 0$, *we can construct the filter using*

$$h'[n] = h_0^{(AF)}[n-m] \tag{2.181}$$

for some m such that

$$h_0^{(AF)}[n-m] \approx 0 \text{ for } n < 0. \tag{2.182}$$

Although the above technique seems to provide an optimum solution, its implementability (as shown by the above simple example) is in question. Furthermore, even if one arrives at a realizable solution, it may very well be an infinite impulse response (IIR) solution, which may not be of interest in some practical scenarios. An alternate approach is to seek an finite impulse response (FIR) solution; that is

$$\hat{y}[n] = \sum_{k=0}^{L-1} x[n-k]\, h_n[k]; \tag{2.183}$$

where L is the number of taps (delays) for the FIR filter, which is set by hardware restrictions, processing delay, etc. This operation uses the past history of the observed signal to render a decision on the value of the signal portion of the signal. Note that $h_n[k]$ will have to be computed each time (n) a new data sample has been observed (that is the reason for using the subscript of n in the definition of impulse response, denoted by $h_n[k]$). One can then benefit from the orthogonality principle to arrive at a practical solution. That is, the coefficients of the filter must satisfy

$$E\left\{\left(y[n] - \hat{y}[n]\right) x^*\left[n-k'\right]\right\} = 0; \text{ for } k' \in [0, L-1] \tag{2.184}$$

This is a set of L equations with L unknowns, and hence it can readily be solved. To do so, we need to simplify the above as follows:

$$\begin{aligned} R_y\left[n-k',n\right] &= \sum_{k=0}^{L-1} R_x\left[n-k',n-k\right] h_n[k] \\ &= \sum_{k=0}^{L-1} R_y\left[n-k',n-k\right] h_n[k] \\ &\quad + \sum_{k=0}^{L-1} R_w\left[n-k',n-k\right] h_n[k], \end{aligned} \tag{2.185}$$

for $k' \in [0, L-1]$, which reduces to

$$\mathbf{A}[n]\,\mathbf{h}[n] = \mathbf{b}[n] \tag{2.186}$$

where $\mathbf{A}[n] = \left\{a_{ij}[n]\right\}$ and $L \times L$ matrix, $\mathbf{b}[n] = \{b_i[n]\}$ is an $L \times 1$ vector, and $\mathbf{h}[n] = \{h_n[i]\}$ is an $L \times 1$ vector with $a_{ij}[n] = R_y\left[n-i, n-j\right] + R_w\left[n-i, n-j\right]$ and $b_i[n] = R_y[n-i, n]$; $i, j \in [0, L-1]$. Hence, assuming that $\mathbf{A}^{-1}[n]$ exists, we have

$$\mathbf{h}[n] = \mathbf{A}^{-1}[n]\,\mathbf{b}[n] \tag{2.187}$$

Note that this solution allows for time variation (it must be computed for all n). Namely, one can utilize this technique over an observation interval and update the solution for the ensuing data frames. Furthermore, WSS is not required! This is particularly of interest in biomedical signal processing as the condition of WSS cannot be always assumed.

Assuming, however, that we have WSS at least over the observation interval, we have the following simplification:

$$\mathbf{h} = \mathbf{A}^{-1}\mathbf{b} \tag{2.188}$$

where $\mathbf{A} = \left\{a_{ij}\right\}$, $\mathbf{b} = \{b_i\}$, and $\mathbf{h} = \{h[i]\}$ with $a_{ij} = R_y\left[i-j\right] + R_w\left[i-j\right]$ and $b_i = R_y[i]$. Note that this estimation process requires a knowledge of the 2nd order statistics of the signal, which may be obtained from the PSD of the signal. In some applications, one can use spectrum analyzer to estimate the spectrum of a signal. In that event, one can invoke the above process for the estimation of $y[n]$, assuming that the additive noise $w[n]$ is independent of the signal.

This solution, although not optimum for all scenarios, provides an implementable strategy which can achieve a quasi-optimum performance for additive Gaussian noise channels.

There is yet another mechanism that one can use to compute $\mathbf{h}[n]$; that is, we can find a solution that maximizes SNR when $w[n]$ is WSS and $y[n]$ is not random and is limited to a length of L samples (this may be an observation interval). This situation may be encountered when the statistical variations to the signal over an observation interval are minute and the signal is corrupted by an additive, WSS noise. Using the definition of SNR provided above, we have

$$\begin{aligned} P_s &= |y_o[n]|^2 \\ &= \left| \sum_{k=0}^{L-1} y[k] h_o[n-k] \right|^2 \qquad (2.189) \\ &= \left| \int_{-\pi}^{\pi} Y[\omega] H_o[\omega] d\omega \right|^2 \qquad (2.190) \end{aligned}$$

$$\begin{aligned} P_n &= E\left\{ |w_o[n]|^2 \right\} \\ &= \sum_{k=0}^{L-1} \sum_{k=0}^{L-1} R_w[k_1 - k_2] h_o[k_1] h_o^*[k_2] \\ &= \frac{1}{2\pi} \int_{-\pi}^{\pi} S_w[\omega] |H_o[\omega]|^2 \, d\omega \qquad (2.191) \end{aligned}$$

where we have assumed that the filter is also L tap long. Hence,

$$SNR_{x_o} = \frac{\left| \int_{-\pi}^{\pi} Y[\omega] H_o[\omega] d\omega \right|^2}{\int_{-\pi}^{\pi} S_w[\omega] |H_o[\omega]|^2 d\omega}$$

A case of interest is when the additive noise is "white", implying that $S_w[\omega] = \sigma^2$; $|\omega| \leq \pi$. In that scenario, we have

$$SNR_{x_o} = \frac{\left| \int_{-\pi}^{\pi} Y[\omega] H_o[\omega] d\omega \right|^2}{\sigma^2 \int_{-\pi}^{\pi} |H_o(\omega)|^2 d\omega} \qquad (2.192)$$

Using the Cauchy-Schwartz (CS) inequality, we have

$$\begin{aligned} SNR_{x_o} &\leq \frac{\int_{-\pi}^{\pi} |Y[\omega]|^2 d\omega \int_{-\pi}^{\pi} |H_o(\omega)|^2 d\omega}{\sigma^2 \int_{-\pi}^{\pi} |H_o(\omega)|^2 d\omega} \\ &= \frac{\int_{-\pi}^{\pi} |Y[\omega]|^2 d\omega}{\sigma^2} = \frac{\sum_{n=0}^{L-1} |y[n]|^2}{\sigma^2} \qquad (2.193) \end{aligned}$$

Interestingly, the maximum possible SNR is the ratio of the power of $y[n]$ to the PSD of the additive noise. More important question is "*How does one achieve the maximum SNR?*" From the CS inequality, we know that the equality is achieved when

$$H_o^{(MF)}(\omega) = Y^*(\omega)\, e^{j\omega(L-1)} \tag{2.194}$$

(the linear phase is added for implementability sake). This simply implies that the filter system function is "matched" to the "shape" of the spectrum of the signal (hence the name matched filter). Using the inverse FT,

$$h_0[n] = y^*[L-1-n] \tag{2.195}$$

If the filter is to be implementable, $h_0[n] = 0$ for $n < 0$, which is insured by the assumption that $y[n] = 0$ for $n \notin [0, L-1]$. If the signal (or observation interval) is finite, this will be an easy condition to satisfy. For infinite observation interval, then one requires an infinite delay (impractical). Hence, a design parameter that needs to be selected carefully is the observation interval, which in turn will define the key parameters of the optimum filtering.

Example 6 For the previous example, find the matched filter and the SNR that it will achieve. Assume that the signal is confined to L taps. That is, $y[n] = A\sin(2\pi n f_c T_s + \theta)$ for $n \in [0, L-1]$, and 0, otherwise.

Solution: *The matched filter is given by $h_0[n] = y^*[L-1-n]$, which leads to the following:*

$$h_0^{(MF)}[n] = y[n] = A\sin(2\pi((L-1)-n)f_c T_s + \theta);\; 0 \le n \le L-1$$

and zero, otherwise. This assumes that one "knows" the phase θ (this is one difficulty in realizing matched filter). The maximum SNR is

$$SNR_o^{(MF)} = \frac{\frac{1}{2\pi}\int_{-\pi}^{\pi}|Y[\omega]|^2\, d\omega}{\sigma^2/2\pi} = \frac{\sum_{n=0}^{L-1}|y[n]|^2}{\sigma^2}$$

which for a relatively large L leads to

$$SNR_o^{(MF)} = \frac{A^2 L/2}{\sigma^2/2\pi}$$

It is interesting to note that the ratio of the matched filter SNR and the SNR of the approximate filter is

$$\frac{SNR_o^{(AF)}}{SNR_o^{(MF)}} = \frac{1}{2\Delta\omega L}$$

Furthermore, if one lets $\Delta\omega \rightarrow 0$, $SNR_o^{(AF)}\infty$, *which is* $SNR_o^{(WH)}$ *for this signal. It is instructive to note that the WH filter provides the minimum MSE for the noise that may not be white. In that sense, it is the most ideal form of filtering, although its implementability is in question. In this problem, however, we deal with white noise. Hence, we are able to find implementable solutions that achieve near ideal performance.*

For the situations where one is comparing the approximate filter (AF), motived by WH, to the MF, one has to consider $\frac{1}{2\Delta\omega L}$ *as the degradation factor. However, for* $\frac{1}{2\Delta\omega L} > 1$, *the use of AF is advantageous over the MF! This is due to the fact that AF for* $\frac{1}{2\Delta\omega L} > 1$ *is approximating the WH filter closely. Furthermore, an ideal filter with a bandwidth of* $2\Delta\omega$ *has infinite taps, whereas MF used here has only L taps. We can, then, use* $\frac{1}{2\Delta\omega L}$ *to assess the number of taps needed to achieve a performance that is better than that of the AF. It is instructive to consider the system function of the matched filter, which is given by*

$$H_o^{(MF)}(\omega) = \frac{A(L-1)e^{j\phi}}{2}\sin c\left(\frac{(\omega-\omega_0)(L-1)}{2}\right)\exp\left\{-j\frac{\omega(L-1)}{2}\right\}$$
$$-\frac{A(L-1)e^{-j\phi}}{2}\sin c\left(\frac{(\omega+\omega_0)(L-1)}{2}\right)\exp\left\{-j\frac{\omega(L-1)}{2}\right\}$$

where $\omega_0 = 2\pi f_c T_s$ and $\phi = \theta + \frac{3}{2}(L-1)\omega_0$. *In this case,* $\omega_0 = \frac{\pi}{4}$ *(note that* $T_s = \frac{1}{8f_c}$*). As can be see, this filter can be implemented in time domain with relative ease, while a frequency domain implementation is cumbersome. The magnitude of this filter is shown in Figure below. The system function reflects a bandpass filter with theoptimum filter point of view, this filter focuses only on a portion of the spectrum that signal is present. Note that this system function, although similar to the approximate optimum filter, in principle, matches the spectrum of the signal and is not constant over the bandwidth of interest. However, one can argue that the most significant portion of the filter response resides in a null-to-null bandwidth of the spectrum. That is, one can use* $\frac{4\pi}{(L-1)} = 0.0982$ *for* $L = 128$ *as the bandwidth of the matched filter. If we set* $\Delta\omega = \frac{4\pi}{(L-1)} \approx \frac{4\pi}{L}$ *for the AF solution, then* $\frac{SNR_o^{(AF)}}{SNR_o^{(MF)}} = \frac{1}{8\pi}$, *which implies that the MF still outperforms the AF solution by a significant margin.*

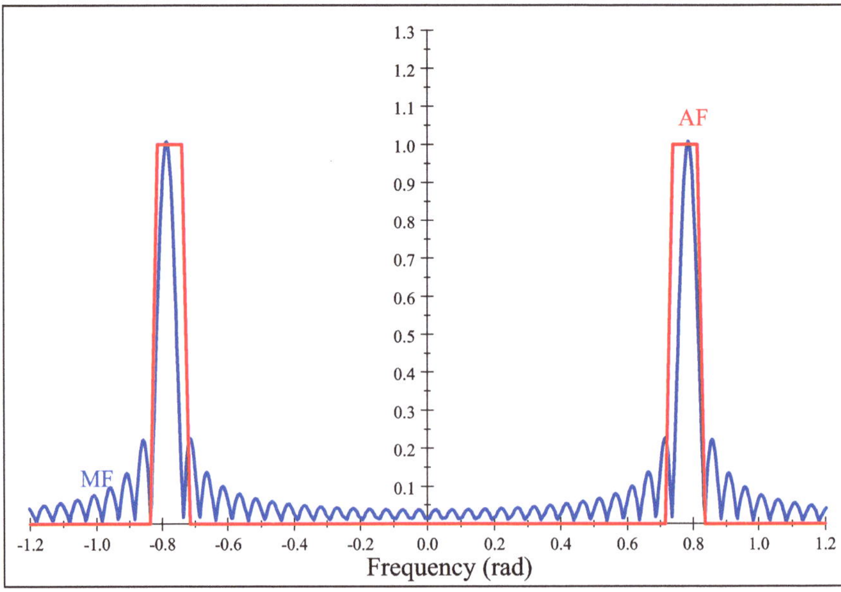

$\frac{\left|H_o^{(\mathrm{MF})}(\omega)\right|}{\max\left(\left|H_o^{(\mathrm{MF})}(\omega)\right|\right)}$ and $\frac{\left|H_o^{(\mathrm{AF})}(\omega)\right|}{\max\left(\left|H_o^{(\mathrm{AF})}(\omega)\right|\right)}$ for the sinusoidal signal with $L = 128$ taps, $\omega_0 = \frac{\pi}{4}$, $A = 1$, and $\theta = 0$.

In the above example, of course, one can reduce $\Delta\omega$ for the AF solution to be competitive with MF solution, with the nontrivial cost of an increased impulse response length (hence, delay). However, caution must be exercised when using this comparison. To elaborate, the AF solution assumes that the power of sinusoid is entirely centered around ω_0, which further assumes $L = \infty$. In reality (and for the sake of fairness), one must compare $SNR_o^{(\mathrm{MF})}$ with the SNR of the AF solution when the AF system is excited by an $L-$tap sinusoid (note that as $\Delta\omega \to 0$, $SNR^{(AF)} \to \infty$ only if the spectrum of the signal is a pair of delta functions centered at $\pm\omega_0$ ($L = \infty$), whereas the signal considered here will render $SNR^{(\mathrm{AF})} \to 0$ as $\Delta\omega \to 0$). In that event, the SNR of the AF solution will be even smaller than that computed above. For these reasons, the use of MF is highly recommended as it offers an implementable architecture (an FIR implementation with filter coefficients (i.e., $h[n]$) which can be stored and used in a DSP platform) and offers an SNR that is superior to that of the other linear filters.

Finally, note that the MF solution resembles a bandpass filter centered around $\pm\omega_0$. This was our "intuitive" answer to the question "*How do you recover the signal here?*" which was posed in the System with Signal + Noise section of this book. We caution the reader about using intuition for all cases of signal detection indiscriminately. One key reason that our intuition leads us to an optimum structure is due to the fact that noise presents itself as a “white” noise with identical power levels at all frequencies in this scenario. If one deviates from this assumption, the resulting optimum device will surely not follow our intuition, and in fact, the optimum solution may very well be counter-intuitive. In that event, one has to rely entirely on the optimum receiver principle presented earlier to arrive at a realizable implementation (or realizable approximation to the possibly unrealizable optimum solution).

Chapter 3
Biological Signals

In the previous chapters, we addressed man-made signals or signals generated by non-biological systems. In this chapter, we switch gears and tackle the type of signals which are observed when a biological system is monitored using electronic systems. In principle, the underlying principles that give rise to the biological signals differ from their non-biological counterparts. This fact results in rendering many signal processing algorithms, which are developed for processing man-made signals, ineffective tools for studying biological signals. For one, most biological signals appear "noise-like," leading one to an erroneous conclusion that biological signals are similar to noise corrupted deterministic signals, such as the ones observed in AWGN case. In fact, noise plays a role in the cases of interest as some electronic noise due to electronic measurement devices is an always-present fixture of biomedical signal processing. However, the real culprits for the noise-like behavior of biological signals are the underlying nonlinear processes at work, which determine the statistical behavior of such signals, along with the transient nature of biological processes, which add to the complexity of the analysis that one can carry out using the conventional signal processing tools. Many of algorithms developed in the signal processing world make an implicit assumption of stationarity (or at least stationarity on a micro time scale). Such an assumption is typically not valid for the case of biological signals. Furthermore, non-additive-noise behavior of biological signals lead to further complication.

The question that must be on the mind of any student of biomedical signal processing subject is "How does one unveil a set of key underlying principles governing the behavior of an observed biological signal through short- or long-term observation in time?" This is, in general, a fairly difficulty question to answer for reasons that will become clear in this chapter. There exists, however, a key difference between biomedical signal processing techniques and their counterparts used in RF communication world; that is, most RF communication signal processing techniques are designed to operate at fairly high speeds using small observation intervals (real time data transmission and reception at Mbps if not Gbps), whereas in the case of biomedical signal processing, one has the benefit of GHz signal processing tools when dealing with sub-kHz (if not sub-Hz) signals! This implies that we have quite a bit

K. Kiasaleh, *Biological Signals Classification and Analysis*,
Lecture Notes in Bioengineering, DOI 10.1007/978-3-642-54879-6_3

of "time" when dealing with biological signals and that large number of observations can be made. Unlike RF signals, however, longer observation does not imply a more effective means of extracting relevant information. This is in large part due to the presence of transient, highly nonlinear phenomena that are responsible for the non-stationary, self-noise component of biological signals. With the presence of the transient behavior, longer observation interval does not aid detection of the desired parameters. Rather, an undesirable time-averaging of the signal is resulted, which may or may not aid the detection of the underlying processes hidden to an untrained observer.

Finally, one has to be aware of the fact that the time-series signals observed in the field of biomedical signal processing are often the outcome (or output) of highly nonlinear, multidimensional control loops which may or may not arrive at a chaotic behavior depending on a variety of physiological markers. This behavior is of particular interest to clinicians who are engaged in the treatment of patients with epileptic seizures (ES) or those with malignant ventricular fibrillation (MVF), for example.

To fully appreciated the intricacies of the problem at hand, it is instructive to consider several classes of signals which are often observed by biomedical engineers and clinicians in the field of medicine. We emphasize that the intent here is not to present the reader with an exhaustive list of biological signals, but rather to introduce the reader to a set of well-known, well-studied biological signals in order to reveal the challenges one would face. To that end, we will examine in some detail electrocardiogram (ECG), electroencephalogram (EEG), and electromyogram (EMG) as three key signals in the field of biomedical engineering, dealing with heart, brain, and muscle-nerve interaction. The signals used in this book originate from PhysioNet ATM, which is provided by the collaboration between Massachusetts Institute of Technology and Boston's Beth Israel Hospital (MIT-BIH). The reference to this data can be found at the end of this textbook in the reference chapter.

3.1 Electrocardiogram (ECG)

ECG signals are perhaps the most recognizable and well-known biological signals to date due to their widespread depictions in countless movies and TV shows to heightened the senses of the viewers and make them sympathetic (or perhaps apathetic) to the plight of the character (patient) under observation! The familiar periodic "beeping" sound in these movies strongly suggests a periodic, rhythmic nature to this signal. Naturally, one expects a similar feature when the electronic version of the signal is captured using any one of the signal processing tools in use today. ECG signals, however, contain a great deal of information which cannot be completely understood by a cursory view of the signal. In general, the analog ECG signals collected by electrodes, which are placed in strategically important spots around human chest, back, abdomen, and sides, are sampled at a high sampling rate (high compared to the heart beat rate (HBR)) that allows an accurate reconstruction of the signal. Typically, 128 or 256 (always a power of 2) samples per second (SPS) are acquired. We note that with the progress in the development of high speed DSP and field-programmable-gate-array (FPGA) devices, larger number of samples can be obtained and processed. However, given that the most interesting features of ECG signals are limited to small bandwidths (<60 Hz), from the Sampling Theorem, a sampling rate of 128 SPS is sufficiently high to reconstruct the signal from its sampled values. In the ensuing sections, we will address the signal processing techniques that one can use to characterize the ECG signal. For this section, the interest is primarily in considering the overall behavior of ECG signals. To that end, we use the MIT-BIH database extensively. This database provides a variety of biological signals which are collected over long periods of time and for subjects ranging from "normal" to those experiencing severe medical conditions, hence allowing for one to build a meaningful baseline for healthy behavior and design algorithms to detect anomalies for clinical diagnosis.

3.1.1 QRS Complex

In Fig. 3.1 a typical ECG signal for a normal patient using MIT-BIH database is depicted over a 4 s interval. The signal is normalized to its peak value (the database provides raw data). One is required to subtract the "base" (0 in this case) from the raw data and divide the result by the "gain" (200 in this case). For the sake of studying the behavior of the signal and not the exact amplitude of the signal, the data has been normalized.

This sample is a small section of a 2-h, 10-min, and 12-s samples of a normal sinus rhythm from the database (file 16265 m), which contains a matrix of 2×10^6 data elements ($128 \times (2 \times 3600 + 10 \times 60 + 12) = 999{,}936$, additional zeros added

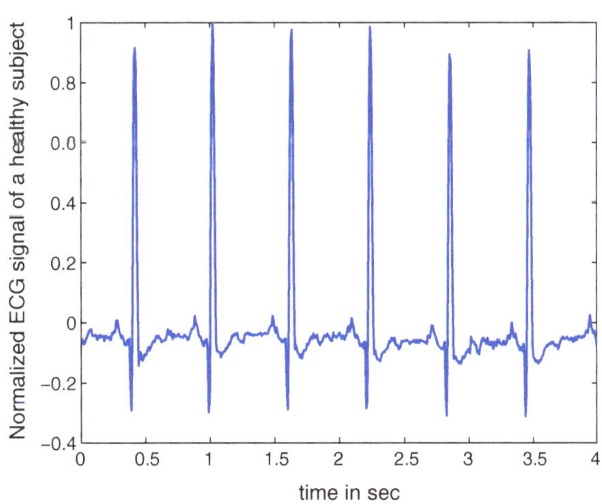

Fig. 3.1 A 4 s sample of the ECG of a healthy patient. The signal is sampled at 128 SPS

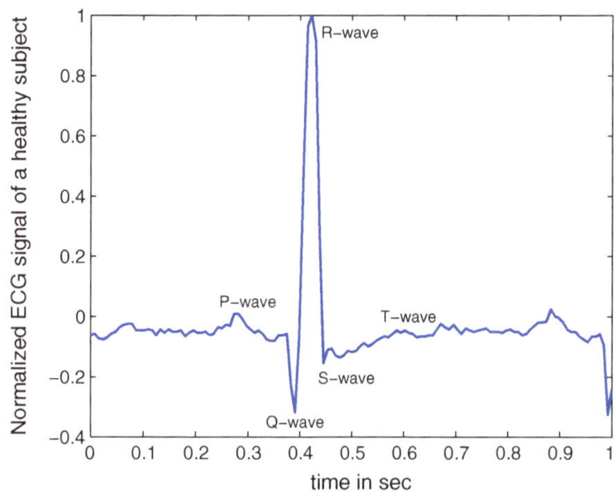

Fig. 3.2 A 1-s sample of the ECG of a healthy patient. QRS complex is depicted with P-, Q-, R-, S-, and T-waves

to round the data set to 10^6 data points per row). The two rows correspond to two distinct electrodes. From the outset, this signal meets some of our expectations, such as having a strong periodic component possessing a fairly regular structure. A closer examination (Fig. 3.2) reveals a somewhat different picture; that is, we clearly see a variation in the peak value of the signal along with a noise-like behavior over the range of time intervals where signal is not significant. The noise may be attributed to the electronic noise present in the acquisition circuitry and to electrical activities on the surface of the body which has been picked up by the electrodes. Note that the small noise-like perturbations are, as expected, wideband as compared with the significant periodic component. Hidden in this picture are other periodic components of the signal, somewhat overshadowed by the dominant peaks. In particular, there are 5 distinct periodic signals that make up an ECG signal; they are P-wave, Q-wave, R-wave, S-wave, and T-wave. Figure 3.2 points out the relative positioning of these waves in relation to one another. The combination of Q, R, and S waves is known as the *QRS complex*. Not only the amplitude but also the relative positioning of these waves in time convey important information regarding the state of the heart, and hence can be used to diagnose various heart anomalies and diseases. Furthermore, there are several intervals of interest here. To elaborate, there are 5 important intervals in time, described in Fig. 3.3. That is, there is PR interval, which refers to the distance in time between the start of the P wave and the start of the QRS complex. QT interval refers to the distance in time between the start of QRS complex and the end of the T wave, and ST interval refers to the distance in time between the end of QRS complex and the end of the T wave. Furthermore, PR segment refers to the distance in time between the end of P wave and the beginning of the QRS complex, and ST segment refers to the distance in time between the end of QRS complex and the beginning of the T wave. Note that the PR and ST segments refer to the resting period where potentially there are no electrical output from the heart.

It will be useful to examine the various waves which make up the QRS complex in some detail.

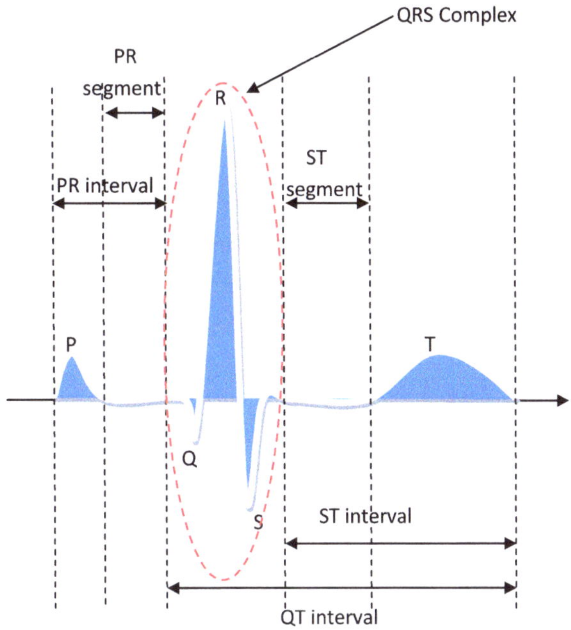

Fig. 3.3 QRS complex

3.1.2 The P Wave

The P wave is a relatively small signal (as compared with QRS wave) which precedes the onset of QRS signal. Typically, less than 0.2 mv, this wave is related to the atrial contraction event. To elaborate, unlike ventricles which have conduction nerves, atria contraction is carried out by muscle cells, leading to a slow wave with a small amplitude.

3.1.3 The PR Segment

This interval enables the completion of the atrial contraction and transfer of blood to the ventricles before the contraction of ventricles is initiated. This may be viewed as a necessary delay to promote stability in a control loop.

3.1.4 The QRS Wave

This wave is the most pronounced (up to 1 mv) wave among the QRS complex and last for 80–100 ms. In short, the QRS wave represents the contraction of ventricle muscles, starting from apex and moving upward.

3.1.5 The ST Segment

Similar to PR segment, the ST segment is a necessary delay in the QRS complex which signifies the flattening of the action potential of the left ventricle. Its duration is in the range of 100–120 ms (up to half of the actual duration of ventricle plateau). Although the ST interval is not a wave, but rather a relaxation interval, some depression or elevation of ST segment may be attributed to heart anomalies, such as myocardial infarction or ischemia. For this reason, ST segment may be used by pathologist to identify heart disease.

3.1.6 The T Wave

The T wave is generated by the ventricle muscle when it is making a transition from its plateau state to its resting state (or relaxation state). The T wave may be viewed as an independent wave (from the QRS signal) that is relatively small in its amplitude (<0.3 mv) with a duration of 120–160 ms (a rather slow wave). In general, the T wave is hardly visible to a naked eye (see Fig. 3.1). Due to the fact that relaxation is in principle a part of the contraction process, a label of "relaxation wave" for the T wave is indeed inaccurate.

Given that these waves are all periodic with the same period, a frequency analysis of the data may not reveal anomalies with the ECG. To illustrate this point, the FT of the signal (assuming the entire length of the signal) is depicted in Fig. 3.4. The FFT output is re-sampled to underscore the frequency range of interest of [0,3] Hz. Surprisingly, this spectrum does not resemble that of a periodic signal, whose spectrum will contain simple spectral lines spaced by the multiples of the

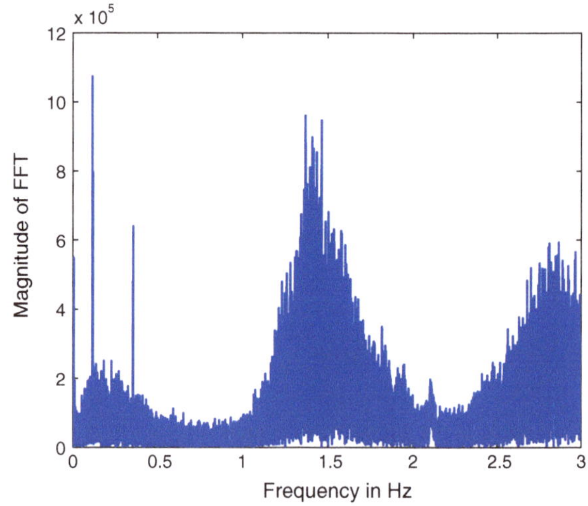

Fig. 3.4 Magnitude of the FFT of the ECG signal observed over 4 min

fundamental frequency (heart beat rate). In fact, the heart rate, which is about 1.5 Hz (or 90 bpm) is not entirely clear from this picture! This is in large part due to variation in the heart rate, which gives rise to a "modulated" signal. The dc value of the signal also gives rise to a large peak at 0 Hz. However, there are other significant components at 0.3 and 0.1 Hz. These components, in the absence of anomalies in the heart function, can be attributed to breathing and other artificial artifacts. The components around 3 Hz corresponds to the harmonics of the spectrum as the QRS complex does not resemble a sinusoidal signal, although the term 'sinus' rhythm has been used in literature to refer to the ECG signal.

This figure clearly depicts the non-stationarity of biomedical signals; over the course of 2 h, the ECG signal has experienced a change in its fundamental frequency! It is important to note that variation in the intervals mentioned above also manifest itself as a "modulation" effect, and hence will result in spectral broadening. Given the abundance of data, one is then compelled to use the "standard" signal processing algorithms to study these signals. To that end, let us use the PSD function using Welch window on the large data sample provided in the database using matlab. The result is depicted in Fig. 3.5. To our disappointment, we observe a spectral depiction that is void of any meaningful information, which cannot even be used to isolate the heart beat rate or other artifacts present in this measurement. Clearly, the standard signal processing tools designed to isolate stationary signals from wideband noise seem to be ineffective in this realm.

If we were to consider a shorter observation interval (say 1 min), we arrive at a more useful result. In Fig. 3.6, the magnitude of FFT is computed for a 1 min (60×128) observation interval. Unlike the results depicted in Fig. 3.4, this figure clearly isolates the beat rate (about 1.5 Hz, or 90 bpm) and underscores the periodic nature of the signal. This is not at all unexpected; in the event of observing signals with inherent time-varying parameters, it is best to limit the observation interval to

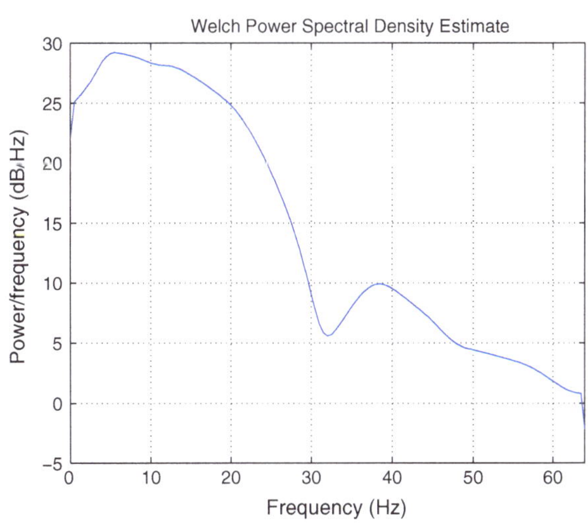

Fig. 3.5 The PSD of the ECG signal

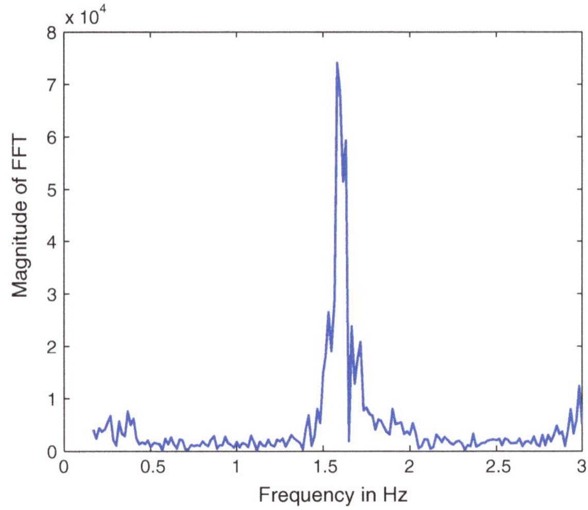

Fig. 3.6 Magnitude of the FFT of the ECG signal observed over 1 min

an interval for which the signal parameters are more or less stationary. This will give the signal a quasi-stationary attribute. Note that we have not established stationarity (as described in the previous chapters) in any sense. This observation also points to a need for computing FFT over short windows of time, instead of the traditional way of computing FT, which is done over an infinite window of time (in theory). We will revisit this problem and study the stationarity of these types of signals in later chapters when signal processing techniques for biomedical applications are presented. The purpose of this initial study is to underscore the inefficacy of the traditional signal processing techniques in isolating important characteristics of biomedical signals.

3.2 Electroencephalogram (EEG)

EEG detection is largely attributed to Hans Berger, who in 1924 was able to capture the EEG signal (in the form of small current, which was collected by a galvanometer) using two small, round, metal disks (electrodes) on a patient's scalp. Since then, we have come a long way in capturing and processing the massive amount of information that can be collected from mammals' brains with modern electronics. EEG signals, unlike ECG, do not exhibit periodicity (at least to the naked eyes) and are subject to large variations due to various internal and external stimuli. In fact, the presence of some periodic signals in EEG may point to underlying physiological anomalies, which in turn may point to the presence of pathological conditions, such as epilepsy. There are, however, periodic components in EEG that are due to normal activities. Furthermore, EEG signals are collected from various parts of scalp, which is in general a hard substance, leading to the presence of various (and often patient-dependent) artifacts in the observed signals.

To gain better insight, it is instructive to examine the EEG signal of a pediatric subject with epilepsy (no seizure) who has been studied over a 48-h period. To do so, we use the MIT-BIH database (see the Reference chapter) and use scalp EEG data (chbmit). This database contains the 23-lead hourly scalp EEG of 24 cases, involving 23 individuals (5 males, ages 3–22, and 17 females, ages 1.5–19, and 1 unidentified-record chb01 and chb21 correspond to the same individual with a gap of 1.5 years), collected over several hours (in most cases, up to 48 h). The data is not contiguous in time as interruptions are embedded to accommodate hardware limitations. We present the data for chb01. The subject chb01 does not experience epileptic seizures during the first hour of observation. Hence, what is depicted here contains seizure-free data. The signals are in μv and are sampled at a rate of 256 Hz (to convert the raw data to μv, one needs to subtract the "base" (in this case 0) form the data and then divide the result by the "gain" (in this case, 2.559375)). Unlike ECG case, where we were focusing on one signal and normalizing the signal was an acceptable practice, for EEG, we are considering 23 different signals where the relationship between the signals is of importance. For this reason, we present the actual data in this case. That is, the data is plotted directly from downloaded record.

In the following 46 Figures shown below, the EEG signals along with their FFT's are depicted for chb01 over the first hour of observation. These figures show the signals collected by the 23 probes, referred to as "channels." A cursory view of the time-series plots indicates that there are little visible correlation between these signals to a naked eye. In fact, large peaks occur rather randomly at various times and at various levels, making an analysis of EEG signals using time-series rather difficult. However, the short term (in this case, one hour) FFT of these signals reveals a consistent picture. Indeed, it is not hard to see the presence of a periodic component with the fundamental frequency of 16 Hz in all channels; clearly, we can see up to 7 harmonics of this fundamental frequency, pointing to a common feature of the subject's EEG that is captured by all the channels. It is also interesting to note that a number of spectral lines at 60 Hz and higher are present in some of the EEG signals. This may be attributed to the 60 Hz AC line interference that may be considered as an environmental noise. It is suggested that an RF isolation chamber be utilized to minimize any environmental impacts when EEG signals are collected. However, for most application of EEG, the signals are collected in the presence of such interferences, requiring one to be cognizant of such interferences.

Since the sampling rate is 256, 128 Hz is the largest frequency that can be measure reliably. That is the reason for not being able to capture any harmonics greater than the 8th harmonic ($8 \times 16 = 128$). Although FFT seems to unravel a key feature of EEG signal, some key information have been suppressed. Furthermore, it appears that the EEG case can be classified as "signal+noise" scenario, although we have not established the characteristic of the noise. Furthermore, we have little information about the dependency of "noise" and "signal" and about whether the "noise" can be classified as AWGN. We will address these problem when signal processing techniques are discussed in later chapters.

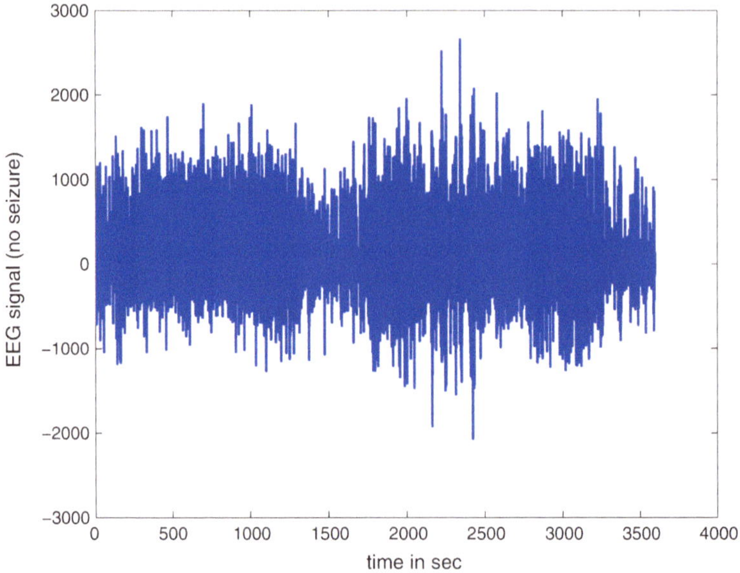

Channel 1

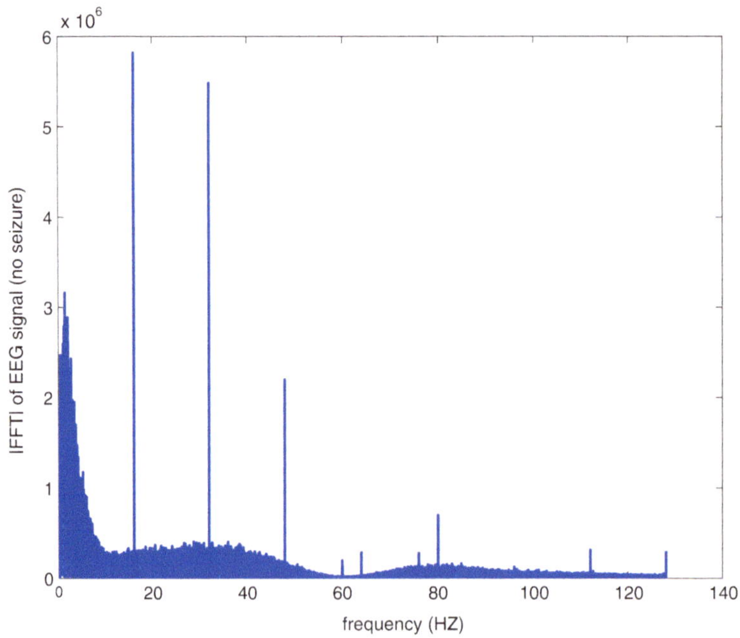

Channel 1

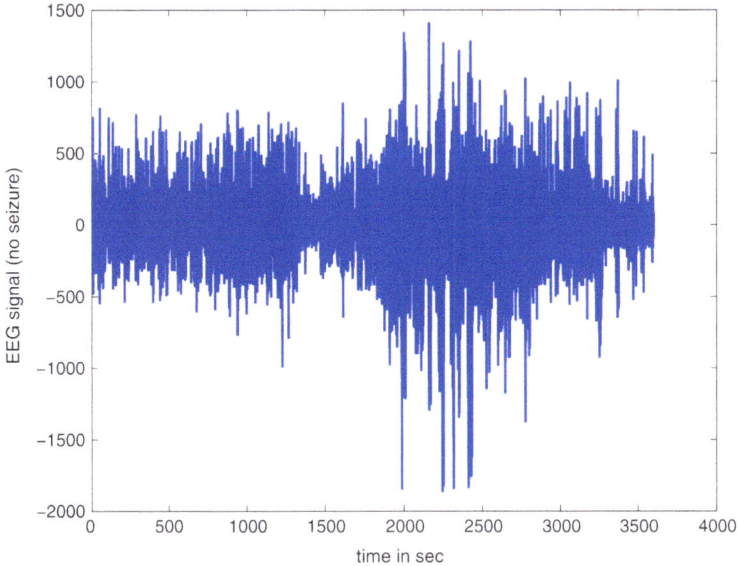

Channel 2

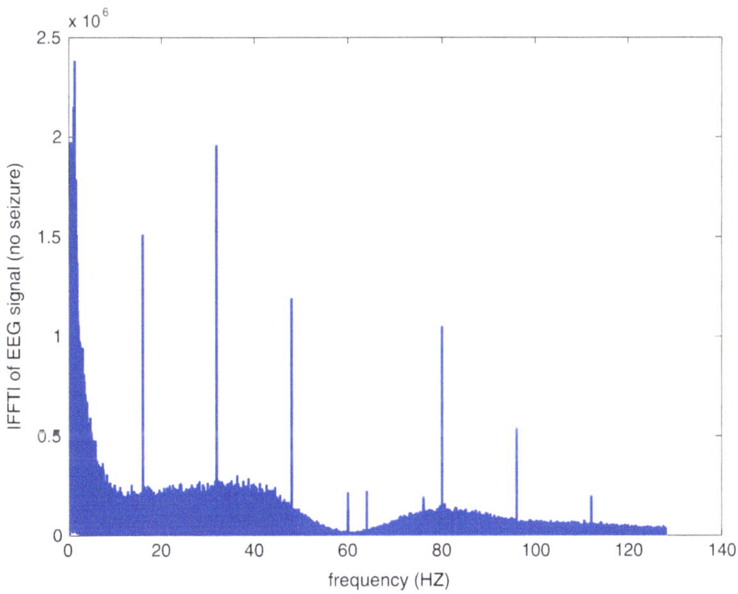

Channel 2

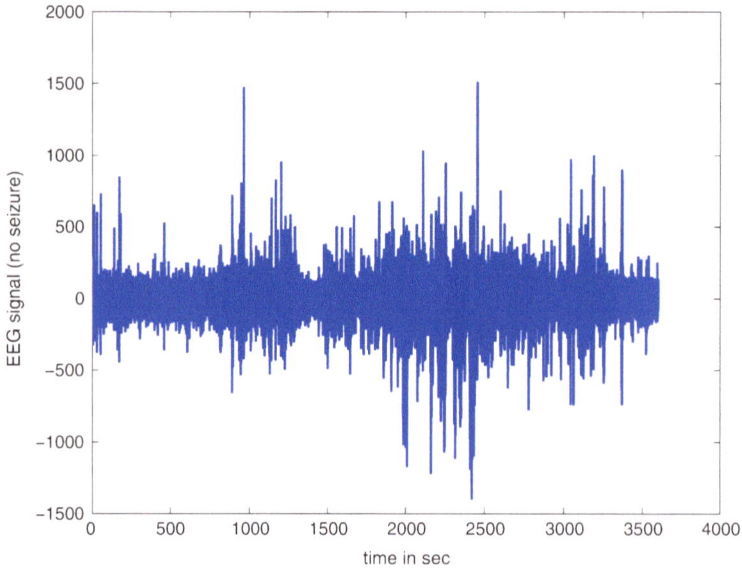

Channel 3

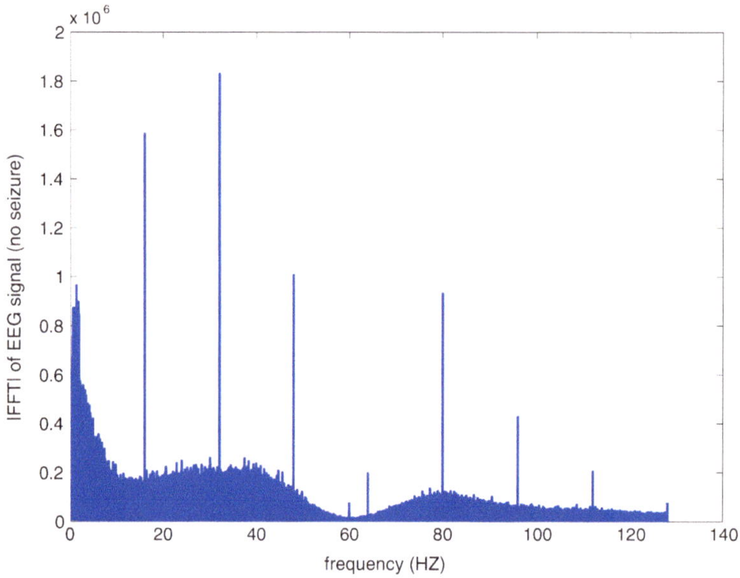

Channel 3

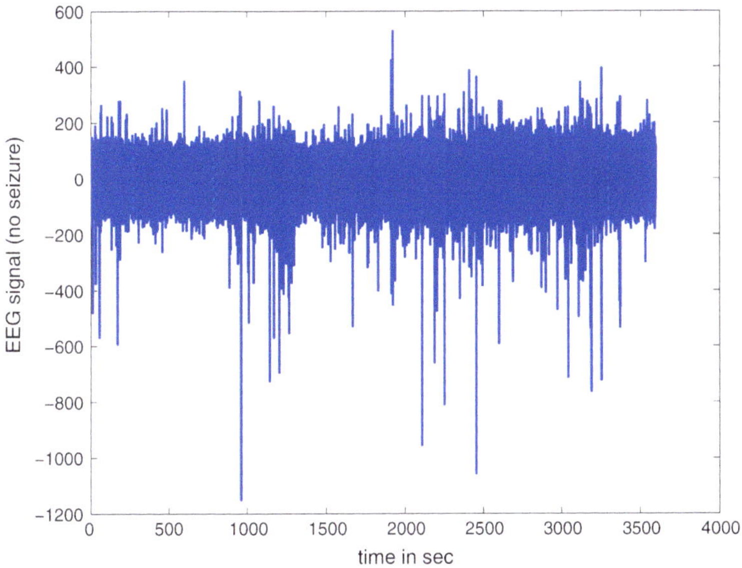

Channel 4

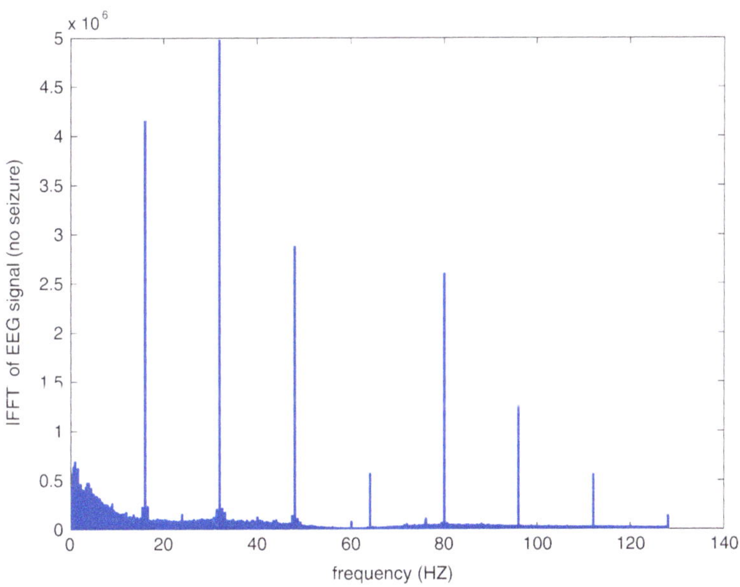

Channel 4

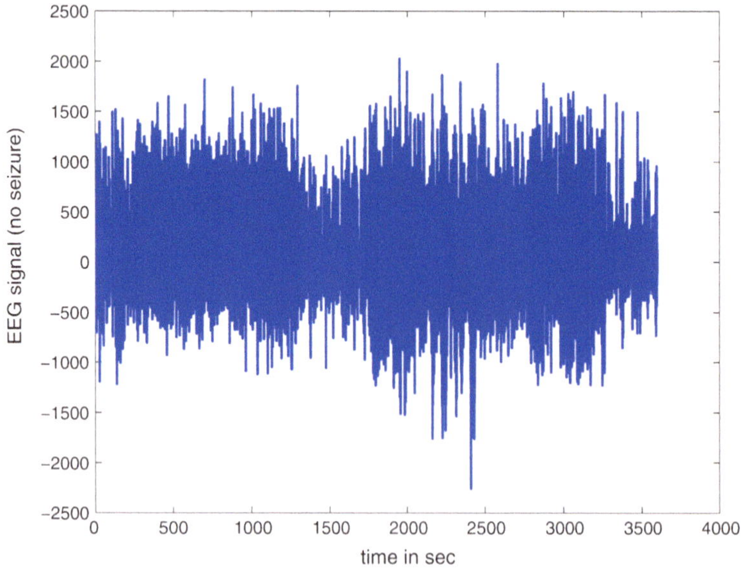

Channel 5

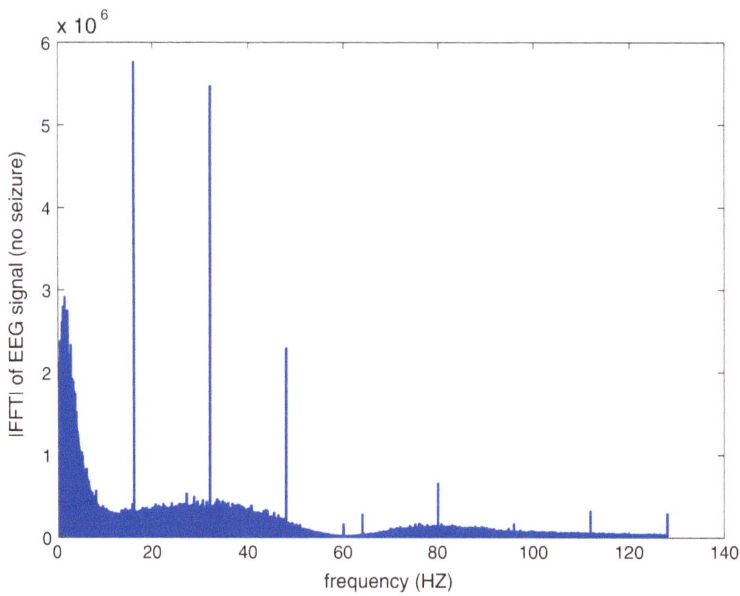

Channel 5

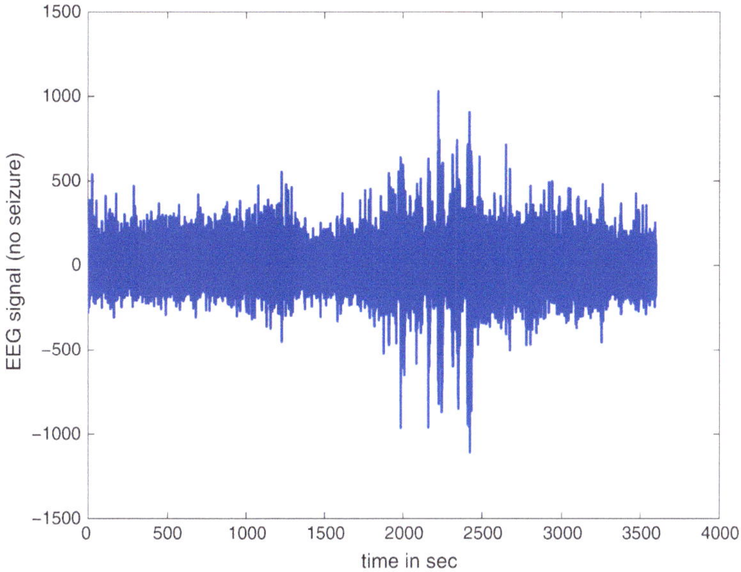

Channel 6

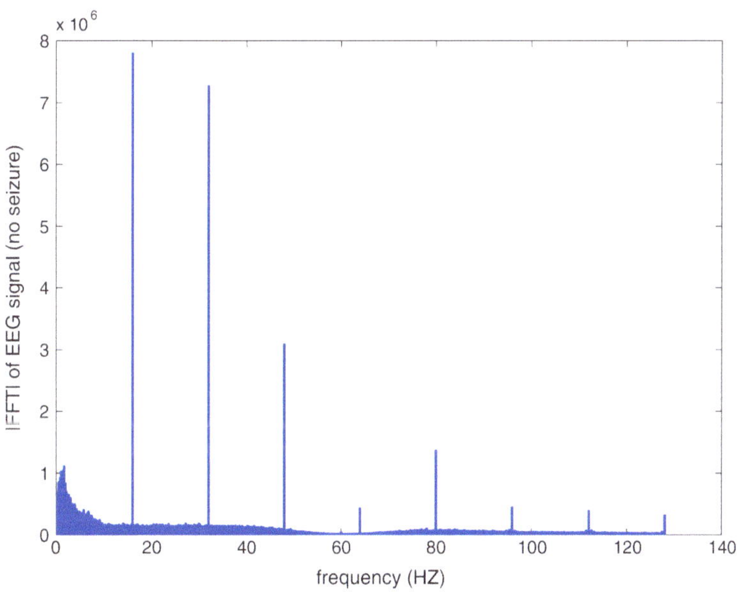

Channel 6

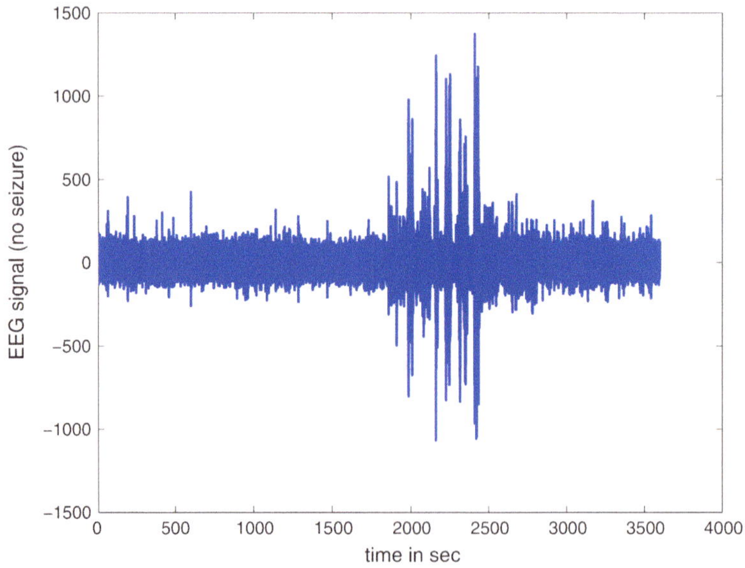

Channel 7

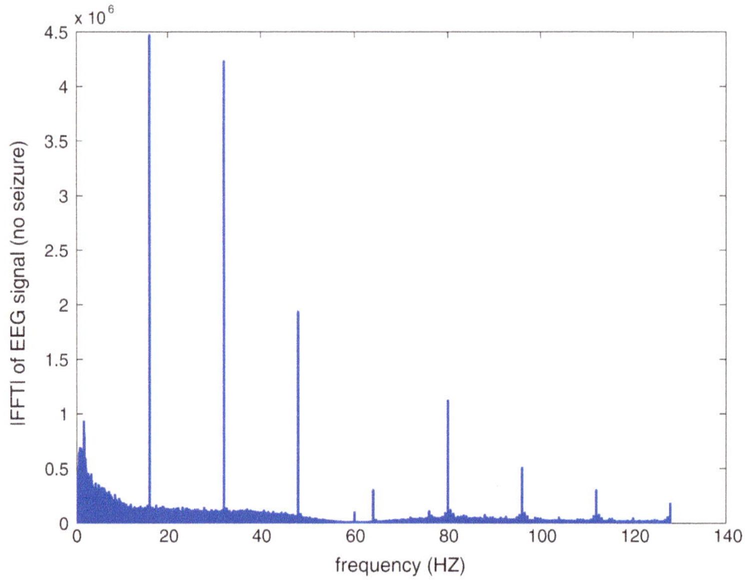

Channel 7

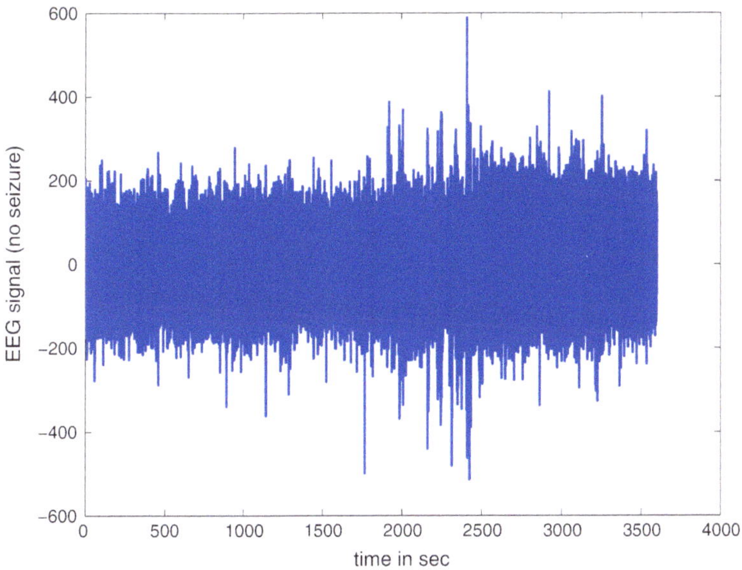

Channel 8

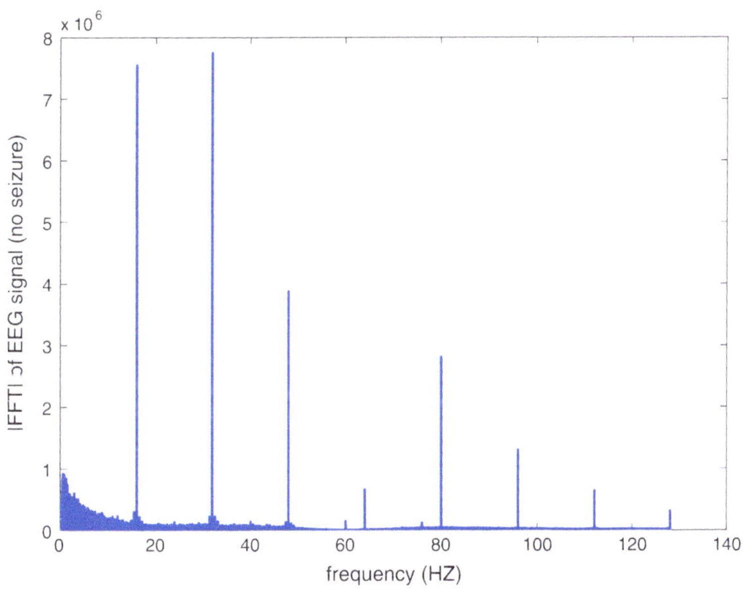

Channel 8

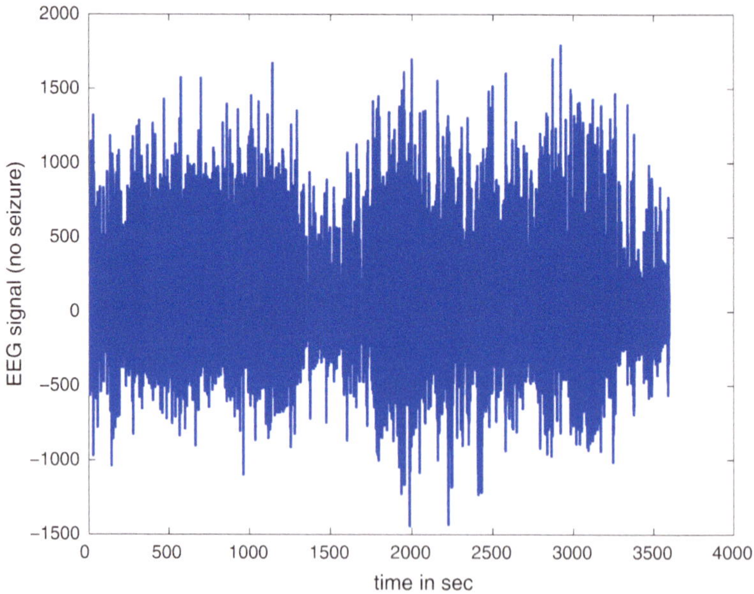

Channel 9

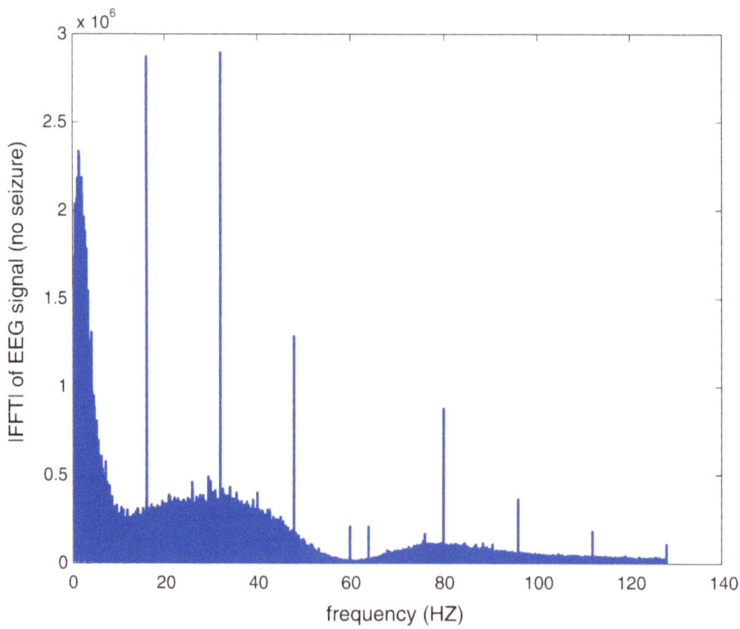

Channel 9

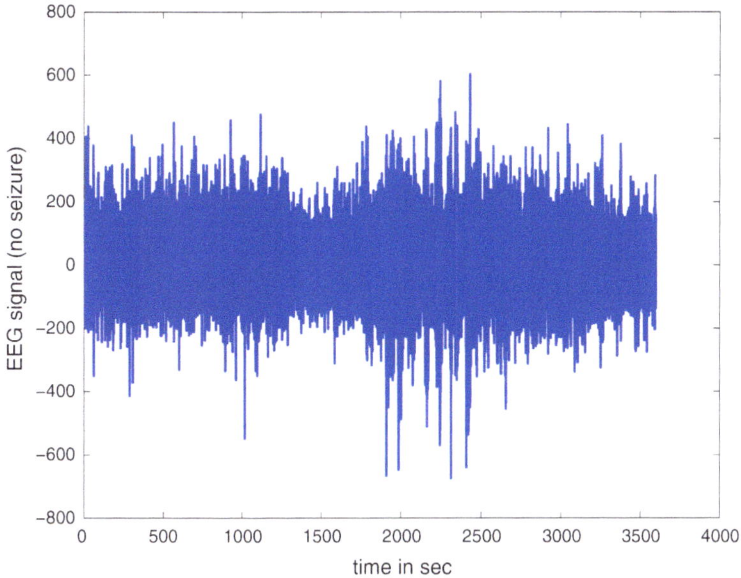

Channel 10

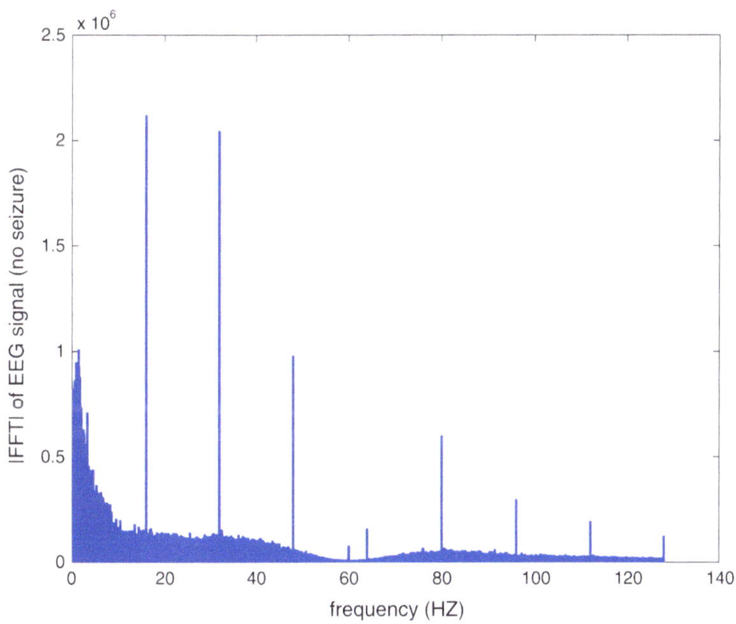

Channel 10

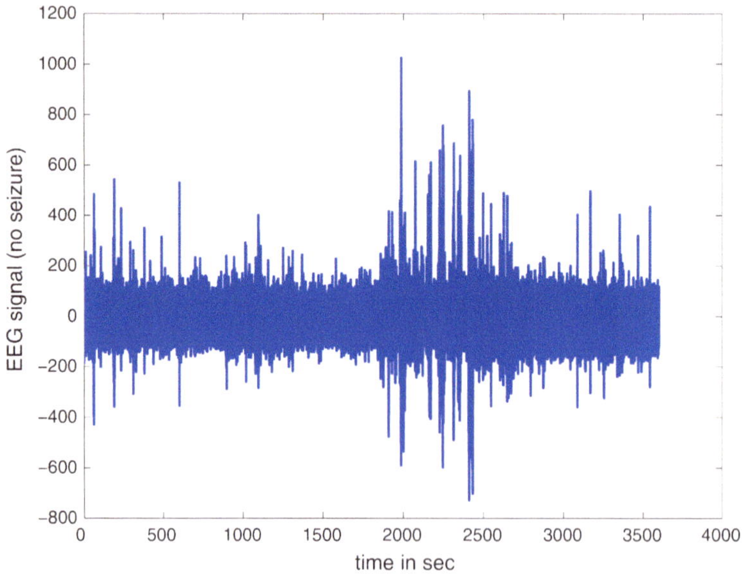

Channel 11

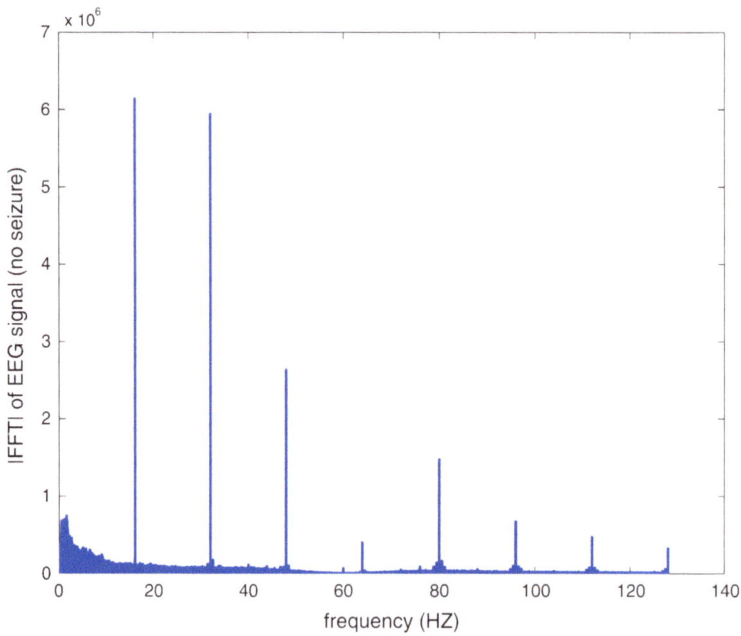

Channel 11

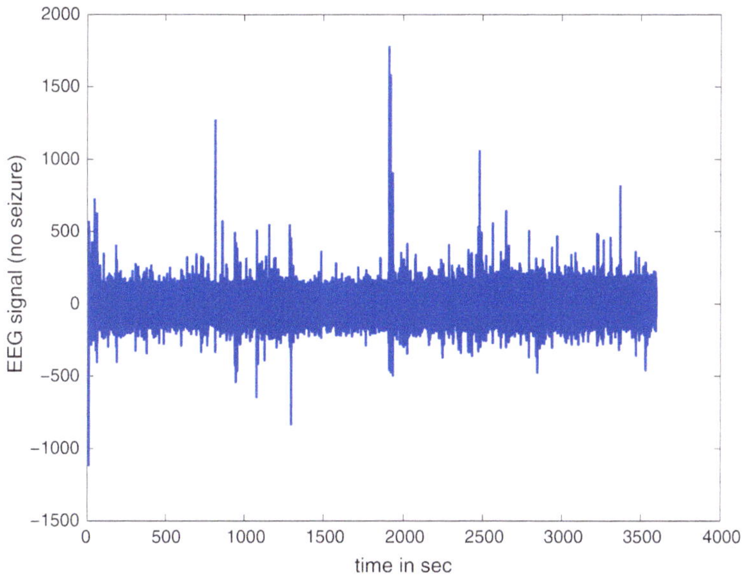

Channel 12

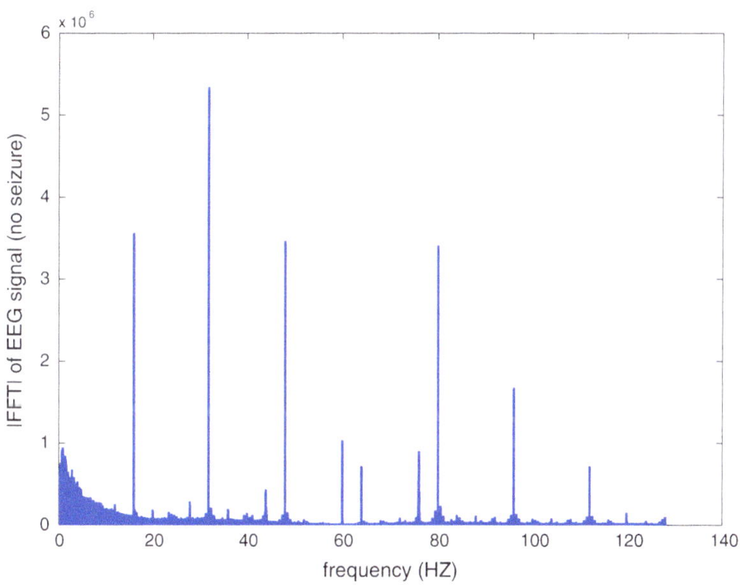

Channel 12

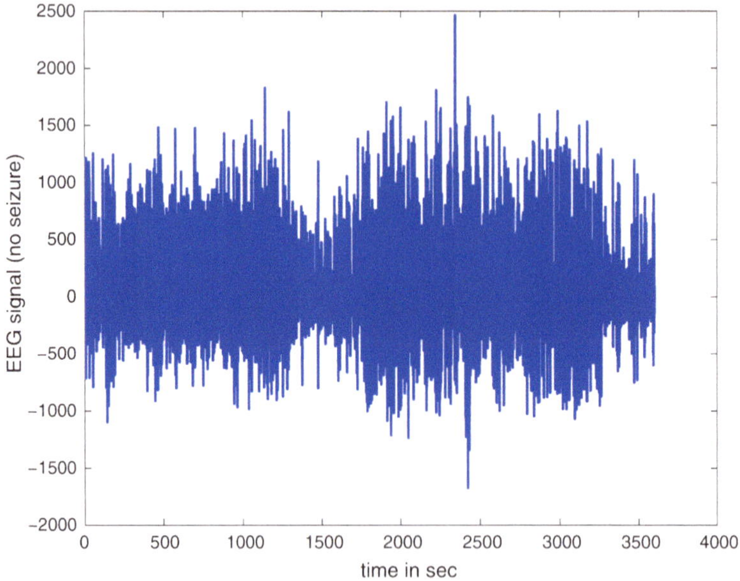

Channel 13

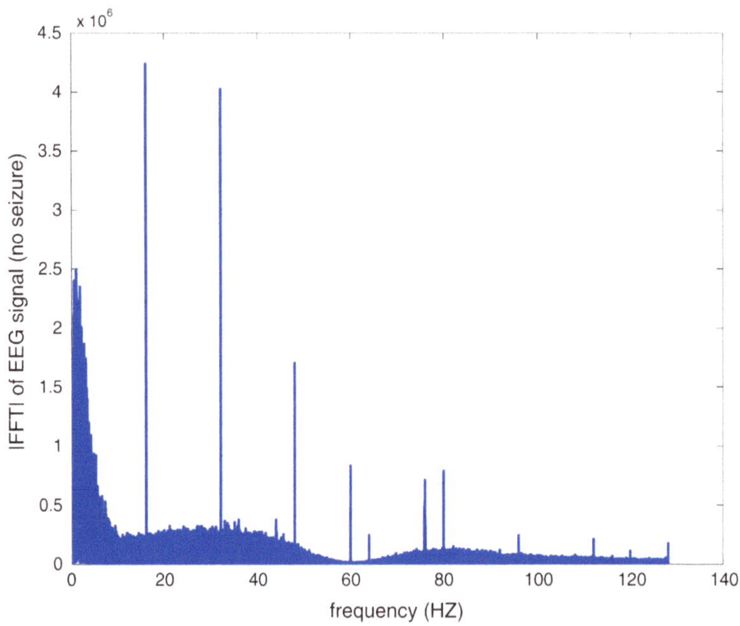

Channel 13

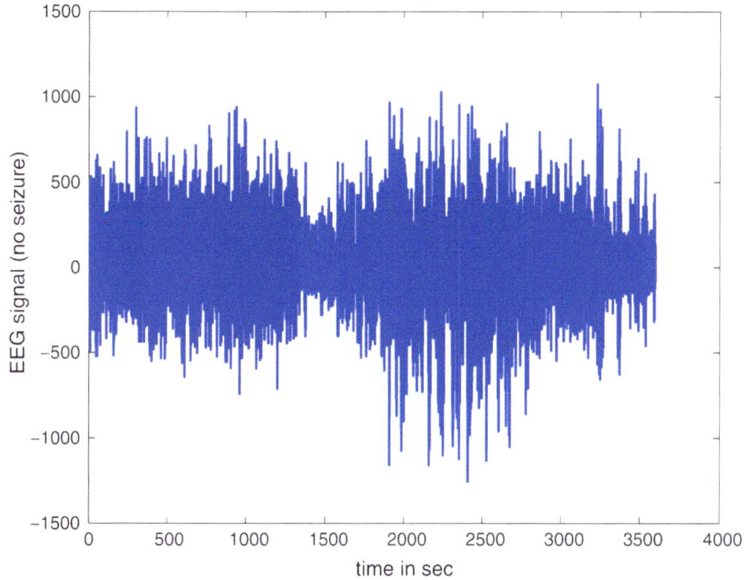

Channel 14

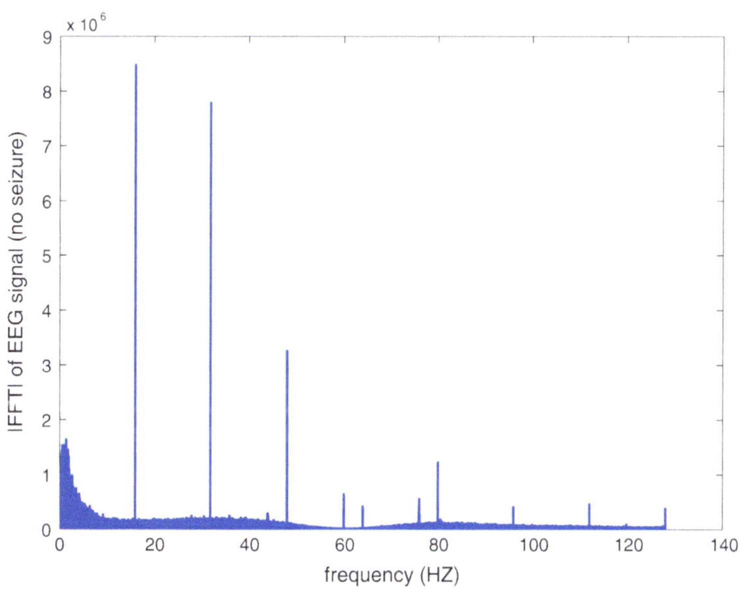

Channel 14

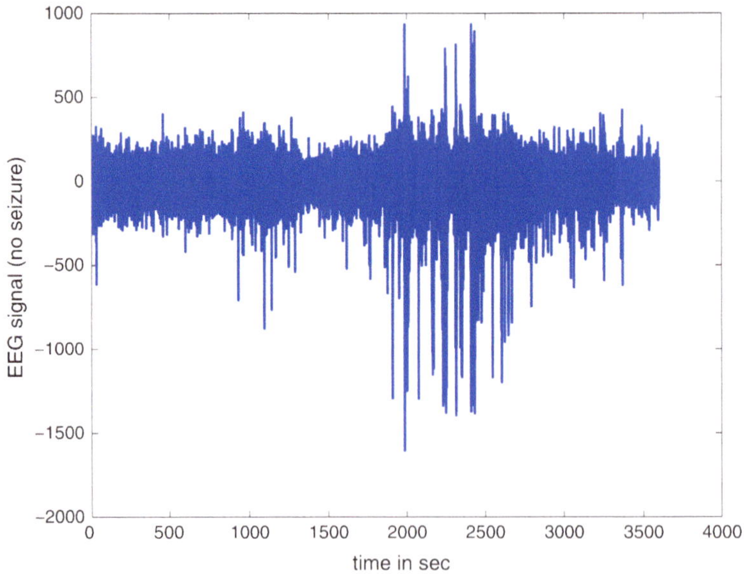

Channel 15

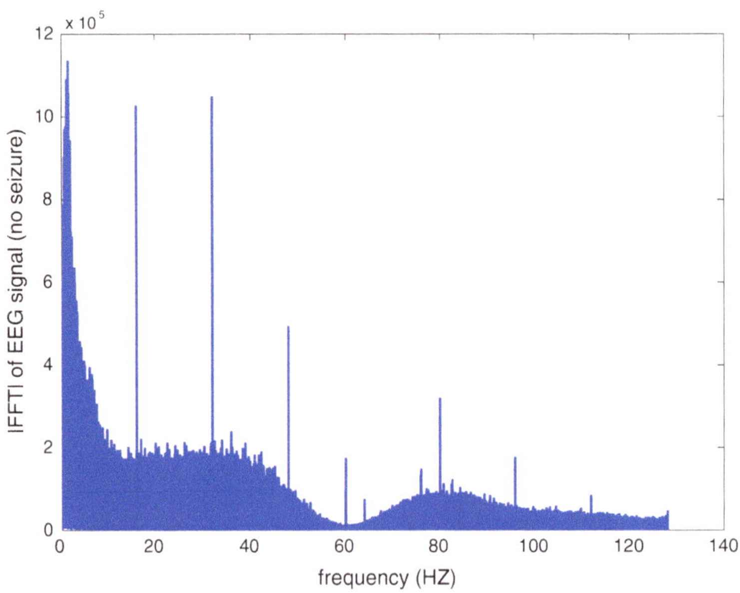

Channel 15

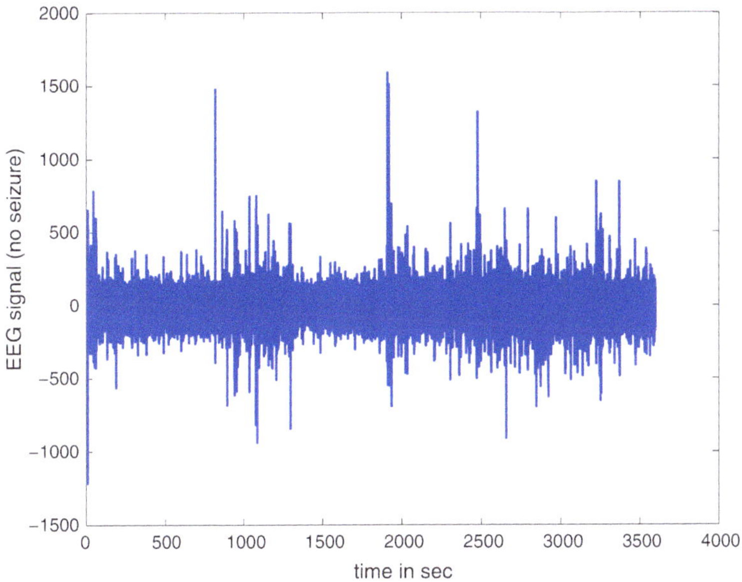

Channel 16

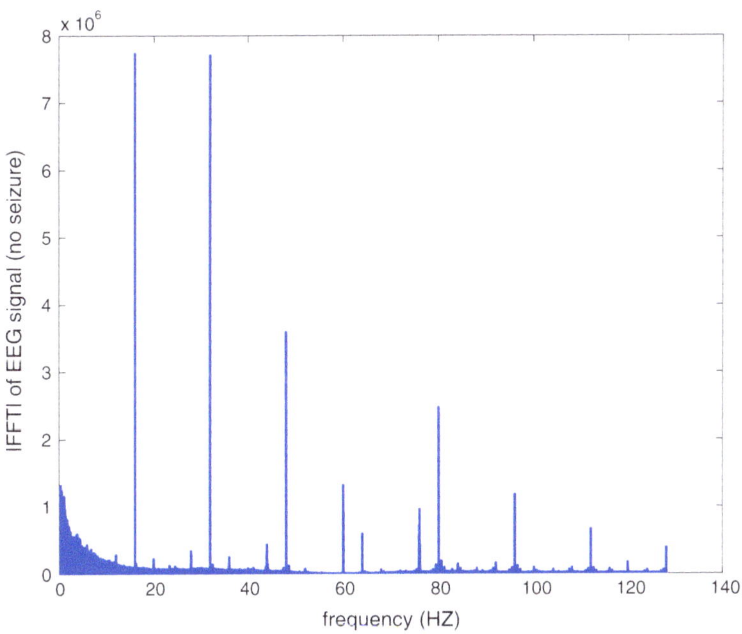

Channel 16

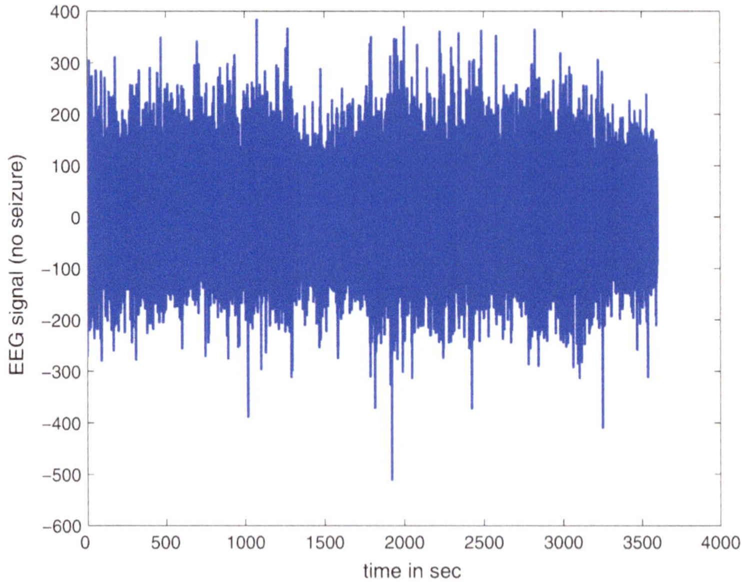

Channel 17

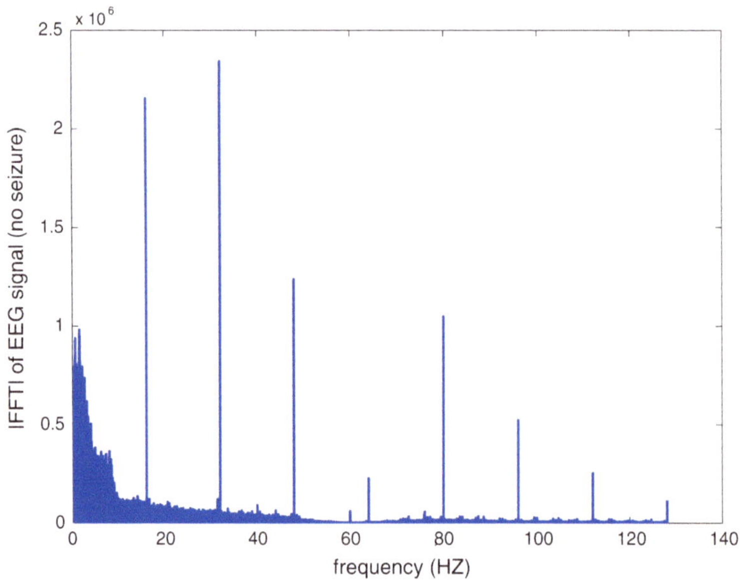

Channel 17

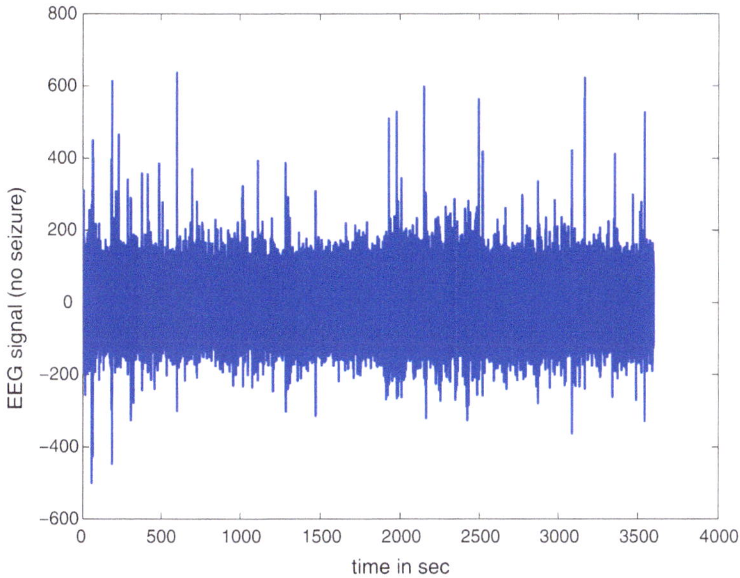

Channel 18

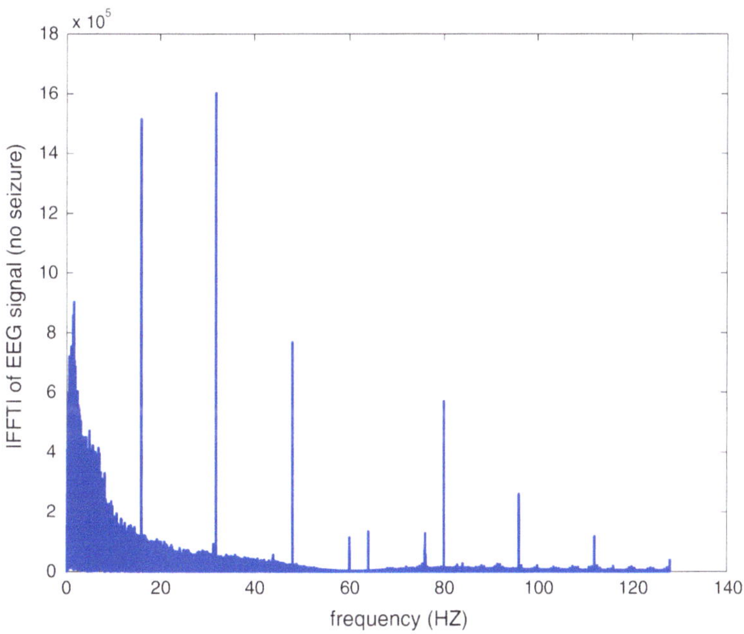

Channel 18

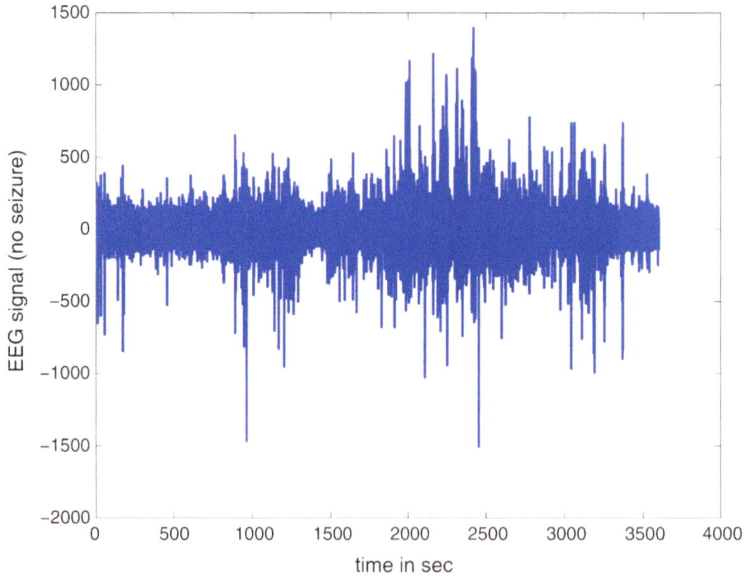

Channel 19

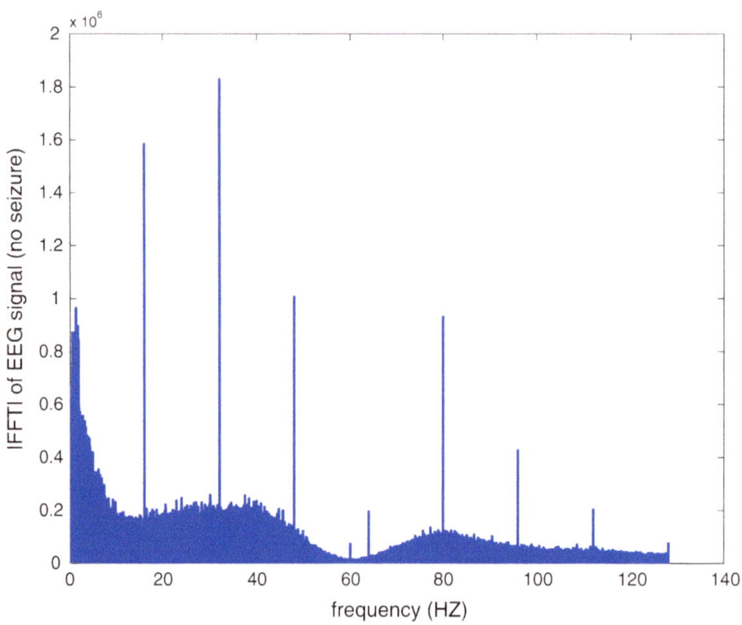

Channel 19

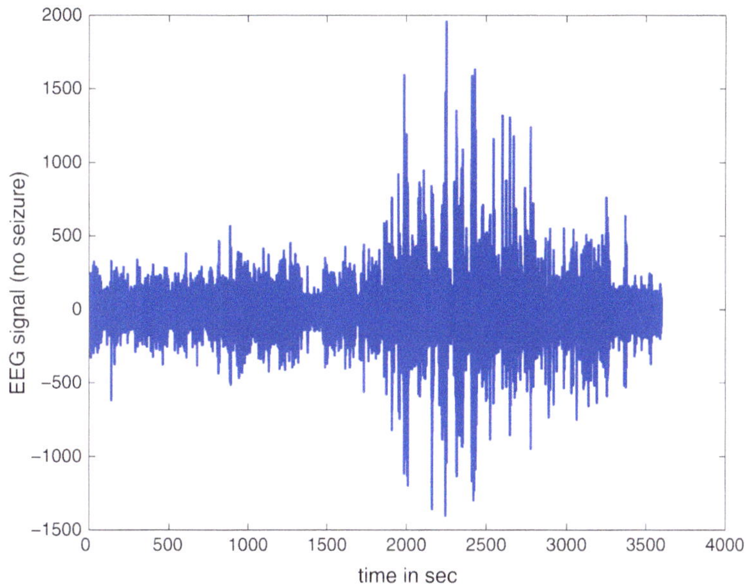

Channel 20

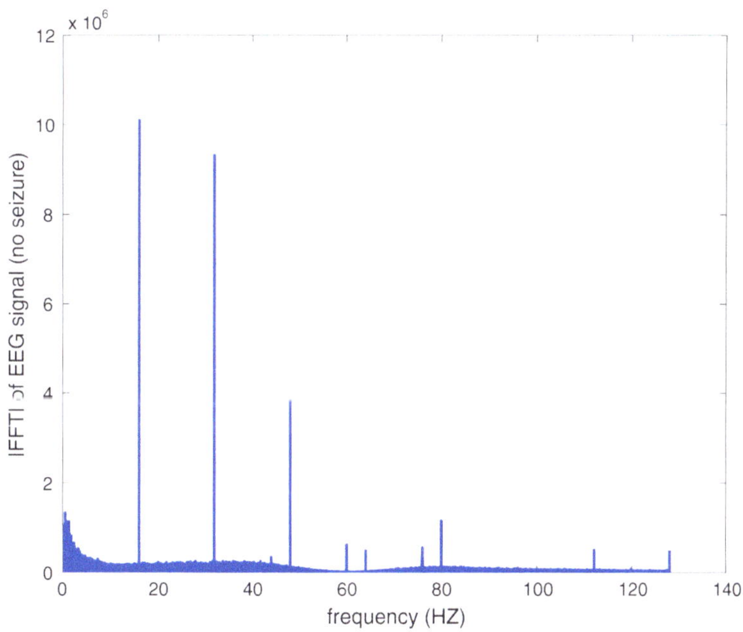

Channel 20

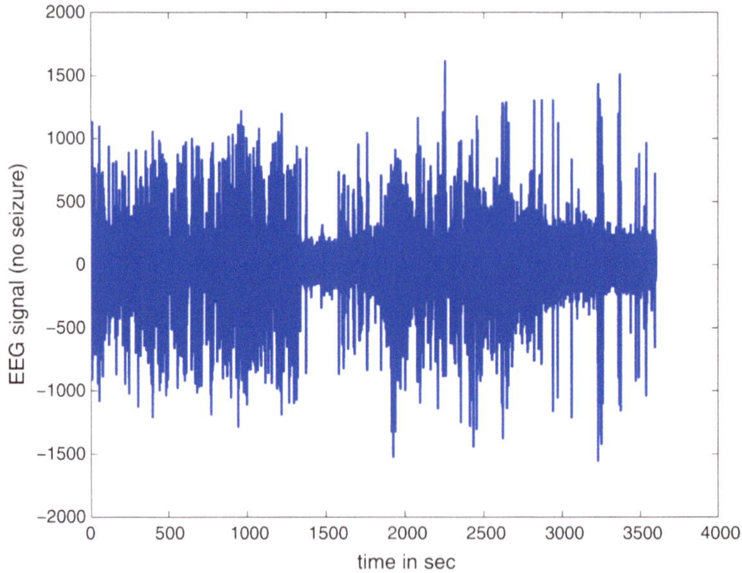

Channel 21

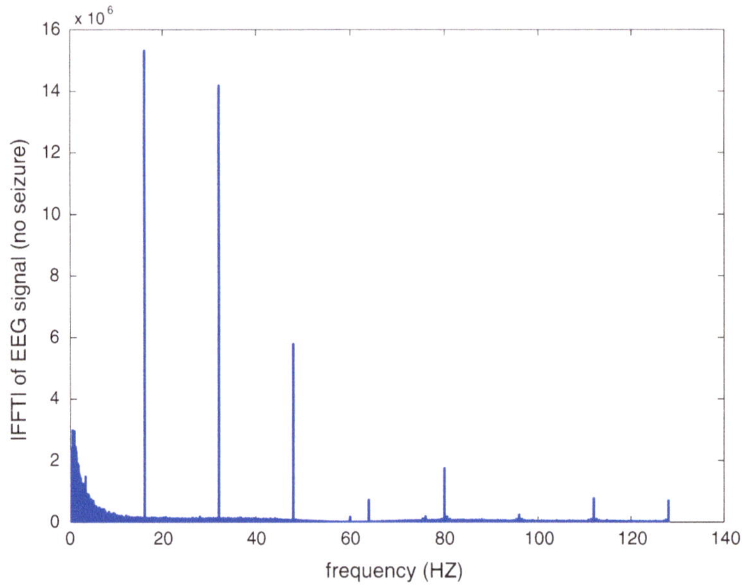

Channel 21

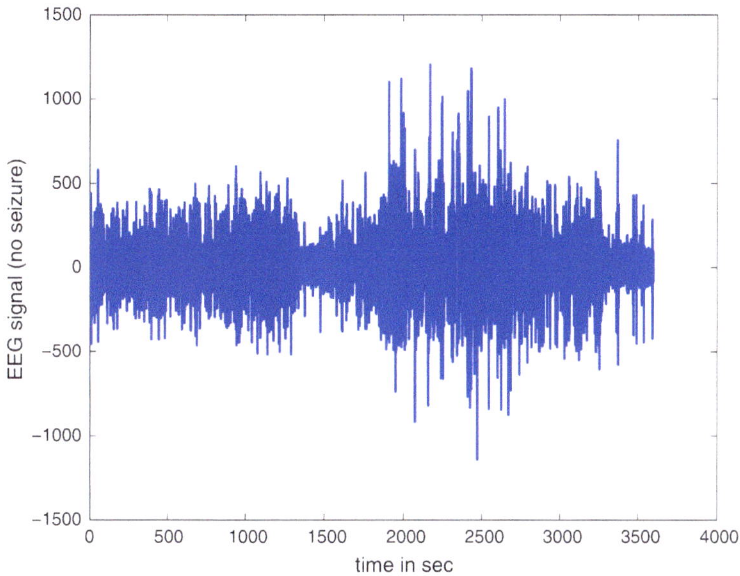

Channel 22

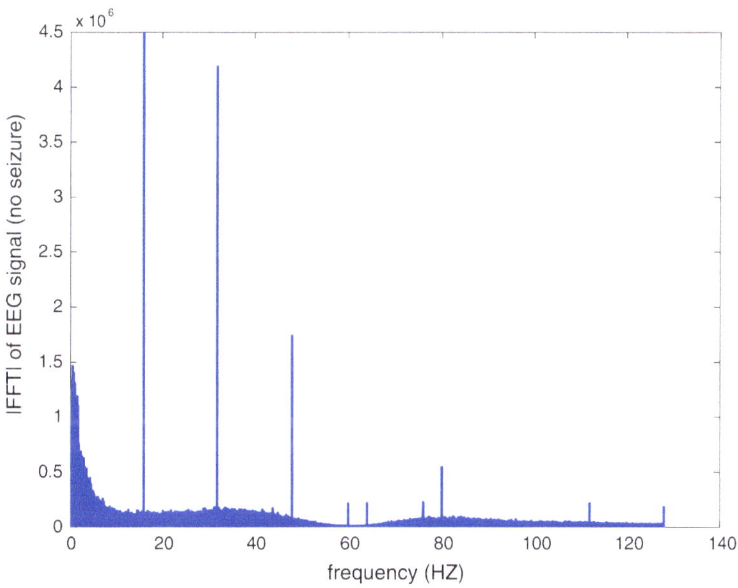

Channel 22

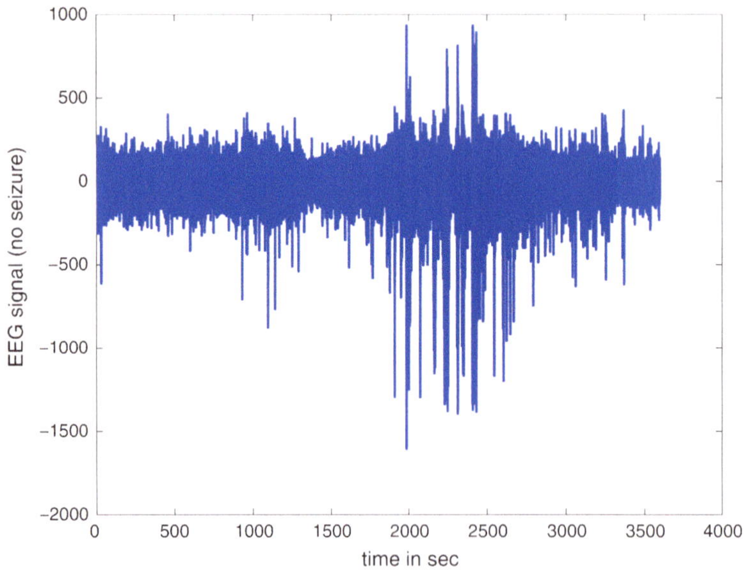

Channel 23

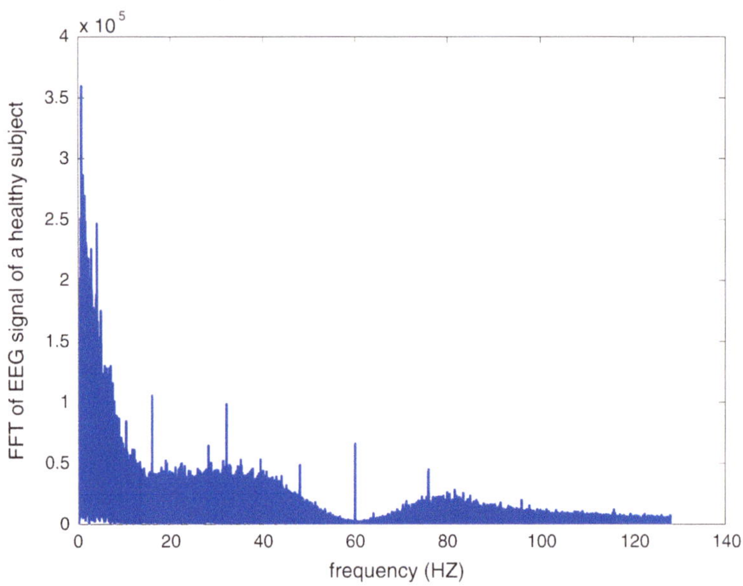

Channel 23

Before we leave this section, let us examine various bands of EEG signals, as they are known in the medical and scientific communities (Figs. 3.7 and 3.8).

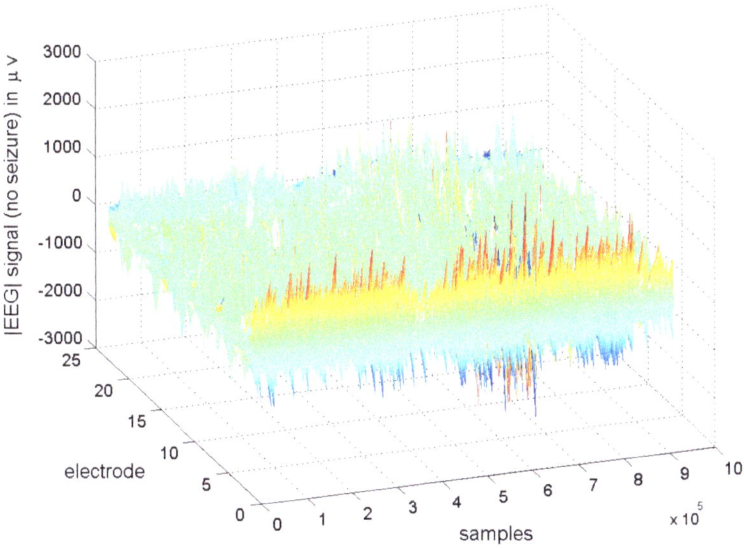

Fig. 3.7 The 3D plot of all 23 ECG signals

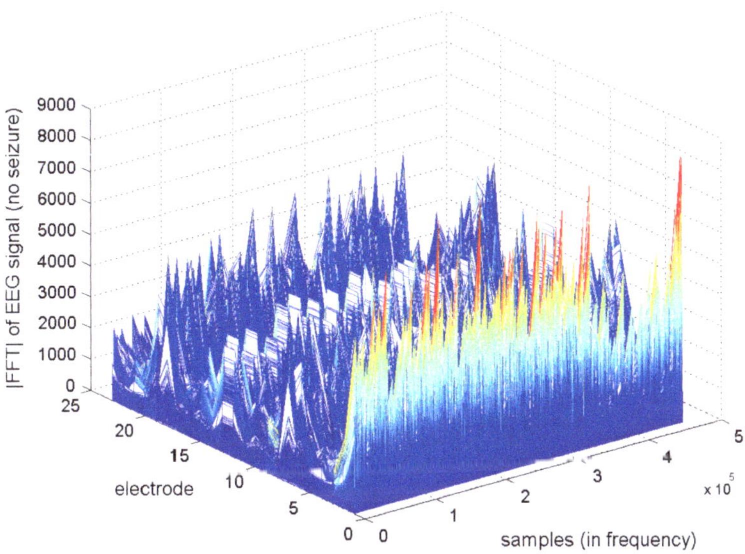

Fig. 3.8 Magnitude of the FFT of all ECG signals

3.2.1 δ Band

The lowest frequency band of an EEG is known as the δ band. This band corresponds to the rage [0, 2] Hz.

3.2.2 θ Band

The section of EEG signal spectrum falling in the range of [4, 8] Hz is referred to as the θ band.

3.2.3 α Band

This is a well-known band that corresponds to deep brain relaxation, which is associated with sleep mode of brain. The section of EEG signal spectrum falling in the range of [8, 12] Hz are referred to as the α band.

3.2.4 β Band

The frequency band [12, 30] Hz is referred to as the β band.

3.2.5 γ Band

Finally, the range of [30, 80] Hz is known as the γ band.

It is important to distinguished between the above bands and their corresponding waves. To illustrate this point, in the above example, we observed the presence of a 16 Hz periodic signal (which obviously is not a pure sinusoid). That is, we observed the presence of a periodic signal whose fundamental frequency is 16 Hz. Such a wave will be classified as a β wave, although its harmonics extend well into the γ band. Hence, any periodic component of EEG whose fundamental frequency is in a band is identified as a wave belonging to that band even though its harmonic may extend to higher bands.

Once again, we refrain from exploring signal processing techniques for analyzing EEG signals at this stage and postpone such a discussion to the ensuing chapters.

So far, we have only the frequency content of EEG and identified specific wave bands and waves associated with EEG. Furthermore, the FFT was computed over the entire range of observation, which as noted earlier, will suppress slow-varying as well as infrequent, but violent changes in the EEG signal. In reality, EEG signals contain rhythms, waves, transients, and spindles which cannot be fully analyzed using FFT, as done above. It is instructive to consider these attributes of EEG before we proceed to the next section. However, in the spirit of this book where all waves, transient or otherwise, are considered as signals, we introduce EEG rhythms, waves, transients, and spindles as EEG signals.

3.2.6 EEG Signals

The word "rhythm" is often used to underscore the present state of the brain, and hence, in general, refers to "slow" changing processes. For instance, we have μ rhythm which is a group of "comb" shape waves in the range 7–11 Hz with amplitudes less than 50 μV. This rhythm is typically attenuated by movement, contralateral movement, tactile simulation, and readiness for movement. Interestingly enough, μ waves are suppressed when the subject observes others' movements. However, the movement of inanimate objects has no bearing on the suppression of μ waves! For this reason, μ waves may be used to move a cursor, for instance, for brain-computer interface (BCI) purposes.

Yet another rhythm is the sleep spindle (or σ wave), with a frequency of 14 Hz, which occurs during sleep and is defined as a short sequence of single-form waves with fusiform appearance, helping one to remain sleep in the presence of external stimuli.

Among the transients, one can refer to spikes (these are large peak signals with durations of 20–30 ms), sharp waves (similar to spikes, but with a longer duration of up to 200 ms), and K-complex (a slow changing transient with sharp components of up to 200 μv, which is generated by brain in response to sudden stimuli during sleep). K-complex has some similarities to the QRS complex of ECG.

λ waves are a group of 50 μv waves occurring in the occipital location dealing primarily with eye movement. Spike-and-wave is yet another rhythm in 2.5–3.5 Hz band, which may contain several sharp spikes of size 1000 μv (in that case, it is called polyspike-and-wave complex). Certain spikes in the spike-and-wave complex are attributed to partial epileptic seizures. Finally, V-wave (vertex-sharp wave) is a sharp signal with 300 μv vertex (maximal at the vertex), which is negative with respect to the EEG in areas other than the vertex. V-wave is often produced spontaneously during sleep or due to sensory stimuli.

As can be seen, EEG may contain many transient signals, which can only be captured by short term signal analysis.

3.3 Elcctromyogram (EMG)

As the name implies, EMG corresponds to the rcsponse of muscles to nerve stimuli generated by sympathetic, parasympathetic, or both nervous systems. Unlike EEG signals, the discovery of this phenomenon dates back to the 16th century. In Fig. 3.9, a sample of a healthy EMG is depicted. Unlike EEG and ECG, the EMG signal may contain high frequency components. For this reason, the MIT-BIH database is created using a sampling rate of 4 kHz. Furthermore, teh raw data must be divided by the gain of 1000 to convert the signal into mv. An FFT (in this event, we only focus on the overall signal), yields a spectrum (see Fig. 3.10) that does not present significant spectral components at high frequencies, with the main components of the signal assuming low frequencies. In contrast, when anomalies are present in

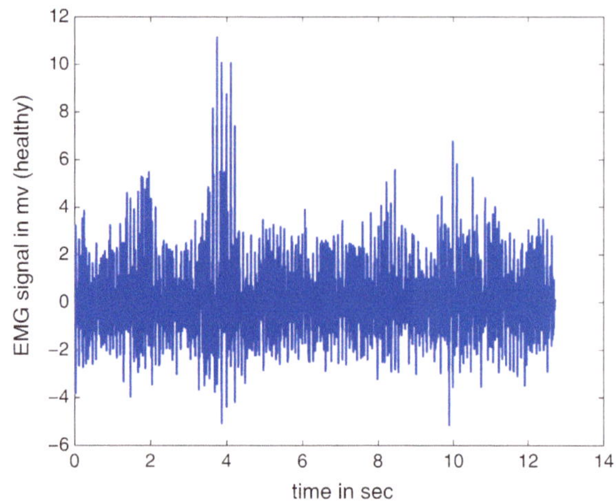

Fig. 3.9 The EMG signal as a function of time

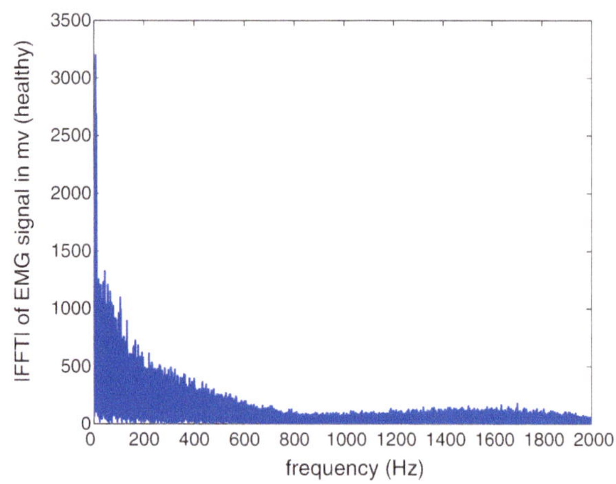

Fig. 3.10 Magnitude of the FFT of the EMG signal

the forms of myopathy (Figs. 3.11 and 3.12) and neuropathy (Figs. 3.13 and 3.14), EMG signals present higher frequency components. In the case of myopathy, the signal has an appearance of being uniform and noise like, although the FFT reveals spectral lines around 100 Hz. For the case of neuropathy, a somewhat different picture emerges. Note that the EMG signal in the presence of neuropathy resembles an amplitude modulated signal, where the signal assumes two distinct and pronounced levels. Furthermore, the low level seems to have a period of 12 s, while the high value is limited to 6 s intervals. This phenomenon may be due to the test subject

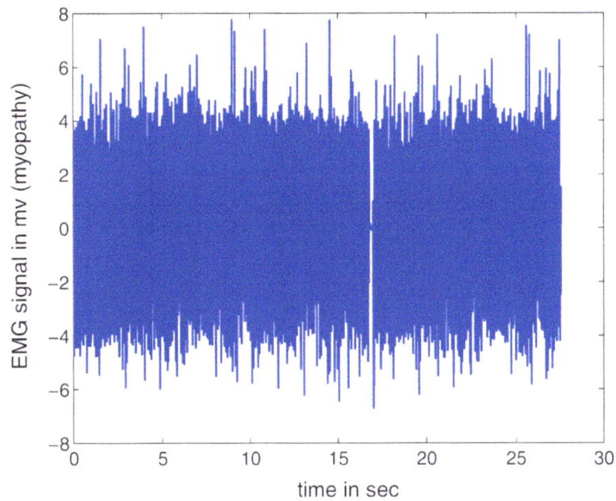

Fig. 3.11 The EMG signal for a patient with myopathy as a function of time

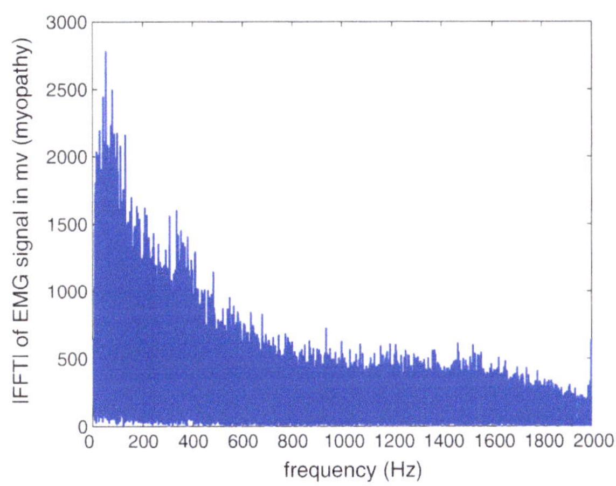

Fig. 3.12 Magnitude of the FFT of the EMG signal for a patient with myopathy

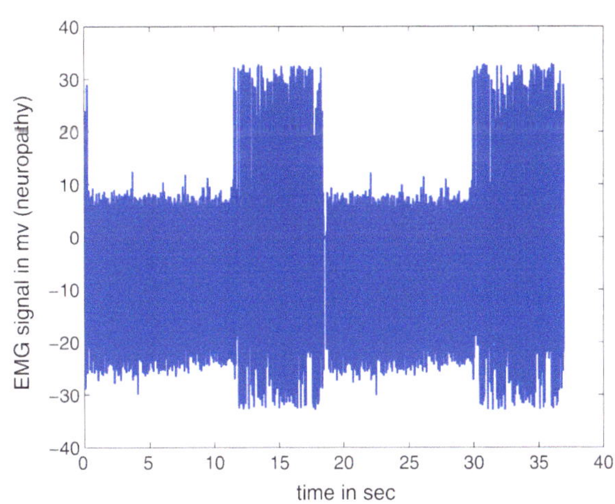

Fig. 3.13 The EMG signal for a patient with neuropathy as a function of time

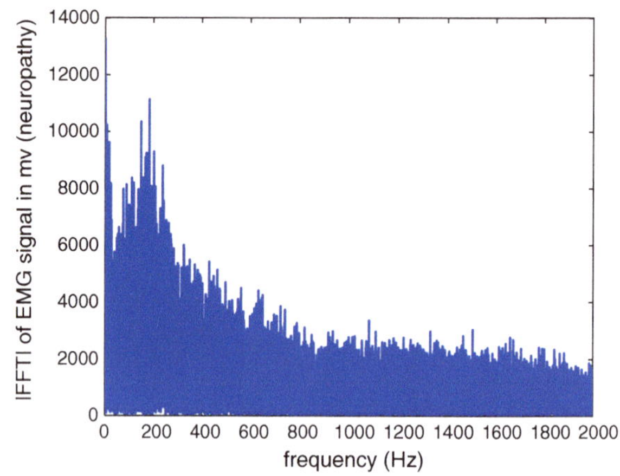

Fig. 3.14 Magnitude of the FFT of the EMG signal for a patient with neuropathy

either undergoing some form of periodic exercise with well-defined temporal characteristics or experiencing a "seizure-like" event, which is a viable possibility for an individual with muscle nerve damage (neuropathy). Furthermore, the EMG signal for this case contains frequency components at higher frequencies ($\geq$200 Hz), pointing to a physiological anomaly.

Chapter 4
Signal Processing Methods for Biological Signals

The previous chapters focused on the signal processing techniques suited for handling stationary (the classification of stationarity was also provided) signals. Such methods become increasingly ineffective in dealing with non-stationary signals to the extent that application of such techniques results in suppressing vital information about the signal or providing misleading information. Once example was the use of FFT in computing the spectrum of ECG signals, where the simple heart beat rate could not be extracted using FFT when FFT is computed over a large window (time) of data (see section on ECG). It, then, becomes exceedingly obvious that a new set of tools need to be utilized in dealing with signals whose fundamental properties are subject to change due to a host of (perhaps independent) events. To that end, in the ensuing discussion, we outline some key methods for analyzing non-stationary biomedical signals. It is imperative to distinguish between (feature) detection mechanisms versus decomposition techniques. The former focuses on detecting the presence of the signal in noise, undesirable disturbances, and/or when the signal features are subject to change over time, whereas the latter is concerned with extracting key features of the signal for pathological or diagnosis purposes. For instance, when dealing with ECG, one may be interested in detecting the QRS complex when signal is corrupted by noise and temporal variations. The detection process in this case may be used for autonomous heart beat rate detection and monitoring. For the same scenario, signal decomposition techniques may be used to isolate an underlying heart condition which may elude a cursory examination of the signal. It is, however, important to note that the boundaries between the methods may be blurred at times. For instance, extracting a feature of the signal, which may also impact the decomposed signal, may be used to diagnose a disease.

Although numerous methods have been proposed in the literature on signal detection, we focus on the most widely accepted techniques. Before doing so, it is of some importance to address the issue of "independence" when observing large volume of data. This is of interest when one is required to identify independent components that are shared by a set of, perhaps very large, observations.

K. Kiasaleh, *Biological Signals Classification and Analysis*,
Lecture Notes in Bioengineering, DOI 10.1007/978-3-642-54879-6_4

4.1 Independence

In the field of biomedical engineering, one often faces a simple question. *"Are the data samples I am acquiring independent?"* A legitimate follow on question is: *"If so, how do I establish that they are truly independent?"* In order for us to address these questions, one has to first establish the concept of independence. Let $\mathbf{x}[n]$ denote a vector of length l_v (as defined in the previous chapters) of sampled data. That is, $\mathbf{x}[n] = [x[n], x[n+1], \ldots, x[n+l_v-1]]^\mathsf{T}$. Then, one can claim the sequence of samples (they need not be consecutive samples in time) that form this vector are independent if and only if

$$f_{\mathbf{x}[n]}(\mathbf{x}) = \prod_{m=n}^{n+l_v-1} f_{\mathbf{x}[m]}(x_m) \tag{4.1}$$

Note that $x[n]$ and $x[n+1]$ do not need to refer to a pair of consecutive samples in time. That is, the samples under investigation can merely be samples arranged in a sequence of l_v observation points in time. Although this is an elegant result, its utility in practice is questionable as its application hinges upon a knowledge of the joint and marginal pdf's of the observed samples. Such a powerful statistical information is often missing in practice or is quite difficult to acquire. In particular, in the field of biomedical signal processing, one has little or no prior knowledge about the statistical characteristics of the signal under investigation.

One may be tempted to establish independence through a number of other incorrect means. In particular, the concepts of "uncorrelated" or "orthogonal" data may be confused with "independence". To gain a better insight, let us define these concepts in some detail.

4.1.1 Uncorrelated

This concept, as the name implies, deals with the correlation between data set and, hence, is concerned with the second order statistics of the data. That is, the set of samples in $\mathbf{x}[n]$ are called mutually uncorrelated if

$$E[x[m]x[q]] = E[x[q]]\,E[x[m]] \text{ for any } q \text{ and } m \in \{n, n+1, \ldots, n+l_v-1\} \tag{4.2}$$

It is tempting to ask whether there is a relationship between (4.2) and (4.1). In particular, can one can establish (4.1), therefore independence, using (4.2). If so, then the task of establishing independence reduces to collecting enough data to obtain a large degree of confidence in the correlation among the data samples (assuming that ergodicity in correlation can be established). Before we abandon the idea of establishing independence in a proper manner (i.e., using (4.1)), let us examine this matter a bit more carefully. Clearly,

$$f_{\mathbf{x}[n]}(\mathbf{x}) = \prod_{m=n}^{n+l_v-1} f_{\mathbf{x}[m]}(x_m) \rightarrow E\left[x[m]\,x[q]\right] = E\left[x[q]\right]E\left[x[m]\right] \tag{4.3}$$

for all q and $m \in \{n, n+1, \ldots, n+l_v-1\}$. However,

$$E\left[x[m]\,x[q]\right] = E\left[x[q]\right]E\left[x[m]\right] \xrightarrow{\text{may or may not}} f_{\mathbf{x}[n]}(\mathbf{x}) = \prod_{m=n}^{n+l_v-1} f_{\mathbf{x}[m]}(x_m) \tag{4.4}$$

In fact, we can show that in most cases

$$E\left[x[m]\,x[q]\right] = E\left[x[q]\right]E\left[x[m]\right] \nrightarrow f_{\mathbf{x}[n]}(\mathbf{x}) = \prod_{m=n}^{n+l_v-1} f_{\mathbf{x}[m]}(x_m) \tag{4.5}$$

for all q and $m \in \{n, n+1, \ldots, n+l_v-1\}$. This is a disappointing result, shattering a misconception held by public at large (and, unfortunately, by some students of statistics) regarding independence. One case, however, that justifies asserting independence when data is uncorrelated is when the data is jointly Gaussian! That is, when $\mathbf{x}[n]$ is a Gaussian vector. In that case,

$$E\left[x[m]\,x[q]\right] = E\left[x[q]\right]E\left[x[m]\right] \xrightarrow{\text{Gaussian Case}} f_{\mathbf{x}[n]}(\mathbf{x}) = \prod_{m=n}^{n+l_v-1} f_{\mathbf{x}[m]}(x_m) \tag{4.6}$$

The proof for this assertion is rather straightforward; to that end, let us consider the covariance matrix, given by (1.44), of an uncorrelated data set. It can be seen that

$$\begin{aligned}\mathbf{K}_x[n] &= E\left\{(\mathbf{x}[n] - \mathbf{m}_x[n])(\mathbf{x}[n] - \mathbf{m}_x[n])^H\right\} \\ &= \begin{bmatrix} \sigma_n^2 & 0 & 0 & \cdots & 0 \\ 0 & \sigma_{n+1}^2 & 0 & \cdots & 0 \\ \vdots & 0 & \ddots & 0 & \vdots \\ 0 & \vdots & \cdots & \ddots & 0 \\ 0 & 0 & 0 & 0 & \sigma_{n+l_v-1}^2 \end{bmatrix}\end{aligned} \tag{4.7}$$

where $\sigma_j^2 = E\left\{x^2[j]\right\} - E^2\{x[j]\}$ denotes the variance of the jth sample of the signal. Hence,

$$\mathbf{K}_x^{-1}[n] == \begin{bmatrix} \sigma_n^{-2} & 0 & 0 & \cdots & 0 \\ 0 & \sigma_{n+1}^{-2} & 0 & \cdots & 0 \\ \vdots & 0 & \ddots & 0 & \vdots \\ 0 & \vdots & \cdots & \ddots & 0 \\ 0 & 0 & 0 & 0 & \sigma_{n+l_v-1}^{-2} \end{bmatrix},$$

and using (1.46), we have

$$
\begin{aligned}
f_{\mathbf{x}[n]}(\mathbf{x}) &= \frac{1}{(2\pi)^{l_v/2}\left\{\prod_{m=n}^{n+l_v-1}\sigma_m^2\right\}^{1/2}} \\
&\quad \times \exp\left\{-\frac{1}{2}\sum_{m=n}^{n+l_v-1}\frac{(x[m]-E\{x[m]\})^2}{2\sigma_m^2}\right\} \\
&= \prod_{m=n}^{n+l_v-1} f_{\mathbf{x}[m]}(x_m)
\end{aligned} \tag{4.8}
$$

This completes the proof. Also, this conclusion underscores an important point. That is, the Gaussian process is a second order process (2nd order statistics fully characterizes the process). Some processes observed in nature (including the additive noise in communication) adhere to the Gaussian assumption, and hence, one is tempted to make the misguided extension of this observation to include biological processes. Unfortunately, nonlinear systems, when excited by Gaussian processes, do not yield Gaussian outputs. Since most biological processes are the result of complex and often nonlinear processes, the Gaussian assumption is essentially unwarranted for such scenarios. This implies that we cannot establish independence using the correlation (or lack there of) among the data samples.

The other metric that is used incorrectly to establish impedance is the concept of orthogonality.

4.1.2 Orthogonal

The samples of a process that form $\mathbf{x}[n]$ are considered to be mutually orthogonal if

$$
E[x[m]\,x[q]] = 0;\ \text{for all } q \text{ and } m \in \{n, n+1, \ldots, n+l_v-1\} \tag{4.9}
$$

Obviously, if a process is mutually uncorrelated and zero mean, it is also orthogonal. However, if a process is orthogonal and non-zero mean, is it also uncorrelated? The answer, obviously, is that one cannot assume that a process is uncorrelated in that case. The issue of independence, once again, cannot be addressed when orthogonality is considered. Once again, the samples of a zero-mean, orthogonal Gaussian process are also independent (since they are uncorrelated).

Now that we have established the criterion for independence, we need to introduce metrics which can be used to assess how "far" the data set is from being declared as independent by only examining the 2nd order statistics of the data. As noted above, if one can establish that the data set obeys a Gaussian statistics, uncorrelated data implies an independent data set. The question is, then, how do we establish that the data is Gaussian? In order to answer this question, we examine several methods that are used in practice to determine whether the data set obeys a Gaussian statistic. There are several methods that can be employed with various degrees of effectiveness and utility. We outline some key metrics used in practice.

4.2 Is It Gaussian?

Given the above discussion, it becomes apparent that it would be of interest to assess whether the data we are examining admits to a Gaussian statistics, allowing one to gain access to all moments of the random process using only the second order statistics as well as establishing independence using the correlation function. To that end, we examine a number of popular ways for establishing whether a random variable meets the Gaussian criterion.

4.2.1 Kurtosis

This metric makes an attempt to measure the higher order (>2) statistics of a random variable and determines if such statistics can be related to the second order statistics. The objective is to assess how "far" the pdf of data samples is from a Gaussian one. This metric is defined as

$$Kur = \frac{E\left\{(x[l] - m[l])^4\right\}}{E^2\left\{(x[l] - m[l])^2\right\}} - 3 \tag{4.10}$$

Note that a Gaussian random variable will yield $Kur = 0$. Obviously, this is a metric that is most suited for pdfs with even symmetry about the mean value (this will insure that the odd central moments are zero, which is the case for Gaussian random variables). The effectiveness of this metric, quite clearly, is limited by its reliance on the moments of the 4th order moment of the random variable and not on the pdf of the random variable, which clearly identifies all moments of the random variable. To gain an insight into the efficacy of this metric, let us consider the following example.

Example 1 We are given a random variable with the pdf (this is a double exponential pdf with a location parameter of 0)

$$f_X(x) = \frac{\alpha}{2} \exp\{-\alpha |x|\}. \tag{4.11}$$

Compute its Kurtosis and comment on its relationship to a Gaussian pdf.

Solution: *The Kurtosis for this random variable is given by*

$$Kur = \frac{\int_{-\infty}^{\infty} x^4 \exp\{-\alpha |x|\} dx}{\left(\int_{-\infty}^{\infty} x^2 \exp\{-\alpha |x|\} dx\right)^2} - 3 = \frac{48/\alpha^5}{\left(4/\alpha^3\right)^2} = 3(\alpha - 1) \tag{4.12}$$

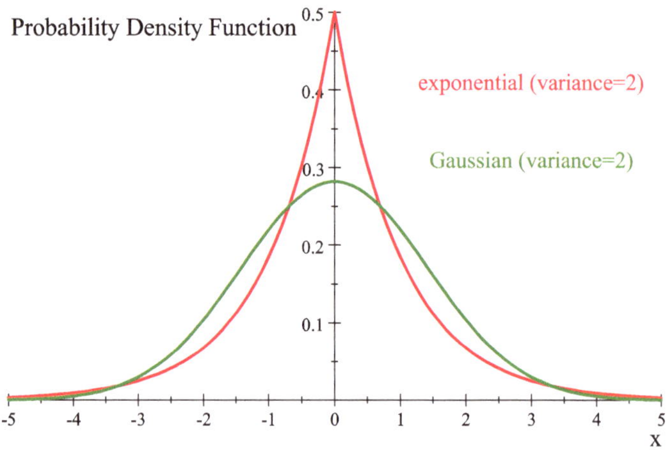

Fig. 4.1 $f_x(x) = \frac{\alpha}{2}\exp(-\alpha|x|)$ for $\alpha = 1$ (*red curve*) and its Gaussian counterpart. $Kur = 0$ for $f_x(x)$. The variances of both random variables are 2

Note that Kur *for this random variable increases linearly with* α. *Increasingly enough, for* $\alpha = 1$, $Kur = 0$, *which is the Kurtosis for a Gaussian random variable! This clearly depicts the inadequacy of this metric in establishing the Gaussian nature of the random variable. In Fig. 4.1, the above pdf for* $\alpha = 1$ *(variance of 2) is plotted along with a Gaussian pdf with the same variance. As can be seen, there exists a marked difference between the two pdfs while their Kurtosis measures agree. From a statistical point of view and when probabilities of events are concerned, the two pdfs are vastly different. Hence, Kurtosis alone cannot be used as a key metric to establish Gaussian behavior for the data set, which in turn can be explored to establish independence using only the second order statistics (namely, correlation among the data set). However, a large Kurtosis can be used to rule out the possibility that the data set obeys a Gaussian statistics. Hence, one may utilize this metric to that end.*

4.2.2 Entropy and Negentropy

Since the concept of independence relies on the joint pdf of data set, one is required to obtain the joint pdf of data to establish independence. One is then compelled to ask the following question: *Is there a mechanism that relies on expectation (rather than the pdf itself) to establish independence?* The answer to this question, fortunately, is "yes." Note that we are always in a better position to obtain the expected values (via experimental studies) rather than acquiring the pdf, which accounts for 'all' events, likely or otherwise. In order to better understand this concept, let us consider, rather birefly, the concept of "entory."

4.2.2.1 Entropy

In short, entropy is a measure of "randomness" of a random variable, which assumes a minimum value of 0 (for a deterministic random variable). For a continuous random variable, it is defined as

$$H(x) = -E\{\ln[f_x(x)]\} = -\int_{-\infty}^{\infty} f_x(x)\ln[f_X(x)]\,dx \tag{4.13}$$

For a discrete random variable,

$$H(x) = -\sum_{n=1}^{N} p_i \ln[p_i] \tag{4.14}$$

where $p_i = p\{x = i\}$. In case where x is a deterministic random variable, $p_i = 0$ except for one value of i, say j. In that case, $p_j = 1$ and $H(x) = 0$. Furthermore, note that one can obtain a maximum value for $H(x)$ by setting $\frac{\partial H(x)}{\partial p_i}$ equal to zero for all i, except for one, since we have the following constraint: $\sum_{i=1}^{N} p_i = 1$. That is,

$$\frac{\partial H(x)}{\partial p_i} = \ln[p_i] + 1 = \lambda, \text{ for} \forall i \text{ (except for } i = j \text{ for some } j) \tag{4.15}$$

where λ is the Lagrange multiplier. The $p_i's$ that satisfy the above equation must all be the same, which in turn results in

$$p_i = \frac{1}{N} \text{ for } \forall\, i \tag{4.16}$$

This results is significant, since it suggest that the above metric is maximized for the most "random" case where the random variable can assume any of the possible values with equal probability. Note that the peak value is

$$H(x) = \ln[N] \tag{4.17}$$

Therefore, one can use this metric to establish the level of randomness in a random variable. Before introducing the concept of negentropy, let us consider a key example.

Example 2 Compute the entropy of a Gaussian random variable with mean and variance of m and σ^2, respectively.

Solution: *From the definition entropy,*

$$\begin{aligned} H(x) &= -E\{\ln[f_x(x)]\} = \frac{1}{2}\ln\left(2\pi\sigma^2\right) - E\left\{-\frac{(x-m)^2}{2\sigma^2}\right\} \\ &= \ln\left(\sqrt{2\pi e\sigma^2}\right) \end{aligned} \tag{4.18}$$

Note that the entropy of a Guassian random variable is independent of its mean, as one would expect, and is an increasing function of its variance.

Finally, note that the measure of entropy is merely an expectation, although it relies on having a priori knowledge of the pdf. Furthermore, the exact value of entropy may not be as important as its "distance" from the entropy of a Gaussian random variable. This subject is tackled next.

4.2.2.2 Negentropy

Given the above formulation, we can then establish negentropy as

$$\begin{aligned} J(x) &= H_G(x) - H(x) \\ &= \ln\left(\sqrt{2\pi e\sigma^2}\right) - H(x) \end{aligned} \tag{4.19}$$

with

$$\sigma^2 = E\left\{(x - E\{x\})^2\right\} \tag{4.20}$$

denotes the variance of the random variable x. In the above, $H_G(x)$ denotes the entropy of a Gaussian random variable with a variance that is identical to that of the random variable x. One may then view negentropy as a measure of difference between the entropy of a random variable and its Gaussian counterpart. To gauge the effectiveness of this metric, let us re-examine the double exponential pdf introduced above.

Example 3 Let us consider a zero-mean double exponential (also known as Laplacian) random variable with the pdf

$$f_X(x) = \frac{\alpha}{2}\exp\{-\alpha|x|\};\ \alpha > 0. \tag{4.21}$$

Compute the negentropy for this random variable.

Solution: *First, it can readily be shown that the variance of x is given by (this is a zero mean random variable)*

$$\sigma^2 = \int_{-\infty}^{\infty} \frac{\alpha x^2}{2}\exp\{-\alpha|x|\}\,dx = \frac{2}{\alpha^2}.$$

Furthermore, for this pdf

$$\begin{aligned} H(x) &= -\int_{-\infty}^{\infty} \frac{\alpha}{2}\exp\{-\alpha|x|\}\ln\left\{\frac{\alpha}{2}\exp\{-\alpha|x|\}\right\}dx \\ &= 1 - \ln\left(\frac{\alpha}{2}\right) = \ln\left(\frac{2e}{\alpha}\right) \end{aligned}$$

Hence, using a variance of $\frac{2}{\alpha^2}$ for the Gaussian pdf,

$$\begin{aligned} J(x) &= \ln\left(\sqrt{\frac{4\pi e}{\alpha^2}}\right) - H(x) \\ &= \ln\left(\sqrt{\frac{4\pi e}{\alpha^2}}\right) - \ln\left(\frac{2e}{\alpha}\right) \\ &= 0.5\ln(\pi/e) = 0.0724 \end{aligned}$$

This is rather remarkable in that it suggest negentropy as a metric that is unaffected by the value of α, allowing one to distinguish the zero-mean double exponential statistics from its zero-mean Gaussian counterpart (with the same variance) for all values of α. It is noteworthy that negentropy was able to distinguish between the two pdfs. However, the value for negentropy is rather small, underscoring the similarity between Gaussian and Lapacian pdfs.

Since we have introduced the concept of entropy, it is instructive to also discuss the concept of "mutual information," which can be used to establish the independence between two random variables.

4.2.3 Mutual Information

As the name implies, mutual information is a metric that can be used to assess how much uncertainty is removed from one random variable by observing yet another random variable. Quite clearly, if two random variables are independent, mutual information, based on its definition, must be 0. Before we introduce the concept of mutual information, it is instructive to discuss the conditional entropy. That is,

$$\begin{aligned} H(y|x) &= -E\{\ln(f(y|x))\} \\ &= -\int_{-\infty}^{\infty}\int_{-\infty}^{\infty} \ln(f(y|x))\, f_{xy}(x, y)\, dxdy \end{aligned} \tag{4.22}$$

where

$$f(y|x) = \frac{f_{xy}(x, y)}{f_x(x)} \tag{4.23}$$

is the conditional pdf of y conditioned on x. One can view this quantity as the level of uncertainty in y by observing x. If x and y are independent (namely, $f(y|x) = fy(y)$), one can see that

$$H(y|x) = -\int_{-\infty}^{\infty}\int_{-\infty}^{\infty} \ln(f(y))\, f_{xy}(x, y)\, dxdy$$

$$= -\int_{-\infty}^{\infty} \ln\left(f\left(y\right)\right) f_y\left(y\right) dy \tag{4.24}$$

$$= H\left(y\right) \tag{4.25}$$

That is, observing x had no bearings on reducing the uncertainty in y (as one would expect from two independent random variables). Given this definition, then one can define mutual information as follows:

$$I\left(x; y\right) = H\left(y\right) - H\left(y|x\right) \tag{4.26}$$

Interestingly enough, this equation can be re-written as

$$\begin{aligned}
I\left(x; y\right) &= -\int_{-\infty}^{\infty} \ln\left(f_y\left(y\right)\right) f_y\left(y\right) dy \\
&\quad + \int_{-\infty}^{\infty}\int_{-\infty}^{\infty} \ln\left(f\left(y|x\right)\right) f_{xy}\left(x, y\right) dxdy \\
&= -\int_{-\infty}^{\infty}\int_{-\infty}^{\infty} \ln\left(fy\left(y\right)\right) f_{xy}\left(x, y\right) dxdy \\
&\quad + \int_{-\infty}^{\infty}\int_{-\infty}^{\infty} \ln\left(f\left(y|x\right)\right) f_{xy}\left(x, y\right) dxdy \\
&= -\int_{-\infty}^{\infty}\int_{-\infty}^{\infty} \ln\left(\frac{f_Y\left(y\right)}{f\left(y|x\right)}\right) f_{xy}\left(x, y\right) dxdy \\
&= -\int_{-\infty}^{\infty}\int_{-\infty}^{\infty} \ln\left(\frac{f_X\left(x\right)}{f\left(x|y\right)}\right) f_{xy}\left(x, y\right) dxdy \\
&= -\int_{-\infty}^{\infty} \ln\left(f_x\left(x\right)\right) f_{xy}\left(x, y\right) dydx \\
&\quad + \int_{-\infty}^{\infty}\int_{-\infty}^{\infty} \ln\left(f\left(x|y\right)\right) f_{xy}\left(x, y\right) dxdy \\
&= -\int_{-\infty}^{\infty} \ln\left(f_x\left(x\right)\right) f_x\left(x\right) dx
\end{aligned}$$

$$+\int_{-\infty}^{\infty}\int_{-\infty}^{\infty} \ln\left(f\left(x|y\right)\right) f_{xy}\left(x,y\right) dxdy$$

which is

$$I\left(x;y\right) = H\left(x\right) - H\left(x|y\right). \tag{4.27}$$

Furthermore, given the definition of the conditional entropy,

$$I\left(x;y\right) = H\left(x\right) + H\left(y\right) - H\left(x,y\right) \tag{4.28}$$

where

$$H\left(x,y\right) = -E\left(\ln\left(f_{xy}\left(x,y\right)\right)\right) \tag{4.29}$$

can be considered as the joint entropy of x and y. Note that when x and y are independent,

$$I\left(x;y\right) = 0.$$

Collectively, the above equations point to the mutual information as a means of assessing the dependency between a pair of random variables. Unlike the previous metrics, when $I\left(x;y\right) = 0$, one can conclusively state that the two random variables are independent. This answers our earlier question about whether one can obtain a method which uses expectation to establish independence. It is imperative to note that the mutual information does require a knowledge of the exact (or perhaps approximate) pdf of data. This may, in turn, be an ineffective tool if pdf information is not available.

Example 4 A pair of zero-mean, Gaussian random variables are correlated with the normalized correlation of ρ. Compute the mutual information assuming that the variances of the two random variables are σ_x^2 and σ_y^2.

Solution: *The joint pdf for the Gaussian pair is given by*

$$f_{xy}\left(x,y\right) = \frac{1}{2\pi\sigma_x\sigma_y\sqrt{\left(1-\rho^2\right)}} \exp\left\{-\frac{x^2}{2\sigma_x^2\left(1-\rho^2\right)} - \frac{y^2}{2\sigma_y^2\left(1-\rho^2\right)} + \frac{\rho xy}{\sigma_x\sigma_y\left(1-\rho^2\right)}\right\} \tag{4.30}$$

where $\rho = \frac{E\{xy\}}{\sigma_x\sigma_y}$, $\sigma_x^2 = E\left\{x^2\right\}$, *and* $\sigma_y^2 = E\left\{y^2\right\}$. *Also, one can readily show that* $f_x\left(x\right) = \frac{1}{\sqrt{2\pi}\sigma_x}\exp\left\{-\frac{x^2}{2\sigma_x^2}\right\}$ *and* $f_y\left(y\right) = \frac{1}{\sqrt{2\pi}\sigma_x}\exp\left\{-\frac{y^2}{2\sigma_y^2}\right\}$. *Now (see previous examples),*

$$H\left(x\right) = \ln\left(\sqrt{2\pi e\sigma_x^2}\right), \tag{4.31}$$

$$H(y) = \ln\left(\sqrt{2\pi e \sigma_y^2}\right), \tag{4.32}$$

and

$$\begin{aligned}
H(x, y) &= -E\left\{-\frac{x^2}{2\sigma_x^2\left(1-\rho^2\right)} - \frac{y^2}{2\sigma_y^2\left(1-\rho^2\right)} + \frac{\rho x y}{\sigma_x \sigma_y\left(1-\rho^2\right)}\right\} \\
&\quad + \ln\left(2\pi\sigma_x\sigma_y\right) + \ln\left(\left(1-\rho^2\right)^{1/2}\right) \\
&= \frac{1}{1-\rho^2} - \frac{\rho^2}{\left(1-\rho^2\right)} + \ln\left(2\pi\sigma_x\sigma_y\right) \\
&\quad + \ln\left(\left(1-\rho^2\right)^{1/2}\right). \\
&= \ln\left(2\pi e\sigma_x\sigma_y\right) + \ln\left(\left(1-\rho^2\right)^{1/2}\right).
\end{aligned} \tag{4.33}$$

Hence,

$$I(x, y) = -0.5\ln\left(1-\rho^2\right) \tag{4.34}$$

Since $\rho^2 \in [0, 1]$, $I(x, y)$ is always non-negative (as one would expect). Note that, when $\rho = 0$ in the above example (i.e., the two random variables are uncorrelated), $I(x, y) = 0$. Of course, we already know that when a air of Gaussian random variables are uncorrelated, they are also independent. Hence, we can say that, for Gaussian random variables,

$$I(x, y) = 0 \Leftrightarrow \rho = 0 \Leftrightarrow \text{ independence} \tag{4.35}$$

When x and y are not Gaussian,

$$\boxed{I(x, y) = 0 \Leftrightarrow \text{independence}} \tag{4.36}$$

The last result is significant in that, if a pair of random variables are not Gaussian, one can use the mutual information to establish independence. Since in the field of biomedical signal processing we are concerned with non-Gaussian signals, the mutual information can play a key role in establishing independence.

4.3 "Distance" Between Two PDFs

The previous section was concerned with establishing independence between random variables through either establishing correlation (or lack thereof) between Gaussian random variables or computing mutual information for two random variables. This is an importance analysis as one may be required to establish independence between a pair of time series or decompose a signal into its independence components. The above analysis, however, hinges upon having a knowledge of the pdf for the observed data at hand. In practical problems, including biomedical signal processing arena, often the pdf of data is not available and, hence, must be estimated using histogram or other techniques. Once such "approximate" pdf is obtained, one can compute the measures discussed in the previous section to establish mutual information between the unknown random variable of interest and a possible known candidate pdf with identical first and second moments (or as many moments needed to match the data to the known pdf). That is, if one obtains a pdf through experimental means, the next task is to see which one of the known pdfs match the observed data. Given that the known pdf may have several parameters, one can match the known pdf to the data via matching the moments.

Ultimately, once a match is performed, one would be interested in establishing the "distance" between the observed pdf and its mathematical (not necessarily Gaussian) counterpart. Hence, we need to define the concept of "distance" between two pdfs in some detail.

4.3.1 Kolmogorov-Smirnov (KS) Distance

A method for measuring the closeness of two pdfs, which is quite intuitive, is the Kolmogorov-Smirnov distance. This distance is defined as follows:

$$D_{KS} = \frac{1}{2} \sup \left\{ \left| \int_{-\infty}^{x} \left[f_x(s) - g_x(s) \right] ds \right| \right\} \tag{4.37}$$

In this equation, sup denotes the supremum. Note that we have included a factor of $\frac{1}{2}$ to make sure that $D_{KS} \in [0, 1]$. One can argue that this distance is merely the peak difference (if one can measure it) between the two probability distribution (not density) functions.

4.3.2 Hellinger Distance (HD)

Yet, other methods that can be used to assess the similarity (or dissimilarity) of the pdf in question to any other pdf (including its Gaussian counterpart) is the Hellinger distance. The Hellinger distance, which is defined over [0, 1], can be used to measure the distance between the two pdfs. This metric is defined as follows:

$$D_{HD}(f, g) = \sqrt{\frac{1}{2}\int_{-\infty}^{\infty}\left(\sqrt{f_X(x)} - \sqrt{g_x(x)}\right)^2 dx}$$
$$= \sqrt{1 - \int_{-\infty}^{\infty}\sqrt{f_x(x)\, g_x(x)}dx} \tag{4.38}$$

which clearly requires that pdf $f_x(x)$ and its candidate pdf $g_x(x)$ be continuous random variables over x. A similar definition can be obtained for discrete case (where the integral is replaced by summation). The advantage of HD is that it requires a match over the entire range of x.

4.3.3 Kullback-Leibler (KL) Divergence

For Kullback-Leibler (KL) divergence when a pair of continuous random variables are considered, we have

$$D_{KL}(f, g) = \int_{-\infty}^{\infty} \ln\left(\frac{f_x(x)}{g_x(x)}\right) f_x(x)\, dx$$
$$= -\int_{-\infty}^{\infty} \ln(g_x(x))\, f_X(x)\, dx + \int_{-\infty}^{\infty} \ln(f_x(x))\, f_x(x)\, dx$$
$$= H_c(f, g) - H(x) \tag{4.39}$$

where

$$H_c(f, g) = -E\{\ln(g_x(x))\} \tag{4.40}$$

is called the cross-entropy. This quantity is different from the joint entropy $H(x, y)$ which was introduced earlier. Unlike the Hellinger distance, the KL divergence is not limited to the range [0,1]. In fact, KL divergence cannot be classified as "distance" between the two pdfs. In general,

$$D_{KL}(f, g) \neq D_{KL}(g, f) \tag{4.41}$$

and the triangle inequality is not satisfied. That is, the triangle inequality, given by

$$D(f, g) \leq D(f, q) + D(q, g) \tag{4.42}$$

can be satisfied for all $f(x)$, $g(x)$, and $q(x)$ pdfs for HD and KS distance. However, the above cannot be established for KL divergence. For this reason, KL is not referred to as the KL "distance," but rather the KL divergence. It can be shown than

$$D_{KL} \geq 2D_{HD}^2 \tag{4.43}$$

This implies that one would obtain, in general, values for KL divergence than are larger than that of Hellinger distance. If the objective is to assess whether two pdfs are similar (or dissimilar), then KL metric is a more useful tool.

Example 5 For the pdfs shown below, compute the KS and HD along with the KL divergences. Determine what value of α will minimize the above distances.

$$f_x(x) = \frac{\alpha}{2} \exp\{-\alpha|x|\}; \alpha > 0 \tag{4.44}$$

$$g_x(x) = \frac{1}{\sqrt{4\pi}} \exp\left\{-\frac{x^2}{4}\right\}$$

Solution: *First, the Gaussian pdf has a variance of 2 (the double exponential pdf with the same variance has a Kurtosis of 0, and hence this example is of importance as Kurtosis fails to distinguish $f_X(x)$ from its Gaussian counterpart). Using straightforward integration, we arrive at*

$$\int_{-\infty}^{x} f_x(s)\,ds = \begin{cases} 1 - \frac{1}{2}e^{-\alpha x} & x \geq 0 \\ \frac{1}{2}e^{\alpha x} & x < 0 \end{cases} \tag{4.45}$$

and

$$\int_{-\infty}^{x} g_x(s)\,ds = \frac{1}{2}\left\{1 + \mathrm{Erf}\left(\frac{x}{2}\right)\right\} \tag{4.46}$$

In this equation, Erf() denotes the Error Function. For these pdfs, we resort to numerical computation to evaluate the maximum of $\left|\int_{-\infty}^{x} f_x(s)\,ds - \int_{-\infty}^{x} g_x(s)\,ds\right|$. *The results is depicted in Fig. 4.2. As can be seen, the minimum distance is achieved at α of nearly 0.75. Note that the two pdfs have identical variances of 2 at $\alpha = 1$, and hence it is logical for the two pdfs to be "close," but not identical near $\alpha = 1$. This metric certainly distinguishes the two pdfs for all values of α, except for α near 1. The HD and KL divergence are given as follows:*

$$D_{HD} = \sqrt{1 - \pi^{1/4}\sqrt{2\alpha}\exp\left(\frac{\alpha^2}{2}\right)\left(1 - \mathrm{erf}\left(\frac{\alpha}{\sqrt{2}}\right)\right)}, \tag{4.47}$$

$$D_{KL}(f, g) = \frac{1 + \ln(4\pi)\,\alpha^2}{2\alpha^2} - \left(1 - \ln\left(\frac{\alpha}{2}\right)\right); \alpha > 0, \tag{4.48}$$

and

$$D_{KL}(g, f) = \frac{2\alpha}{\sqrt{\pi}} - \ln\left(\sqrt{\pi e \alpha^2}\right); \alpha > 0. \tag{4.49}$$

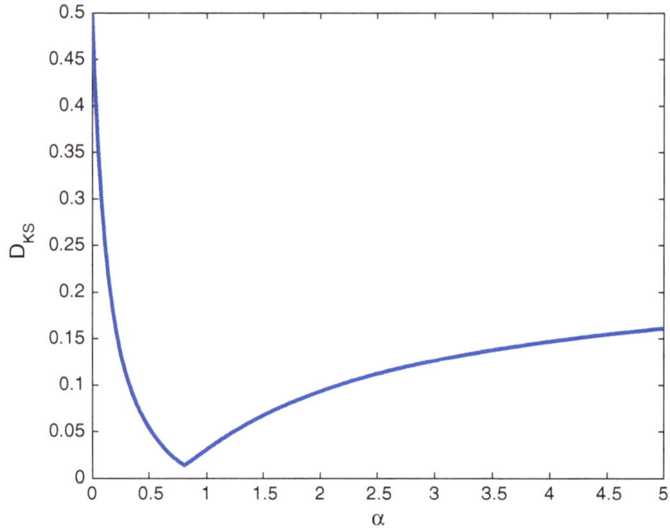

Fig. 4.2 D_{KS} as a function of α for the Gaussian pdf with variance of 2 and $f_x(x) = \frac{\alpha}{2}\exp(-\alpha|x|)$

The HD and KL metrics are shown in Fig. 4.3 *where, as expected, HD metric is confined to* [0, 1]. *Interestingly enough, similar to KS distance, both HD and KL metrics produce distances of nearly 0 in the vicinity of* $\alpha = 1$. *Nonetheless, these metrics, unlike Kurtosis, successfully distinguish (albeit by a small margin) the double exponential pdf from its Gaussian counterpart for* $\alpha = 1$. *Also, note that all metrics increase with values of* α *other than* 1, *whereas negentropy remains at a small value*

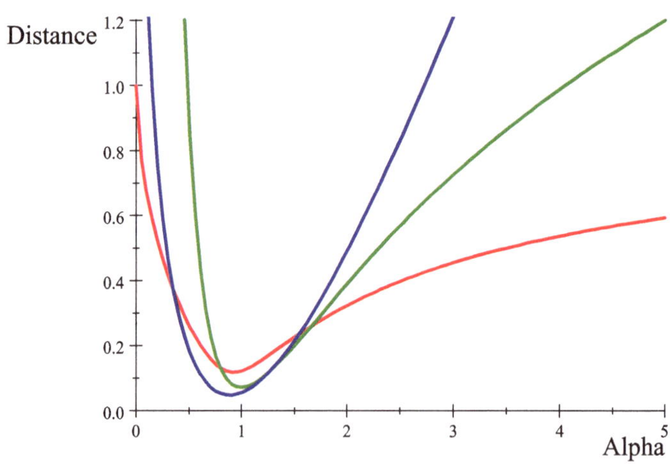

Fig. 4.3 Hellinger (*red*) distance along with KL divergences $D_{KL}(f, g)$ (*green*) and $D_{KL}(g, f)$ (*blue*) for $f_x(x) = \frac{\alpha}{2}\exp\{-\alpha|x|\}$ and its Gaussian counterpart as a function of α

(non-selective). Finally, $D_{KL}(f, g) - 2D_{HD}^2$ *and* $D_{KL}(g, f) - 2D_{HD}^2$ *are plotted for various values of* $\alpha > 0$ *in Fig. 4.4. One can see that the difference remains positive for either of the two KL divergences (as expected, since* $D_{KL} - 2D_{HD}^2 \geq 0$*) for all values of* α*. Finally, note that the HD provides a larger minimum value (about 0.125) at* $\alpha = 1$ *than that produced by the KS distance (about 0.025). Since Gaussian is not a good match to the Lapacian pdf, one can see that HD is a better metric than its KS counterpart in establishing closeness of the two pdfs.*

Example 6 We are interested in modeling the raw signal collected by an EEG probe over a period of approximately 40 h at the rate of 256 samples/s. The data is obtained from www.physionet.org (CHB-MIT Scalp EEG Database (chbmit) chb01/chb01_1.edf-chb01 to chb01_44.edf). The database contains the record of EEG for 24 children (one case is the same child over a different periods of time) with histories of epileptic seizures over a period of approximately two days. Without the loss of generality, we consider the data for the fist child (hence, chb01) and the first probe of the 23-probe EEG data collection mechanism (that is, FP1-F7). We understand that the collected signal is corrupted by AWGN and that the signal may contain periodic components (perhaps multiple of them). A cursory examination of the raw data histogram has suggested that a double exponential pdf can perhaps model the long term behavior of the signal. Find the Kurtosis to establish that data is not Gaussian and then compute HD and KS metrics along with KL divergence for the double exponential pdf to validate or dispute the double exponential pdf assertion. Also, depict the probability distribution function of data and its double exponential fit (compute α using the variance of the observed data).

Solution: *In the absence of an analytical model, we resort to the histogram of the data set. The histograms are generated for the entire data set, decimated by 16,*

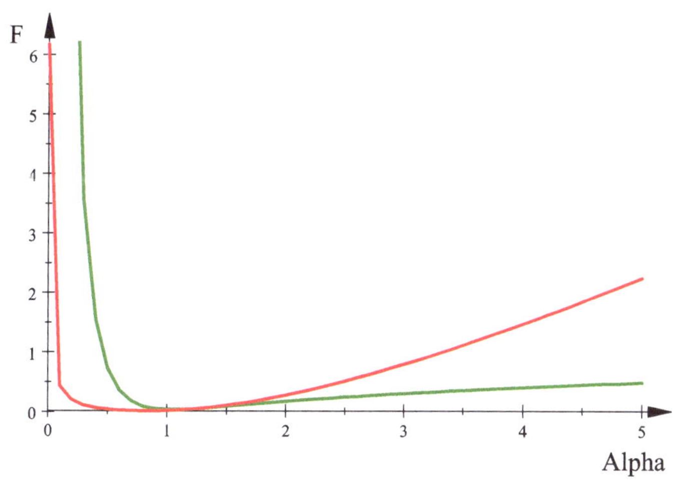

Fig. 4.4 The functions $F_1(\alpha) = D_{KL}(f, g) - 2D_{HD}^2$ (*green*) and $F_2(\alpha) = D_{KL}(g, f) - 2D_{HD}^2$ (*red*). In both cases, the functions remain positive

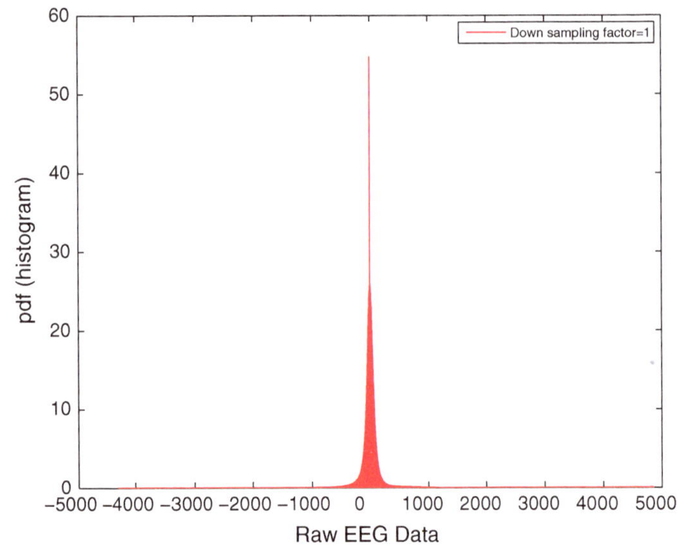

Fig. 4.5 Pdf of raw EEG signal

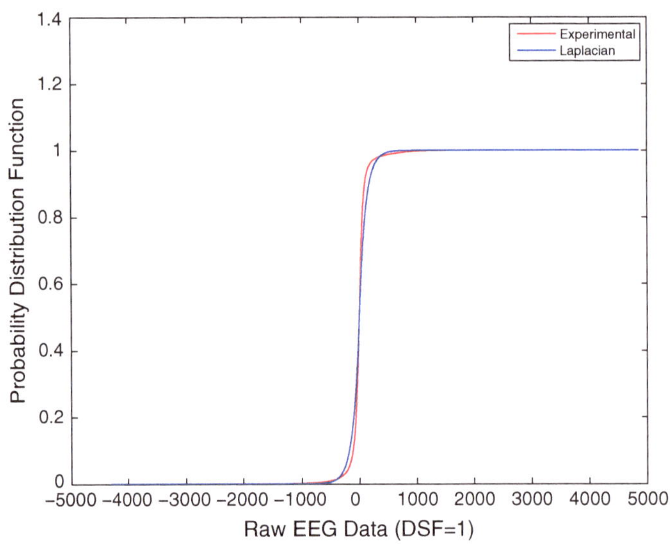

Fig. 4.6 PDF of the raw EEG signal along with its Laplacian counterpart

128, and 256 (decimation of 256 results in data samples which are spaced by 1 s). Figures 4.5, 4.6, 4.7, 4.8, 4.9, 4.10, 4.11 and 4.12 *depict the resulting histograms (pdf), probability distribution function (PDF) of data along with their associated double exponential pdf. That is, we compare the experimental PDFs to the PDF of a double exponential random variable, given by*

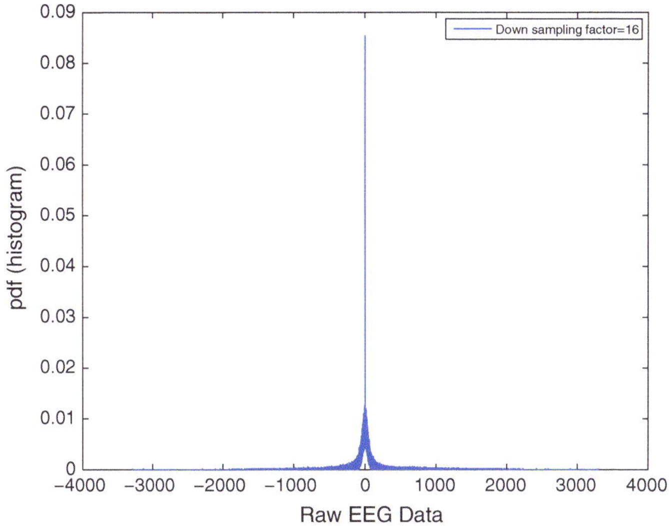

Fig. 4.7 Pdf of the raw EEG signal

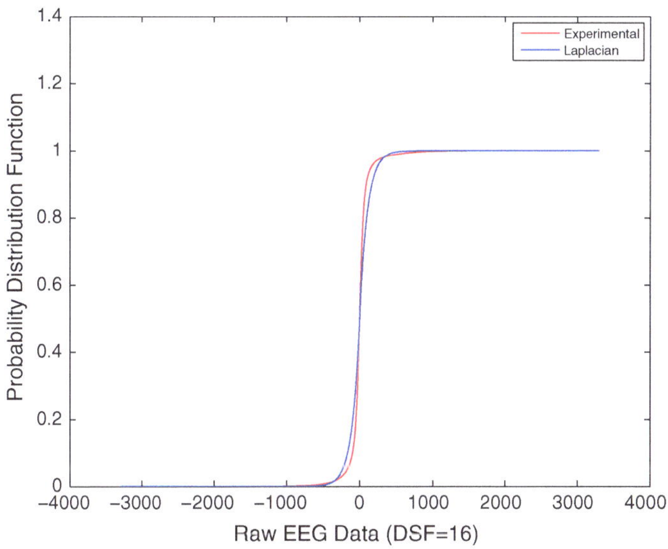

Fig. 4.8 PDF of the raw EEG signal along with its Laplacian counterpart

$$f_X(x) = \frac{\alpha}{2} \exp\left(-\alpha |x|\right),$$

for values of α *which yield the same variances as those of the samples of the observed signal. Specifically,* $\alpha = \sqrt{\frac{2}{\sigma^2}}$ *where* σ^2 *is the numerical variance of the data. A*

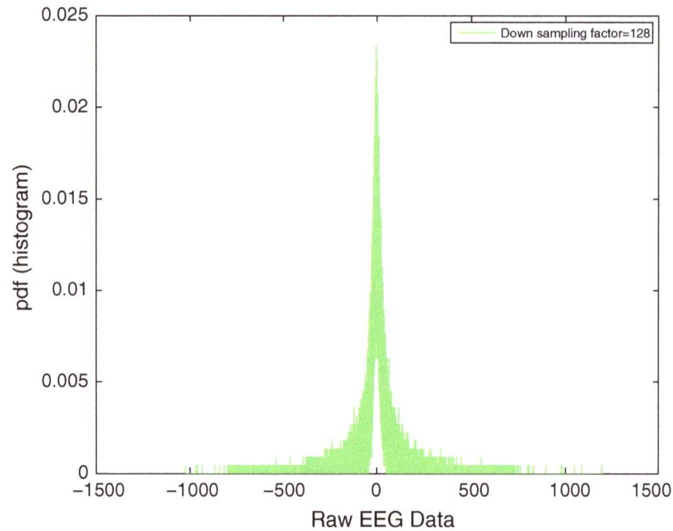

Fig. 4.9 Pdf of the raw EEG signal

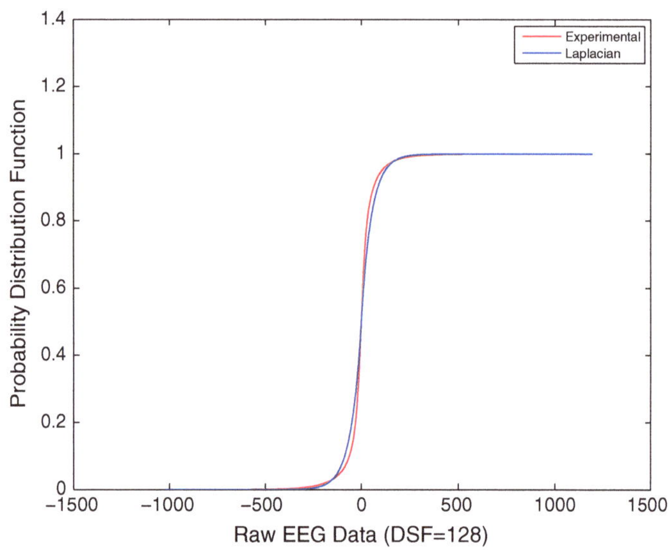

Fig. 4.10 PDF of the raw EEG signal along with its Laplacian counterpart

cursory view confirms the double exponential pdf assertion. To gain a better insight, the Kurtosis is depicted in Fig. 4.13*. Quite clearly, the Kurtosis reaffirms our initial observation that the signal does not obey Gaussian statistics. However, as the data samples are separated in time, the Kurtosis begins to drop, although not by a significant degree. Next, in Fig.* 4.14*, the Hellinger and KS distances are plotted. For*

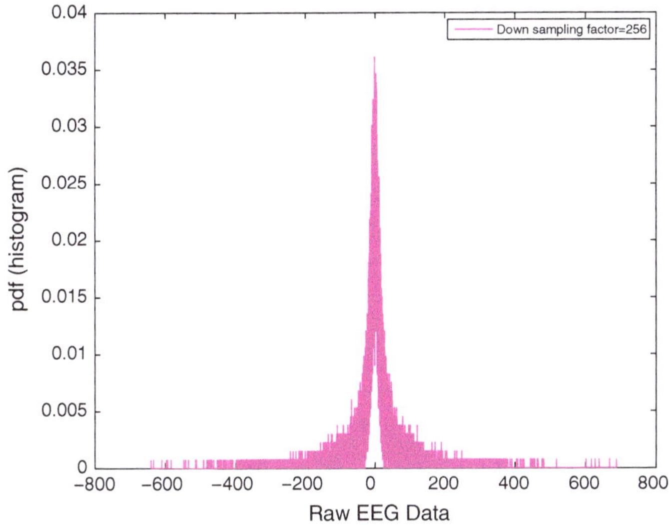

Fig. 4.11 Pdf of the raw EEG signal

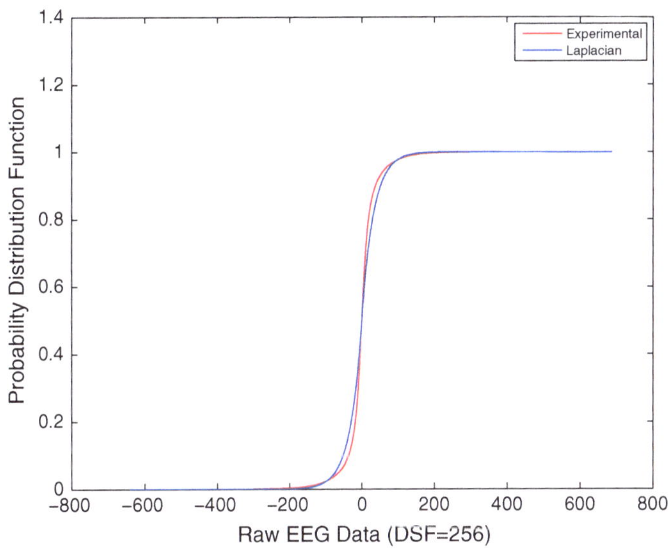

Fig. 4.12 PDF of the raw EEG signal along with its Laplacian counterpart

no decimation (down sampling rate of 1), the Hellinger distance is at its maximum, signifying a substantial departure from a double exponential pdf. However, as the decimation rate increases, the Hellinger distance drops dramatically, giving some validation to the double exponential assertion. KS distance, however, shows a consistent similarity between the observed statistics and those of a double exponential pdf. Similarly, the KL divergences (g describes the histogram) are plotted in Fig. 4.15.

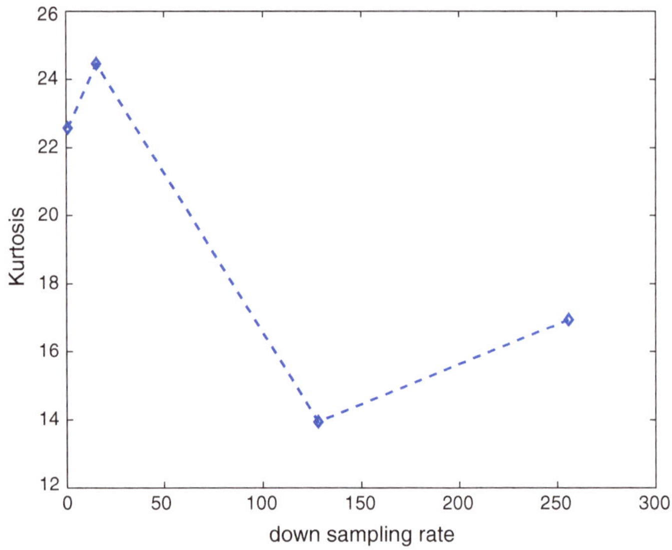

Fig. 4.13 Kurtosis for the raw EEG signal

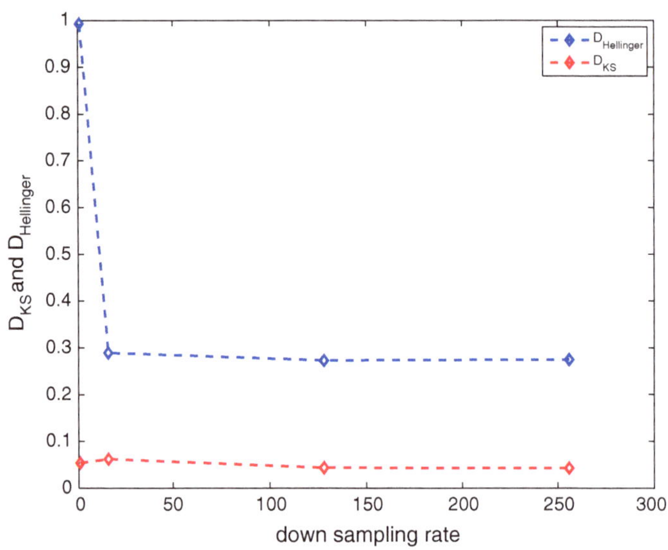

Fig. 4.14 KS and Hellinger distances for the pdf of the raw EEG signal and its Laplacian counterpart

KL divergences also arrive at a similar conclusions; that is, once data is decimated, the double exponential pdf assertion becomes increasingly accurate.

In the above example, we observed that decimation of data resulted in the data assuming a somewhat double exponential statistics. One can argue that a fair (statistical) comparison has not been made in this example since no decimation case offers

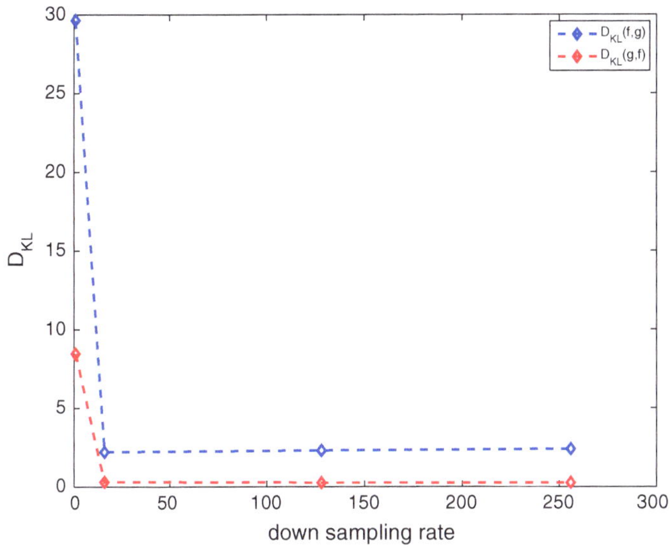

Fig. 4.15 KL divergence for the pdf of the raw EEG signal and its Laplacian counterpart

the largest number of samples (37221888), and hence is statistically more significant than other observations. However, a decimation of data by a factor of only 16 results in a dramatic reduction in the HD and KL divergence. The trend stays consistent for larger decimation factors, underscoring that spacing the data samples perhaps provides a more meaningful measurement (namely, allows for data to become independent or uncorrelated). To give this observation some consideration, let us compute the correlation and mutual information to establish independence and correlation.

Example 7 For the signal in the previous example, compute the correlation function and the mutual information for samples spaced by 16, 128, and 256. To compute the mutual information, use the pdfs fond using histogram method. Comment on your observations. In particular, comment on the statistical dependency of the samples obtain in this manner.

Solution: *The normalized (peak value of 1) autocorrelation for the data of the previous example is depicted in Fig.* 4.16 *for various values of down sampling factor (DSF). The x-axis shows the normalized separation between data samples. To elaborate, for* $DSF = 256$ *and* $n = 256$ *(the actual separation between the original data samples),* $\frac{n}{DFG} = 1$ *and one has to read the blue curve for* $x = 1$. *It becomes immediately obvious that when samples are separated by 16 samples or more, the data assumes an uncorrelated attribute. In fact, for DSF of 128 or 256, the data exhibits a "white" noise characteristics, implying that samples obtained in these manners are uncorrelated. To the contrary, without down sampling, the data exhibits a high level of correlation among adjacent samples. This further reaffirms our earlier assertion that a histogram based on no decimation is most likely inaccurate due to the presence*

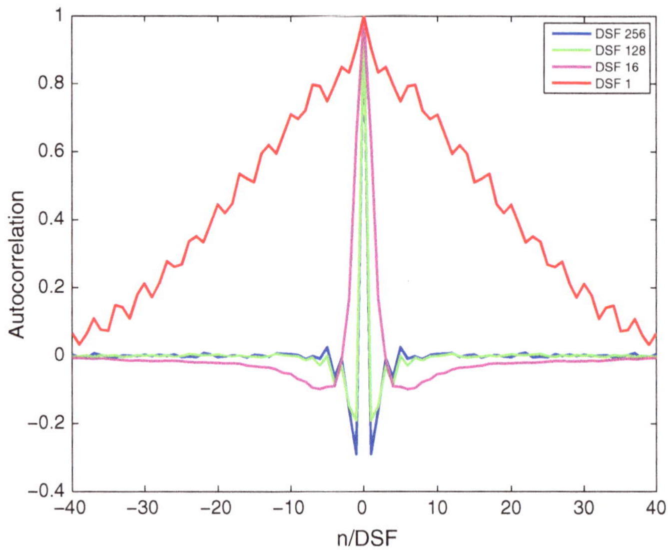

Fig. 4.16 Autocorrelation as a function of separation in time for various down sampling factors (DSF). In this figure, *n* denotes the separation between the original data set

of high correlation among the samples. This also explains the wide variations in the HD as well as the KL divergence calculations for no decimation case and those for the decimated cases. Next, we move to compute mutual information between the "adjacent" samples. Note, once again, adjacent sample for DSF of 256 are data samples that are separated by 256 samples. The mutual information requires the joint pdf (histogram), which is computed for adjacent data pairs. We also do not consider $DSF = 1$ *as it is quite obvious from the correlation function that the true adjacent samples are indeed correlated and, hence, dependent. Figure* 4.17 *depicts the mutual information as a function of DSF. Quite predictable, mutual information becomes a decreasing function of DSF. Furthermore, unlike the correlation function, mutual information points to an important fact; samples remain dependent (statistically) albeit the dependence 'weakens' with an increase in DSF. By extrapolation, one may conclude that samples must be separated by more than 1200 samples to achieve nearly uncorrelated samples (this implies, roughly speaking, samples which are separated by 5 s).*

From Figs. 4.16 and 4.17, one may conclude that the observed signal does not enjoy a Gaussian statistics as the uncorrelated samples do not become independent. Furthermore, we can see that one cannot merely rely on the correlation function to render meaningful decisions regarding the observed signal. As will be seen in the next section, the concept of hypothesis testing will require a knowledge of pdf in order to allow one to make decisions with significant statistical significance. For this reason, joint pdfs will be needed, and a knowledge of independence is critical to forming the joint pdfs from marginal ones.

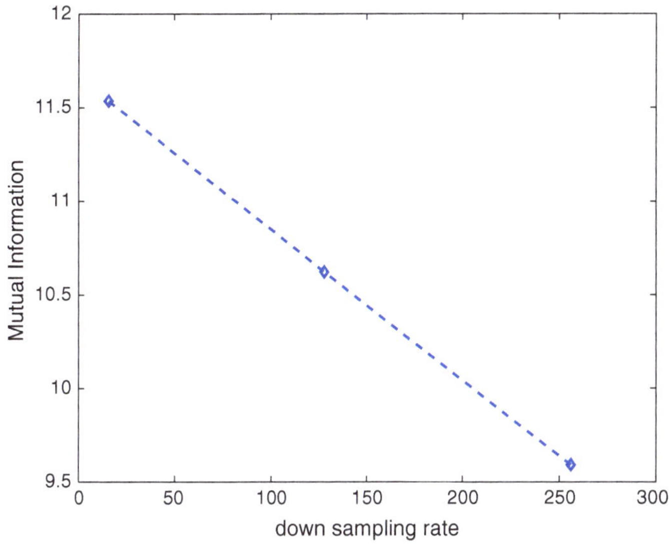

Fig. 4.17 The mutual information between samples separate in time

It is imperative to note that the above examples assume that the pdf of the signal can be estimated using the histogram method. Such an assertion is valid when one can establish independence, as noted in the previous chapter, among the samples which are separated in time by a large amount. That is, one has to establish ergodicity in distribution. In the absence of further information regarding the underlying processes which dictate the statistical behavior of the EEG signal, an assumption of independence among samples may be flawed. However, given the large variations in the EEG signal over time, which may be due to independent influencing factors (external stimuli), it is safe to assume that signals separated in time become increasingly independent with an increase in the separation distance. This assumption may give credence to the histogram method as a means of estimating the pdf of the EEG signal. Interestingly enough, the mutual information in the above example also points to this fact; that is, even with the downsampling of 256, the mutual information remains high. However, the downward trend of the mutual information with downsampling factor reaffirms the above observation regarding the independence of samples with large separations in time.

Regardless of independence, one can establish the correlation length (the time separation where the samples become uncorrelated) of EEG signal of about 5 s according to the above example. In many studies, this number has been used as the correlation length of the EEG signals in general. However, one can establish the correlation function for any data set and determine the correlation length independent of the above observation.

In conclusion, given the availability of data and the knowledge that a histogram can be used to estimate pdf, one can obtain the mutual information. Furthermore, one can match the histogram of the data to a known pdf. The accuracy of the "matching" process, then, can be established using a number of well-known metrics discussed above. If one can successfully argue in favor of the pdf estimation, such as histogram, and then match the pdf to a known pdf, which will yield a relatively small HD, KS, or KL metrics, a pdf for the observed data can be obtained. Such a pdf is of critical importance to the process of detecting the desired signal when corrupted by additive noise or other interferences. This subject will be tackled next.

4.4 Detection and Estimation Methods

We are now in a position to tackle the "detection" process. In the world of signal processing, detection refers to the extraction of a signal from an AWGN-corrupted signal In the biomedical world, there is often no significant propagation, and the signal is not only corrupted by AWGN (due to electronic devices) but also is impacted by interference, self-noise, and nonlinear effects. In the electronic communications, we often know what we are looking for (namely, the presence or absence of a know signal-or a known signal with some unknown parameters). In the biomedical domain, we may not even know the exact structure of the signal we are looking to extract. Instead, we may be interested in determining the presence or absence of a class (or collection) of signals or a transient event. In any event, in all cases, one has to make a "decision" using a "criterion" that operates on an "observed" signal or signals. Hence, one has to resort to hypothesis testing to go any further. We tackle this concept next.

4.4.1 Signal Detection Using Hypothesis Testing (HT)

The concept of HT, as the name implies, attempts to verify the validity of a hypothesis (or hypotheses) through observing a signal (or signals). It becomes immediately obvious that this concept has an important place in the field of biomedical signal processing as a key aspect of this field is to determine the presence of a signal, which may indicate a physiological condition, in an observed signal (time series).

4.4.1.1 Bayes Mixed Mode Hypothesis Testing (B-HT)

To begin, let us elaborate on the terminology of "mixed mode." As it will become clear shortly, we are often concerned with the problems that involve discrete events and continuous/discrete random variables. This involves mixing pdfs and event probabilities. For this reason, we use the term "mixed mode" to emphasize this fact.

To illustrate this point, let us consider the scenario where M hypotheses, H_0, ..., and H_{M-1}, are under consideration. That is, our experiment is designed to distinguish between M hypotheses. We are also provided (observing) a vector of length l_v, given by

$$\mathbf{x} = [x[0], x[n+1], \ldots, x[l_v - 1]]^\mathsf{T} \tag{4.50}$$

Note that the observation may be discrete in time, but may be a continuous or discrete random variable. Furthermore, for the sake of simplification, we use x_i and $x[j]$ interchangeably. Hence,

$$\mathbf{x} = \left[x_0, x_1, \ldots, x_{l_v-1}\right]^\mathsf{T} \tag{4.51}$$

For the sake of generality, we make the assumption that the observed interval l_v defines a number of samples (period of time) for which one of the aforementioned hypotheses is true. For signal processing, this is typically a "symbol" duration, whereas for a biomedical signal, this may be a transient event, such as epilepsy or the onset of malignant ventricular fibrillation (MVF). More precisely, let E_0, E_1, ..., and E_{M-1} be M events which are associated with the M hypotheses. That is,

$$\begin{array}{ccc} H_0 & \Leftrightarrow & E_0 \\ H_1 & \Leftrightarrow & E_1 \\ \vdots & & \vdots \\ H_{M-1} & \Leftrightarrow & E_{M-1} \end{array} \tag{4.52}$$

These events may be a collection of underlying events that are to be detected by observing $\mathbf{x}[n]$.

In hypothesis testing, we are concerned with the following questions:

1. *In what manner the M events are mapped to x?*
2. *More importantly, given that we have observed x, can we map this observation back to one of the M events?*
3. *What are the probabilities (known as a-priori probabilities) of occurrence of the M events?*
4. *Finally, realizing that the observation x is almost surely impaired by random events, what are the cost (or costs) of making a wrong decision?*

As you gather, the above questions are important to whoever is interested in conducting any experiments, whose results are of any significance. In the field of biomedical signal processing, this may be a time-sensitive or time-insensitive decision making process. For instance, if the experiment is designed to identify time-sensitive events, such as epileptic seizure or MVF, then the decision making process may have a significant impact on a patient who is using the hypothesis testing to take a corrective action to remedy a life-threatening medical condition. In other cases, the experiment is simply tries to determine the presence of time-insensitive anomalies.

In such a case, one has the benefit of time to make a decision or even repeat the experiment. Hence, the "cost" of making a "decision" is situation dependent and may vary drastically in the filed of biomedical signal processing.

We will tackle the four major questions raised above separately. First, *how does one map an event to* $x[n]$*?* This requires a knowledge of the underlying structure behind your observation. For electronic communication, where you have binary hypothesis testing ($M = 2$), you are concerned about sending a "1" or a "0." In that event, the mapping is impacted by how one represents these hypotheses using waveforms and how the "channel" impacts the signal. In this case, $\mathbf{x}$ will be the signal observed at the receiver. For biomedical applications, $\mathbf{x}$ represents EEG, ECG, EMG, etc. while the hypotheses are various states of a biological system (epilepsy, cardiac arrest, MVF, etc.). Unlike electronics communications, where channels and signals are well-defined, one does not always have access to accurate models for how events map to their corresponding $\mathbf{x}[n]'s$ in the biomedical cases. Hence, we consider probabilistic models, which also take into account the impacts of channel (measurement devices). That is, let us assume that we have access to the following information:

$$\begin{array}{ll} H_0 \Leftrightarrow & f_{\mathbf{x}}(\mathbf{x}|E_0) = f_0(\mathbf{x}) \\ H_1 \Leftrightarrow & f_{\mathbf{x}}(\mathbf{x}|E_1) = f_1(\mathbf{x}) \\ \vdots & \vdots \\ H_{M-1} \Leftrightarrow & f_{\mathbf{x}}(\mathbf{x}|E_{M-1}) = f_{M-1}(\mathbf{x}) \end{array} \tag{4.53}$$

where $f_j(\mathbf{x}) = f_{\mathbf{x}}\left(\mathbf{x}|E_j\right)$ is the conditional probability density function of $\mathbf{x}$ given that event E_j has occurred. These pdfs may be viewed as the probabilistic mapping functions between the events and the observed $\mathbf{x}$. It is noteworthy that for biological signals one may have to estimate these pdfs, whereas in electronic communications problem these pdfs are known a-priori. For instance, when $\mathbf{x}[n]$ is a received signal vector in a digital transceiver, $f_{\mathbf{x}}\left(\mathbf{x}|E_j\right)$ is solely determined by the characteristics of the communications channel, which most often is an AWGN channel. Next, one can define the "cost" of making an erroneous decision. That is, let

$$\begin{aligned} C_{ij} = {} & \text{cost of making a decision in favor of the } i\text{th event } (H_i) \\ & \text{given that the } j\text{th event } (H_j) \text{ has actually occurred} \end{aligned} \tag{4.54}$$

Note that C_{ii} is the cost of making a correct decision, and hence it may be set to 0 for practical problems. As noted above, these cost functions vary drastically form situation to situation. For instance, for the case of MVF, where one is interested in a binary hypothesis testing scenario (*Is MVF present or absent?*), C_{ij}; $i \neq j$, is the cost of either having the patient entering a phase of MVF and the test that is being conducted missing the onset of the MVF, or the patient is free of MVF and the test indicates otherwise. In the field of biomedical signal processing, the costs stated above have important consequences.

Given the above formulation, then the average cost of making a decision in favor of event E_i is

$$C_i = \sum_{j=0}^{M-1} C_{ij} P_{ij} P_j \tag{4.55}$$

where $P_{ij} = P\left\{E_i | E_j\right\}$ denotes the probability of making a decision in favor of H_i given that H_j has occurred (namely, the probability that we state $\mathbf{x}[n]$ is generated by E_i where in reality E_j has occurred). The above equation also uses what is known as the a-priori probabilities, given by

$$\begin{array}{ll} H_0 \Leftrightarrow & P_0 = P\{E_0\} \\ H_1 \Leftrightarrow & P_1 = P\{E_1\} \\ \vdots & \quad\vdots \\ H_{M-1} \Leftrightarrow & P_{M-1} = P\{E_{M-1}\} \end{array} \tag{4.56}$$

which are subject to the constraint

$$\sum_{j=0}^{M-1} P_j = 1. \tag{4.57}$$

This formulation implies that one of the hypothesis always occurs. These probabilities, which determine the chances of various events occurring, are usually known in the world of electronic communications (equally likely symbols are often used in digital communications, leading to $p_j = \frac{1}{M}$ for all j, etc.), whereas in the biomedical world, these probabilities are hard to assess as they are time and/or event dependent. For instance, if we consider the epileptic seizure case, the a-priori probabilities (if the hypothesis is to determine whether a patient is in pre-epileptic period or not) may vary depending on the patient's external stimuli, which clearly can be a random, time-dependent event. Hence, one who is engaged in biomedical signal processing endeavors usually faces a challenging obstacle in determining a-priori probabilities.

Now, when $C_{ii} = 0$ for all i, C_{ij} may be viewed as the cost of making an error in favor of E_i (when H_j is true). Therefore, the average cost of making a decision for all events would be

$$\begin{aligned} C &= \sum_{i=0}^{M-1} \sum_{j=0}^{M-1} C_{ij} P\left\{E_i | E_j\right\} P_j \\ &= \sum_{i=0}^{M-1} \sum_{j=0}^{M-1} C_{ij} P\left\{E_i, E_j\right\}. \end{aligned} \tag{4.58}$$

Assuming that

$$C_{ii} < C_{ji}; \text{ for any } j = 0, 1, \ldots, M-1; \; j \neq i \tag{4.59}$$

(which is true for almost all practical problems), then one must seek to minimize C. To elaborate, the above condition simply implies that the cost of making a correct decision in favor of H_i is smaller than the cost of making an erroneous decision. Consequently, C may be viewed as the cost of making an erroneous decision, and hence, one has to seek to minimize it.

Given that *a*-priori probabilities are known or defined by the problem and $C'_{ij}s$ are determined by factors (misdiagnosis cost, false detection cost, etc.) often beyond our control, the only remaining factor that can be used to minimize the cost are $P\left\{E_i|E_j\right\}$ for all $i \neq j$. One obvious solution is a strategy that leads to

$$P_{ij} = P\left\{E_i|E_j\right\} = \begin{cases} 1 & i = j \\ 0 & i \neq j \end{cases}$$

In this case, one can achieve $C = 0$ for $C_{ii} = 0$. The problem is that $\mathbf{x}[n]$ is a random vector, and hence, one cannot arrive at the above scenario for majority of problems.

The task before us then is to find a decision "rule" that minimizes C. To elaborate, given that $\mathbf{x}[n]$ is a random vector,

$$P_{ij} = \int_{DR_i} f_{\mathbf{x}}\left(\mathbf{x}|E_j\right) d\mathbf{x} \tag{4.60}$$

where DR_i is a region in the n-dimensional domain of the random vector $\mathbf{x}[n]$. Note that the above implies that we declare E_i if $\mathbf{x}$ falls in DR_i. That is,

$$\hat{E} = E_i \text{ if } \mathbf{x} \in DR_i. \tag{4.61}$$

This would seem to be a simple task if DR'_is were known. However, for the problem at hand, we are to find DR'_is. In fact, finding DR'_is is the ultimate goal here; that is, once you know these regions, then one can decide which event has occurred for a given observed $\mathbf{x}[n] = \mathbf{x}$. In fact, that would be the decision rule. *How do we, then, find DR'_is?* This question is answered by the Bayes rule.

In Bayes rule, then, for a given set of $\left\{p_j;\ j = 0, 1, \ldots, M-1\right\}$ and $\left\{C_{ij}; i, j = 0, 1, \ldots, M-1\right\}$, we are tasked with finding a set of decision regions $\left\{DR_j;\ j = 0, 1, \ldots, M-1\right\}$ to minimize C.

It should be clear to the reader that the above is a minimization over the domain of the random vector $\mathbf{x}$. To put this concisely,

$$\underset{j=0,1,\ldots,M-1}{DR_j} = \arg\min C \tag{4.62}$$

subject to the constraint

$$\sum_{j=0}^{M-1} P_j = 1, \tag{4.63}$$

which implies that one of the hypothesis always occurs.

We also point out that the main objective of this decision making policy is to identify M mutually exclusive (to avoid ambiguity in the decision making process) decision regions that are exhaustive. That is, if DR denote the entire decision region (defined by the range of random vector $\mathbf{x}$), then

$$DR = DR_0 \cup DR_1 \cup \ldots DR_{M-1} \tag{4.64}$$

where

$$DR_i \cap DR_j = \varnothing \text{ for all } i \neq j \tag{4.65}$$

with $\varnothing$ denoting the empty set. This further implies

$$\sum_{i=0}^{M-1} P_{ij} = 1 \text{ for all } j \tag{4.66}$$

We can make further simplification of the decision making process by noting the following:

$$C = \sum_{i=0}^{M-1} \int_{DR_i} C_{ij} \sum_{j=0}^{M-1} P_j f_{\mathbf{x}} \left(\mathbf{x}|E_j\right) d\mathbf{x} \tag{4.67}$$

Let

$$\Gamma_j(\mathbf{x}) = f_{\mathbf{x}} \left(\mathbf{x}|E_j\right) P_j \tag{4.68}$$

Then,

$$\begin{aligned} C &= \sum_{i=0}^{M-1} \int_{DR_i} \sum_{j=0}^{M-1} C_{ij} \Gamma_j(\mathbf{x})\, d\mathbf{x} \\ &= \sum_{i=0}^{M-1} C_{ii} P_i + \sum_{i=0}^{M-1} \int_{DR_i} \sum_{j=0; j\neq i}^{M-1} \left(C_{ij} - C_{jj}\right) \Gamma_j(\mathbf{x})\, d\mathbf{x} \end{aligned} \tag{4.69}$$

Notice that

$$C_i(\mathbf{x}) = \sum_{j=0; j\neq i}^{M-1} \left(C_{ij} - C_{jj}\right) \Gamma_j(\mathbf{x}) \tag{4.70}$$

is the cost of making a decision in favor of H_i for a given $\mathbf{x}$. We then have

$$C = \sum_{i=0}^{M-1} C_{ii} P_i + \sum_{i=0}^{M-1} \int_{DR_i} C_i(\mathbf{x})\, d\mathbf{x} \tag{4.71}$$

The first term is a constant and does not impact the decision making process. The second term, however, is a function of the decision regions. Now, if H_k has occurred, the cost of making a decision in favor of H_k is the smallest. That is, $\int_{DR_k} C_k(\mathbf{x})\, dx$ must be the smallest among all $\int_{DRi} C_i(\mathbf{x})\, dx; i = 0, 1, \ldots, M-1; i \neq k$. Realizing that $\left(C_{ij} - C_{jj}\right) > 0$ for all i and j, which in turn implies that $C_i(x) > 0$ for all i, instead of minimizing the integral, we can minimize the integrand. This leads to

$$\boxed{\hat{H}(\mathbf{x}) = H_k \text{ iff } \mathrm{C}_k(\mathbf{x}) = \min\left\{C_i(x); i = 0, 1, \ldots, M-1\right\}}$$

Since the set $\left\{f_{\mathbf{x}}\left(\mathbf{x}|E_j\right); \; j = 0, 1, \ldots, M-1\right\}$ indeed is needed to go any further, we resort to an example to illustrate the Bayes rule.

Example 8 A biological signal is observed and a large number of samples are taken. The samples are averaged to remove the impact of transient noise, additive white noise, etc. The final outcome is a single random variable x. It is believed that the observed signals are generated by 2 distinct events E_0 and E_1 with probabilities p and $(1-p)$, respectively. By observing x, we are hoping to determine which event has occurred. Under each event, x is shown to be exponentially distributed with parameters λ_0 or λ_1 corresponding to E_0 or E_1, respectively. The cost of making decision is as follows: $C_{ij} = \alpha\,|i-j|, i, j = 0, 1$. Find the optimum decision rule and obtain the resulting cost of making a decision using the optimum rule. Assume that your decision is based on observing x only.

<u>**Solution**</u>: *From the description of the problem, there are two hypotheses and $P_0 = p$ and $P_1 = 1 - p$. Also, realizing that $x[n] = x$,*

$$f_{\mathbf{x}}\left(\mathbf{x}|E_j\right) = \begin{cases} \lambda_0 \exp(-\lambda_0 x) & j = 0 \\ \lambda_1 \exp(-\lambda_0 x) & j = 1 \end{cases}. \tag{4.72}$$

Given the cost function,

$$C = \alpha\left\{pP_{10} + (1-p)\,P_{01}\right\}, \tag{4.73}$$

the optimum decision rule is the one that minimizes C. Since the observed signal is a single random variable, the decision region is 1-dimensional. Let η be the boundary between the two decision regions; namely,

$$DR_j \in \{[0, \eta] \; or \; [\eta, \infty]\}. \tag{4.74}$$

Without loss of generality, let us assume that $DR_0 = [0, \eta]$ and $DR_1 = [\eta, \infty]$ (the case of $x = \eta$ has a zero probability of occurrence since x is a continuous random *variable, and hence we have considered* $DR_0 = [0, \eta]$ *and not* $DR_0 = [0, \eta))$*. Quite clearly, finding* η *is the goal here (the optimum rule). This further implies that*

$$\eta = \arg\min C \tag{4.75}$$

Now,

$$C = \alpha \{p \exp(-\eta\lambda_0) + (1 - p)(1 - \exp(-\eta\lambda_1))\}. \tag{4.76}$$

Next, setting $\frac{\partial C}{\partial \eta}$ *to* 0 *leads to*

$$\lambda_1 (1 - p) \exp(-\eta_o\lambda_1) = p\lambda_0 \exp(-\eta_o\lambda_0) \Rightarrow$$

$$\eta_o = \frac{\ln\left(\frac{\lambda_1(1-p)}{\lambda_0 p}\right)}{\lambda_1 - \lambda_0}; \lambda_1 \neq \lambda_0 \tag{4.77}$$

where η_0 *is the optimum* η *and it is assumed that* $\frac{\ln\left(\frac{\lambda_1(1-p)}{\lambda_0 p}\right)}{\lambda_1-\lambda_0} \geq 0$*. If* $\frac{\ln\left(\frac{\lambda_1(1-p)}{\lambda_0 p}\right)}{\lambda_1-\lambda_0} < 0$*, then we need to consider* $DR_1 = [0, \eta_0]$ *and* $DR_0 = [\eta_0, \infty]$ *and repeat the above step (reader can complete this step). Assuming that* $\frac{\ln\left(\frac{\lambda_1(1-p)}{\lambda_0 p}\right)}{\lambda_1-\lambda_0} > 0$*, the decision rule reduces to*

$$x \underset{H_0}{\overset{H_1}{\gtrless}} \eta_0 \tag{4.78}$$

Finally, the cost of this decision making process is

$$C_{\min} = \alpha \{p \exp(-\eta_0\lambda_0) + (1 - p)(1 - \exp(-\eta_0\lambda_1))\}. \tag{4.79}$$

It is of interest to consider a case where $C_{ii} = 0$ for all i and $C_{ij} = 1$ for all $i \neq j$. This implies an experiment where the cost of making an error in favor of any other hypothesis is the same and that there is no cost for a correct decision. One quickly realizes that this special case is of significant practical interest. In that case, we have

$$C_i(\mathbf{x}) = \sum_{j=0; j\neq i}^{M-1} \Gamma_j(\mathbf{x}) \tag{4.80}$$

This further implies that

$$C_k - C_i = \Gamma_i(\mathbf{x}) - \Gamma_k(\mathbf{x}) \tag{4.81}$$

$$= P_i f_{\mathbf{x}}(\mathbf{x}|E_i) - P_k f_{\mathbf{x}}(\mathbf{x}|E_k) \tag{4.82}$$

If H_k is true, then

$$C_k - C_i < 0 \text{ for all } i$$

This leads to

$$\hat{H} = H_k \text{ iff } \Lambda_{ki}(\mathbf{x}) = \frac{f_{\mathbf{x}}(\mathbf{x}|E_k)}{f_{\mathbf{x}}(\mathbf{x}|E_i)} > \frac{P_i}{P_k} \text{ for all } i \neq k \tag{4.83}$$

where now $\Lambda_{ki}(\mathbf{x})$, the likelihood function, which is dependent on the data $\mathbf{x}$ and its pdf under H_k and H_i. Note that the right side of the equation is known and can be computed in advance. It is beneficial, from a computing resource point of view, that binary decisions are made. To that end, the decision making process reduces to this binary test: Declare H_k iff

$$\hat{H} = H_k \text{ iff } \Lambda_{ki}(\mathbf{x}) = \frac{f_{\mathbf{x}}(\mathbf{x}|E_k)}{f_{\mathbf{x}}(\mathbf{x}|E_i)} > \frac{P_i}{P_k} = \eta_{ki} \text{ for all } i \neq k \tag{4.84}$$

It is of importance to note that the decision making process, which required mapping the random vector $\mathbf{x}$ to an l_v—dimensional decision region, now involves comparing random variables $\Lambda_{ki}(\mathbf{x})$ to the known constant η_{ki}. This is technically single dimensional decision making as the likelihood function $\Lambda_{ki}(\mathbf{x})$ for a given k and i is a single random variable. Next, we consider the Neyman-Pearson Hypothesis testing (NP-HT), where the decision costs are not known. This particular hypothesis testing is of significant importance to biomedical signal processing. Before we do so, we need to define some key parameters that are used in the field of biomedical signal analysis involving experimental or test procedures.

4.4.2 Specificity and Sensitivity

In the field of biomedical signal processing, where one is interested in detecting or diagnosing diseases using an observed time-series or signal, one is interested in two key parameters to assess whether an experiment or procedure is effective. The efficacy of the test/experiment is often measured in terms of its specificity and sensitivity, which implicitly suggest a binary hypothesis testing. That is, a test/experiment is conducted and the following hypotheses are examined:

$$\begin{aligned} H_0 &\Leftrightarrow \text{ A disease or a condition is absent} \\ H_1 &\Leftrightarrow \text{ A disease or a condition is present} \end{aligned} \tag{4.85}$$

In this case, P_0 is known, but there are no cost functions. Assuming a decision making process (whether optimum or not), we arrive at P_{10} and P_{01}. By definition, we have

$$\begin{array}{l} \text{Sensitivity} \Leftrightarrow P_{11} = 1 - P_{01} \\ \text{Specificity} \Leftrightarrow P_{00} = 1 - P_{10} \end{array} . \tag{4.86}$$

In the field of signal processing, P_{01} is referred to as the probability of "missed" detection, and hence, P_{11} may be viewed as the probability of detection. Similarly, P_{10} is known as the probability of "false" alarm, and hence, specificity may be viewed as the probability of correctly identifying the absence of a condition. Hence, *one strives to maximize sensitivity and specificity in an experiment/test*. Intuition dictates that one cannot increase both simultaneously. In fact, as we will see, often one has to fix one and maximize the other. That is, one may have to design the experiment to achieve an acceptable level of *specificity* and maximize *sensitivity*. As one can imagine, missing the detection of a disease (small *sensitivity)* can have severe consequences (death), while a small *specificity* may result in unnecessary further testing, and hence, increased financial and emotional cost! In order to examine the above binary hypothesis testing, let us consider NP-HT.

4.4.2.1 Neyman-Pearson Hypothesis Testing (NP-HT)

NP-HT is a special case of the Bayes-HT scheme. Namely, it is a binary hypothesis testing with $C_{ii} = 0$ for all i and $C_{ij} = 1$ for all $i \neq j$. As noted above, this may occur when the decision costs are unknown or hard to assess. The cost of decision making, however, is assessed in terms of P_{ij} $i \neq j$. Namely, we measure the effectiveness of the hypothesis testing in terms of sensitivity and specificity. Since it is a binary decision making, we only have to concern ourselves with P_{01} and P_{10}. NP-HT is often concerned with validating the presence of phenomenon. Hence, H_0 and H_1 are often associated with the absence or presence, respectively, of a phenomenon. That is,

$$\begin{array}{l} H_0 \Leftrightarrow f_{\mathbf{x}}(\mathbf{x}|E_0) = f_0(\mathbf{x}) \\ H_1 \Leftrightarrow f_{\mathbf{x}}(\mathbf{x}|E_1) = f_1(\mathbf{x}) \end{array} \tag{4.87}$$

Note that $1 - P_{01}$ and $1 - P_{10}$ are often of importance in the field of biomedical signal processing, although in the field radar signal processing one is concerned with P_{10} (probability of false alarm) and P_{01} (probability of missed detection). Depending upon the problem at hand, one can fix one of these probabilities and obtain the other. If we denote the entire decision space as DR, then

$$DR = DR_0 \cup DR_1 \tag{4.88}$$

where

$$DR_0 \cap DR_1 = \varnothing. \tag{4.89}$$

The last condition is needed to establish unambiguous decision making policy. Now, let us define

$$\begin{aligned} P_{FA} &= P_{10} \\ P_{MD} &= P_{01} \end{aligned} \tag{4.90}$$

as the false alarm and missed detection probabilities. One can then readily convert these parameters to specificity and selectivity metrics. In this case, the objective is to minimize P_{MD} given a maximum value for P_{FA}, say $P_{FA}^{\max}$. In this case, it is assumed that the maximum "tolerable" probability of false alarm (lowest acceptable specificity) is known to us. We first form a cost function for NP-HT, which is given by

$$C_{NP} = P_{MD} + \lambda \left[P_F - P_{FA}^{\max} \right] \tag{4.91}$$

where λ is the Lagrangian multiplier. Note that $P_F = P_{FA}^{\max}$ results in $C_{NP} = P_{MD}$. Now, minimizing P_{MD} (maximizing selectivity) will minimize C_{NP}, and vice versa. We also have

$$C_{NP} = 1 - P_D + \lambda \left[P_F - P_{FA}^{\max} \right] \tag{4.92}$$

$$\begin{aligned} C_{NP} &= 1 - \int_{DR_1} f_{\mathbf{x}} (\mathbf{x}|E_1)\, d\mathbf{x} + \lambda \left[\int_{DR_1} f_{\mathbf{x}} (\mathbf{x}|E_0)\, d\mathbf{x} - P_{FA}^{\max} \right] \\ &= 1 - \lambda P_{FA}^{\max} - \int_{DR_1} \left\{ f_{\mathbf{x}} (\mathbf{x}|E_1) - \lambda f_{\mathbf{x}} (\mathbf{x}|E_0) \right\} d\mathbf{x}. \end{aligned} \tag{4.93}$$

In order to minimize C_{NP}, one has to to choose DR_1 such that $f_{\mathbf{x}} (\mathbf{x}|E_1) - \lambda f_{\mathbf{x}} (\mathbf{x}|E_0) > 0$ (this in turn minimizes C_{NP} for a fixed $P_{FA}^{\max}$). It is clear from this observation that the decision rule can be expressed as

$$\boxed{\Lambda (\mathbf{x}) = \frac{f_{\mathbf{x}}(\mathbf{x}|E_1)}{f_{\mathbf{x}}(\mathbf{x}|E_0)} \underset{H_0}{\overset{H_1}{\gtrless}} \lambda,} \tag{4.94}$$

which also implies

$$P_{FA} = \int_{\lambda}^{\infty} f_{\Lambda} (s|E_0)\, ds \tag{4.95}$$

and

$$P_{DM} = \int_{-\infty}^{\lambda} f_{\Lambda} (s|E_1)\, ds. \tag{4.96}$$

In these equations, $f_\Lambda\left(s|E_j\right)$ is the pdf of the random variable (likelihood ratio) $\Lambda(\mathbf{x})$ when E_j has occurred. Note that, even though $\mathbf{x}$ is a vector, Λ is a single random variable, and hence the decision region (when Λ is the decision variable) is a 1-dimensional space. From this formulation, and realizing that P_{FA} is a decreasing function of λ while P_{MD} is an increasing functions of λ, the optimum threshold λ_0 can be found by assuming the smallest λ that can be allowed in view of $P_{FA} \leq P_{FA}^{\max}$ condition. Hence, setting $P_{FA} = P_{FA}^{\max}$ will yield the smallest λ and hence the smallest P_{MD}. That is, we can compute λ_0 by solving

$$P_{FA}^{\max} = 1 - P_{SP}^{\min} = \int_{\lambda_0}^{\infty} f_\Lambda\left(\mathbf{s}|E_0\right) d\mathbf{s}, \tag{4.97}$$

where $P_{SP}^{\min}$ is the minimum acceptable specificity. This, in turn, leads to the following P_{MD}

$$P_{DM} = 1 - P_{SE} = \int_{-\infty}^{\lambda_0} f_\Lambda\left(\mathbf{s}|E_1\right) d\mathbf{s} \tag{4.98}$$

where P_{SE} is the resulting selectivity, and the cost function $C_{NP} = P_{MD} = 1 - P_{SE}$. Furthermore, the decision rule is

$$\Lambda(\mathbf{x}) \underset{H_0}{\overset{H_1}{\gtrless}} \lambda_0 \tag{4.99}$$

Example 9 An experiment is conducted to determine the presence of anomaly in a biological sample. The observed sample is a vector of 1000 i.i.d. double exponential random variables. It is known that in the absence of the anomaly, the observed sample obeys a double exponential statistics with zero mean and variance $\frac{2}{\alpha^2}$ ($\alpha > 1$). In the presence of the anomaly, the observed samples form 1000 i.i.d. zero-mean random variables with a double exponential pdf whose variance is $\frac{2}{\beta^2}$ ($\beta > \alpha > 1$). We are also required to keep the false decision in favor of anomaly be kept less than 1 % (specificity of 90 %). Find the decision rule and find the probability of miss detection. You are allowed to invoke the central limit theorem (CLT).

Solution: *It is immediately obvious that*

$$f_{\mathbf{x}}\left(\mathbf{x}|H_0\right) = \left(\frac{\alpha}{2}\right)^{1000} \exp\left\{-\alpha \sum_{i=0}^{999} |x_i|\right\} \tag{4.100}$$

$$f_{\mathbf{x}}\left(\mathbf{x}|H_1\right) = \left(\frac{\beta}{2}\right)^{1000} \exp\left\{-\beta \sum_{i=0}^{999} |x_i|\right\} \tag{4.101}$$

The likelihood function is then

$$\Lambda(\mathbf{x}) = \frac{f_{\mathbf{x}}(\mathbf{x}|H_1)}{f_{\mathbf{x}}(\mathbf{x}|H_0)} = \left(\frac{\beta}{\alpha}\right)^{1000} \exp\left\{-(\beta-\alpha)\sum_{i=0}^{999}|x_i|\right\}. \tag{4.102}$$

Since the log function is a monotonically increasing function, we can consider the log of the likelihood function, known as the log-likelihood (LLH) function, given by

$$\ln[\Lambda(\mathbf{x})] = -(\beta-\alpha)\sum_{i=0}^{999}|x_i| + 1000\ln\left(\frac{\beta}{\alpha}\right). \tag{4.103}$$

Given that the second term is a constant and noting that the coefficient of $\sum_{i=1}^{1000}|x_i|$ *is negative, we can use the following simplified decision variable*

$$\Theta = \frac{1}{1000}\sum_{i=0}^{999}|x_i| \underset{H_1}{\overset{H_0}{\gtrless}} \lambda_0.$$

As we noted earlier, we need to obtain λ_0 *to achieve a desired* P_{FA}. *Since 1000 samples of i.i.d. random variables are observed, one can invoke the central limit theorem, which implies that* Θ *may be viewed as a Gaussian random variable. Since* $|x_i|'s$ *are all zero-mean,* Θ *will be zero-mean. Also, given the independence of* $|x_i|'s$, *the variance of* Θ *is given by*

$$E\left\{\left(\Theta - E^2\{\Theta|H_0\}\right)^2 |H_0\right\} = \frac{1}{1000}\frac{2}{\alpha^2} = \sigma_0^2 \tag{4.104}$$

while

$$E\left\{\left(\Theta - E^2\{\Theta|H_1\}\right)^2 |H_1\right\} = \frac{1}{1000}\frac{2}{\beta^2} = \sigma_1^2. \tag{4.105}$$

Hence,

$$f_\Theta(\theta|E_0) \approx \frac{1}{\sqrt{2\pi\sigma_0^2}}\exp\left\{-\frac{\theta^2}{2\sigma_0^2}\right\} \tag{4.106}$$

and

$$f_\Theta(\theta|E_1) \approx \frac{1}{\sqrt{2\pi\sigma_1^2}}\exp\left\{-\frac{\theta^2}{2\sigma_1^2}\right\}. \tag{4.107}$$

We, then, have the following P_{FA} and P_{MD}:

$$P_{FA} \approx \int_{0}^{\lambda_0} \frac{1}{\sqrt{2\pi\sigma_0^2}} \exp\left\{-\frac{\theta^2}{2\sigma_0^2}\right\} d\theta = 0.5\,\mathrm{erf}\left(\frac{\lambda_0}{\sqrt{2\sigma_0^2}}\right) \tag{4.108}$$

$$P_{MD} \approx \int_{\lambda_0}^{\infty} \frac{1}{\sqrt{2\pi\sigma_1^2}} \exp\left\{-\frac{\theta^2}{2\sigma_1^2}\right\} d\theta = 0.5 - 0.5\,\mathrm{erf}\left(\frac{\lambda_0}{\sqrt{2\sigma_1^2}}\right) \tag{4.109}$$

$$= 0.5\,\mathrm{erf\,c}\left(\frac{\lambda_0}{\sqrt{2\sigma_1^2}}\right) \tag{4.110}$$

where $\mathrm{erf}\,(x)$ *is the standard error function, while* $\mathrm{erf}c(x) = 1 - \mathrm{erf}(x)$ *denotes the complementary error function. There are two important observations that need to be made. First,* Θ *is a positive random variable. However, since* σ_0^2 *is a small number for* $\alpha > 1$, *the value of* $\lambda_0 > 0$ *will be large enough so that the tail of the approximate Gaussian pdf will vanish for* $\Theta < 0$. *Second, the fact that the decision variable is in terms of* $\sum_{i=0}^{999} |x_i|$ *(and not* $-\sum_{i=0}^{999} |x_i|$*), the false alarm probability is an increasing function of* λ_0 *for the above decision variable while* P_{MD} *is a decreasing one. This is not a contradiction to the earlier statement regarding* P_{MD} *and* P_{FA} *and* λ_0. *Hence, we set the threshold at the maximum possible value. That is, we set* λ_0 *such that* $P_{FA} = P_{FA}^{\max} = 0.1$ $(P_{SP} = 0.9)$ *which leads to*

$$\lambda_0 = \sqrt{2\sigma_0^2}\,\mathrm{erf}^{-1}\,(0.1/0.5) \tag{4.111}$$

while

$$P_{MD} \approx 0.5\,\mathrm{erf\,c}\left(\sqrt{\frac{\sigma_0^2}{\sigma_1^2}}\,\mathrm{erf}^{-1}\,(0.1/0.5)\right) \tag{4.112}$$

$$= 0.5\,\mathrm{erf\,c}\left(\frac{\beta}{\alpha}\,\mathrm{erf}^{-1}\,(0.1/0.5)\right) \tag{4.113}$$

In this equation erf^{-1} *denotes the inverse error function. We can also state that*

$$P_{MD} = 0.5\,\mathrm{erf\,c}\left(\frac{\beta}{\alpha}\,\mathrm{erf}^{-1}\,\left(2P_{FA}^{\max}\right)\right)$$

P_{MD} *as a function of* $\frac{\beta}{\alpha}$ *is plotted in Fig.* 4.18 *for various values of* $P_{FA}^{\max}$. *Note that* $P_{SE} = 1 - P_{MD}$. *The red curve depicts* P_{MD} *for the problem at hand. Note that, as expected,* P_{MD} *is a decreasing function of* $\frac{\beta}{\alpha}$. *To elaborate, as* $\frac{\beta}{\alpha}$ *increases, the*

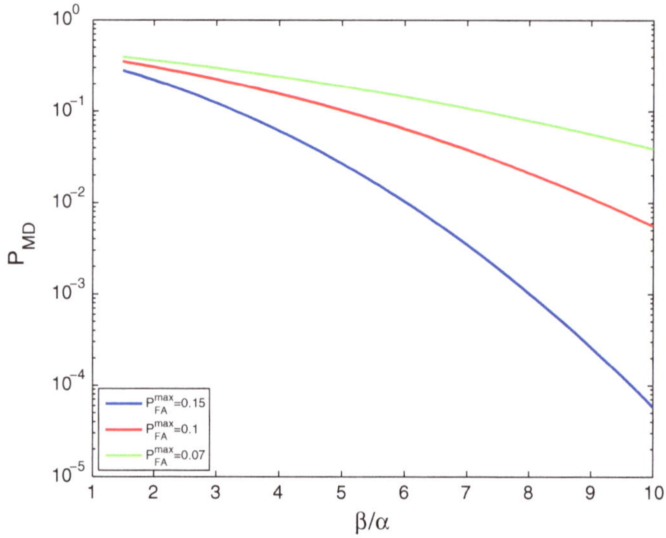

Fig. 4.18 The probability of missed detection

"separation" between the two pdfs increases, resulting in a smaller P_{MD} *for a given* $P_{FA}^{\max}$*. To gain a better insight, we have plotted the pdfs for the two hypotheses for* $\alpha = 2$ *and* $\beta = 3$ *and* $\beta = 4$ *(corresponding to* $\beta/\alpha = 1.5$ *and* 2*, respectively) in Fig.* 4.19*.*

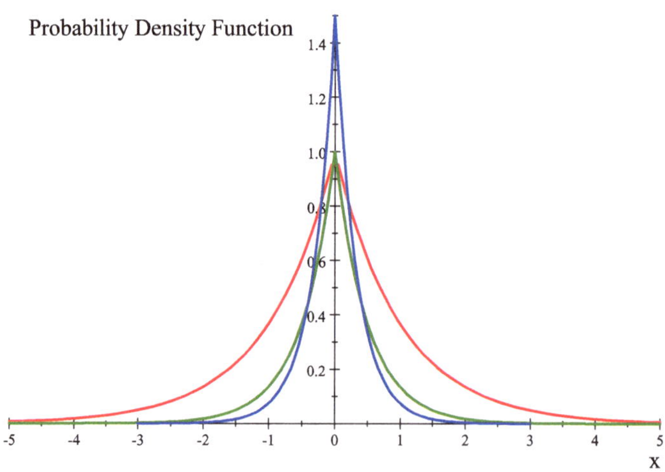

Fig. 4.19 $f_x(x) = \frac{\alpha}{2}\exp(-\alpha|x|)$ for $\alpha = 2$ (*red curve*), $f_x(x) = \frac{\beta}{2}\exp(-\beta|x|)$ for $\beta = 3$ (*green*) and $f_x(x) = \frac{\beta}{2}\exp(-\beta|x|)$ for $\beta = 4$ (*blue*)

Note that the two pdfs are both zero-mean, double exponential, and the separation between the two pdfs comes from a change in the variances, which are related to α *and* β*. Furthermore, larger* $P_{FA}^{\max}$ *(smaller* P_{SP}*) results in a larger* λ_0*, which in turn results in a smaller* P_{MD} *(larger* P_{SE}*). This is apparent from the green and the blue curves.*

4.4.2.2 Maximum A-Posteriori Hypothesis Testing (MAP-HT)

In the previous sections, we examined the hypothesis testing under Bayes and Newman-Pearson rules. In this section, we focus on the MAP-HT rule, where the focus is on maximizing the a posteriori probability (APP). The process is aimed at minimizing the probability of "error." Once again, we assume an M-ary hypothesis testing scenario where now APP probabilities are used to render decision. The cost functions are also unknown. To go any further, we need to define APP. Assuming that the sets $\left\{f_{\mathbf{x}}\left(\mathbf{x}|E_j\right);\ j=0,1,\ldots,M-1\right\}$ and $\left\{P_j;\ j=0,1,\ldots,M-1\right\}$ are available, APP's are defined as

$$P\left\{E_j|x\right\} = \frac{f_{\mathbf{x}}\left(\mathbf{x}|E_j\right)P_j}{\sum_{k=0}^{M-1} f_{\mathbf{x}}\left(\mathbf{x}|E_k\right)P_k};\ j=0,1,..,M-1. \tag{4.114}$$

MAP-HT is then defined as

$$\boxed{\hat{H} = \max \arg P\left\{E_j|\mathbf{x}\right\}} \tag{4.115}$$

Similar to the previous cases, such a decision making cases divides the l_v dimensional space of $\mathbf{x}$ into M distinct decision regions $DR_j;\ j=0,1,\ldots,M-1$. A decision rendered in this manner minimizes the probability of error $P\{ER\}$, defines as

$$P\{ER\} = \sum_{j=0}^{M-1} P\left\{ER|E_j\right\}P_j \tag{4.116}$$

where

$$\begin{aligned} P\left\{ER|E_j\right\} &= 1 - P\left\{C|E_j\right\} \\ &= 1 - \int_{DR_j} f_{\mathbf{x}}\left(\mathbf{x}|E_j\right)d\mathbf{x} \end{aligned} \tag{4.117}$$

with $P\left\{ER|E_j\right\}$ denoting the probability of error given that event E_j has occurred. $P\left\{ER|E_j\right\}$ is the probability of random vector $\mathbf{x}$ falling into any decision region other than DR_j. Note that this is identical to one minus the probability of making a correct decision given E_j, i.e., $P\left\{C|E_j\right\}$.

We also note that the denominator of APP, as defined by (4.114), is independent of j. Hence, maximizing APP is identical to maximizing $f_{\mathbf{x}}\left(\mathbf{x}|E_j\right)$, which maximizes $\int_{DR_j} f_{\mathbf{x}}\left(\mathbf{x}|E_j\right) d\mathbf{x}$. This in turn minimizes $P\left\{ER|E_j\right\}$.

A quick examination of this decision making process reveals an interesting fact. That is, MAP-HT becomes identical to the Bayes rule for the case $C_{ii} = 0$ for all i and $C_{ij} = 1$ for any $i \neq j$. Note that for these conditions, (4.58) and (4.60) reduce to (4.116) and (4.117), respectively. Since we proved that Bayes rule minimizes the cost function, we can also argue that this special case of Bayes rule, which is MAP-HT rule, minimizes its respective cost function, which is the probability of error. MAP-HT has applications in a variety of fields. From which, biomedical signal processing is of interest here. Hence, we consider the following example.

Example 10 In an experiment, we are collecting data with a biological origin. The signal is corrupted by AWGN due to electronic noise $n(t)$. We expect to either get no signal, a 45 Hz, a_1 volt sine signal (phase is important), or a 45 Hz, a volt sine wave with 180 degrees phase shift. Each scenario corresponds to a specific condition that may be present in the subject under examination. Design a detector that minimizes the probability of making a wrong diagnosis. The decision must be made after observing a 1 s sample of the signal. Assume that the sampling rate is 128 samples/s and that the variance of the samples of the AWGN is σ^2. For the decision rule that was obtained above, give an expression for the probability of error. Assume that the three scenarios are equally-likely.

Solution: *Without the loss of generality, we consider the following observation interval*

$$\Gamma = [0, t_s, 2t_s, \ldots, 127t_s]^{\mathsf{T}} \tag{4.118}$$

where $t_s = \frac{1}{128}$. *Since the observed interval is 1 s and the sampling rate is 128, all vectors are now 128 in length. Assume that the observed signal is* $x(t)$. *We also have the corresponding signals in time domain as*

$$s_k(t) = \begin{cases} 0 & k = 0 \\ a\sin(90\pi t) & k = 1 \\ -a\sin(90\pi t) & k = 2 \end{cases}.$$

Clearly, we see three scenarios as follows:

$$\begin{aligned} H_0 &\Leftrightarrow \mathbf{x} = \mathbf{n} + \mathbf{s}_0 \\ H_1 &\Leftrightarrow \mathbf{x} = \mathbf{n} + \mathbf{s}_1 \\ H_{M-1} &\Leftrightarrow \mathbf{x} = \mathbf{n} + \mathbf{s}_2 \end{aligned} \tag{4.119}$$

where

$$\begin{aligned} \mathbf{x} &= [x(0), x(t_s), x(2t_s), \ldots, x(127t_s)]^{\mathsf{T}} \\ &= [x_0, x_1, \ldots, x_{127}]^{\mathsf{T}} \end{aligned} \tag{4.120}$$

$$\mathbf{n} = [n(0), n(t_s), n(2t_s), \ldots, n(127t_s)]^{\mathsf{T}}$$
$$= [n_0, n_1, \ldots, n_{127}]^{\mathsf{T}}, \tag{4.121}$$

$$\mathbf{s}_0 = [0, 0, \ldots, 0]^{\mathsf{T}}, \tag{4.122}$$

$$\mathbf{s}_1 = \left[0, a\sin\left(2\pi\frac{45}{128}\right), a\sin\left(4\pi\frac{45}{128}\right), \ldots, a\sin\left(254\pi\frac{45}{128}\right)\right]^{\mathsf{T}}$$
$$= \left[s_{1,0}, s_{1,1}, \ldots, s_{1,127}\right]^{\mathsf{T}}, \tag{4.123}$$

and

$$\mathbf{s}_2 = -\mathbf{s}_1 = -\left[s_{1,0}, s_{1,1}, \ldots, s_{1,127}\right]^{\mathsf{T}} \tag{4.124}$$

In order to go any further, we need $\left\{f_{\mathbf{x}}\left(\mathbf{x}|E_j\right);\ j = 0, 1, 2\right\}$. *We note that under all three hypotheses, the observed vector x is Gaussian (a linear transformation, which is addition of signals) of a Gaussian vector results in a Gaussian vector. Hence, the mean and covariance matrices in the presence of signal are needed. Since adding a deterministic value (signals) to a random vector does not change its covariance matrix, the covariance matrices of x under all three hypotheses are identical. Given the description of the signal (independent samples as the noise is AWGN), the covariance matrix for x is $\sigma^2 I$. Also,* $E\{\mathbf{x}|E_0\} = 0$, $E\{\mathbf{x}|E_1\} = s_1$, *and* $E\{\mathbf{x}|E_2\} = s_2$. *Hence,*

$$\begin{aligned}
H_0 \Leftrightarrow \quad & f_{\mathbf{x}}(\mathbf{x}|E_0) = \frac{1}{(2\pi\sigma^2)^{64}}\exp\left\{-\sum_{j=0}^{127}\frac{x_j^2}{2\sigma^2}\right\}\\
H_1 \Leftrightarrow \quad & f_{\mathbf{x}}(\mathbf{x}|E_1) = \frac{1}{(2\pi\sigma^2)^{64}}\exp\left\{-\sum_{j=0}^{127}\frac{(x_j - s_{1,j})^2}{2\sigma^2}\right\}\\
H_2 \Leftrightarrow \quad & f_{\mathbf{x}}(\mathbf{x}|E_2) = \frac{1}{(2\pi\sigma^2)^{64}}\exp\left\{-\sum_{j=0}^{127}\frac{(x_j + s_{1,j})^2}{2\sigma^2}\right\}
\end{aligned} \tag{4.125}$$

This can be equivalently written as

$$\begin{aligned}
H_0 \Leftrightarrow f_{\mathbf{x}}(\mathbf{x}|E_0) &= \frac{1}{(2\pi\sigma^2)^{64}}\exp\left\{-\frac{\|\mathbf{x}-\mathbf{s}_0\|^2}{2\sigma^2}\right\}\\
H_1 \Leftrightarrow f_{\mathbf{x}}(\mathbf{x}|E_1) &= \frac{1}{(2\pi\sigma^2)^{64}}\exp\left\{-\frac{\|\mathbf{x}-\mathbf{s}_1\|^2}{2\sigma^2}\right\}\\
H_2 \Leftrightarrow f_{\mathbf{x}}(\mathbf{x}|E_2) &= \frac{1}{(2\pi\sigma^2)^{64}}\exp\left\{-\frac{\|\mathbf{x}+\mathbf{s}_1\|^2}{2\sigma^2}\right\}
\end{aligned} \tag{4.126}$$

where

$$\|\mathbf{x}\|^2 = \sum_{j=0}^{127} x_j^2 \tag{4.127}$$

with $\|\mathbf{x}\|^2$ *denoting the norm of the vector* x. *Since* $\ln(.)$ *is a monotonically increasing function of its argument and pdfs are all non-negative, we can take the* ln *of the pdfs. That is,*

$$\begin{aligned} H_0 &\Leftrightarrow \ln f_{\mathbf{x}}(\mathbf{x}|E_0) = -\tfrac{\|\mathbf{x}-\mathbf{s}_0\|^2}{2\sigma^2} + c \\ H_1 &\Leftrightarrow \ln f_{\mathbf{x}}(\mathbf{x}|E_1) = -\tfrac{\|\mathbf{x}-\mathbf{s}_1\|^2}{2\sigma^2} + c \\ H_2 &\Leftrightarrow \ln f_{\mathbf{x}}(\mathbf{x}|E_2) = -\tfrac{\|\mathbf{x}-\mathbf{s}_2\|^2}{2\sigma^2} + c \end{aligned} \tag{4.128}$$

The constant c does not enter into the decision making process, and hence can be ignored. Using $\hat{H} = H_k$ *iff* $\Lambda_{ki}(\mathbf{x}) = \frac{f_{\mathbf{x}}(\mathbf{x}|E_k)}{f_{\mathbf{x}}(\mathbf{x}|E_i)} > \frac{P_i}{P_k}$ *for all* $i \neq k$*, the decision rule reduces to*

$$\hat{H} = H_k \text{ iff } \ln f_{\mathbf{x}}(\mathbf{x}|E_k) > \ln f_{\mathbf{x}}(\mathbf{x}|E_i) \text{ for all } i \neq k \tag{4.129}$$

Since $\|\mathbf{x}\|^2$ *offers spherical symmetry, one can conclude that*

$$\hat{H} = H_k \text{ iff } \mathbf{x} \text{ is "closest" to } \mathbf{s}_k \tag{4.130}$$

The closeness concept implies that one must define the concept of distance. The distance is defined here as the Euclidean distance for the vector space in which x *is defined. Note that we need to compute* $\|\mathbf{x}-\mathbf{s}_k\|^2$; $k = 0, 1, 2$ *to render a decision. However,*

$$\|\mathbf{x}-\mathbf{s}_k\|^2 = \|\mathbf{x}\|^2 + \|\mathbf{s}_k\|^2 - \langle \mathbf{x}, \mathbf{s}_k \rangle - \langle \mathbf{s}_k, \mathbf{x} \rangle \tag{4.131}$$

where

$$\langle \mathbf{x}, \mathbf{s}_k \rangle = \sum_{j=0}^{127} x_j s_{k,j}^* \tag{4.132}$$

denotes the inner-product between the two vectors. Realizing that $\langle x, s_k \rangle = \langle s_k, x \rangle^*$

$$\|\mathbf{x}-\mathbf{s}_k\|^2 = \|\mathbf{x}\|^2 - \left(2\,\mathrm{Re}\{\langle \mathbf{x}, \mathbf{s}_k \rangle\} - \|\mathbf{s}_k\|^2\right)$$

Hence, in order to maximize $\ln f_{\mathbf{x}}(\mathbf{x}|E_k)$, *one has to minimize* $\|\mathbf{x}-\mathbf{s}_k\|^2$, *which in turn implies maximizing*

$$\mathrm{Re}\{\langle \mathbf{x}, \mathbf{s}_k \rangle\} - \frac{1}{2}\|\mathbf{s}_k\|^2 \tag{4.133}$$

as $\|\mathbf{x}\|^2$ *is not a function of* k *and hence can be omitted from the decision making process. Therefore,*

$$\hat{H} = \arg\max \left\{ \mathrm{Re}\{\langle \mathbf{x}, \mathbf{s}_k \rangle\} - \frac{1}{2}\|\mathbf{s}_k\|^2 \right\} \tag{4.134}$$

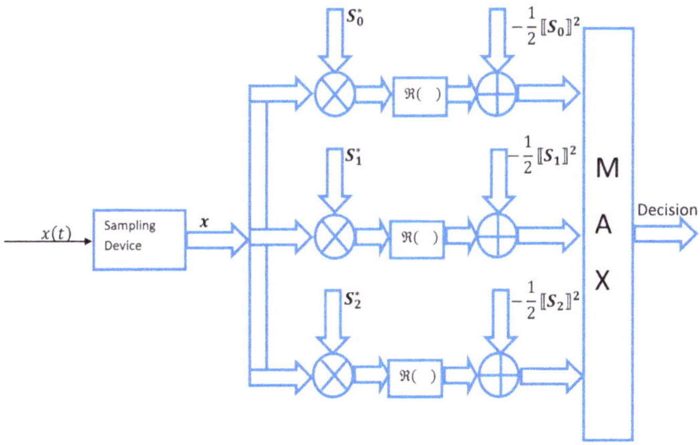

Fig. 4.20 The structure of the optimum receiver

or (since $\|x\|^2$ does not impact the decision making process)

$$\hat{H} = \arg\min \left\{ \|x - \mathbf{s}_k\|^2 \right\} \tag{4.135}$$

The functional block diagram for this detector is depicted in Fig. 4.20 where double lines indicate complex operations. This decision making process requires "correlating" the observed signal x with replicas of the expected signals (s_k; $k = 0, 1, 2$), taking the real parts, adjusting the results by half the "energy" ($\frac{1}{2}\|\mathbf{s}_k\|^2$) of the expected signals, and using the resulting variables to make a decision. Since the signals are real,

$$\hat{H} = \arg\max \left\{ \langle \mathbf{x}, \mathbf{s}_k \rangle - \frac{1}{2} \|\mathbf{s}_k\|^2 \right\} \tag{4.136}$$

To compute the probability of error, we need to characterize the decision variables U_k; $k = 0, 1, 2$ given as follows

$$\begin{cases} U_0 = 0 & H_0 \\ U_1 = \langle \mathbf{x}, \mathbf{s}_1 \rangle - 32a^2 & H_1 \\ U_2 = -\langle \mathbf{x}, \mathbf{s}_1 \rangle - 32a^2 & H_2 \end{cases} \tag{4.137}$$

where we have assumed that

$$\begin{aligned} \|\mathbf{s}_k\|^2 &= \sum_{j=0}^{127} \left| s_{k,j} \right|^2 \approx \frac{1}{t_s} \int_0^1 |s_k(t)|^2 \, dt \\ &= \begin{cases} \frac{a^2}{2} \times 128 = 64a^2 & k = 1 \\ \frac{a^2}{2} \times 128 = 64a^2 & k = 2 \end{cases} \end{aligned} \tag{4.138}$$

We then have

$$P(ER) = \frac{1}{3}\sum_{j=0}^{2} P\left(ER|E_j\right)$$
$$= 1 - \frac{1}{3}\sum_{j=0}^{2} P\left(C|E_j\right) \tag{4.139}$$

where $P\left(C|E_j\right)$ *denotes the probability of making a correct decision given that* E_j *has occurred.* $P\left(C|E_j\right)$ *d is given by*

$$\begin{aligned} P(C|E_0) &= P(\{U_1 < U_0\} \cap \{U_2 < U_0\} |E_0) \\ &= P\left(\left\{\langle \mathbf{x}, \mathbf{s}_1\rangle < 32a^2\right\} \cap \left\{-\langle \mathbf{x}, \mathbf{s}_1\rangle < 32a^2\right\} |E_0\right) \\ &= P\left(\left\{\langle \mathbf{n}, \mathbf{s}_1\rangle < 32a^2\right\} \cap \left\{-\langle \mathbf{n}, \mathbf{s}_1\rangle < 32a^2\right\}\right) \\ &= P\left(-32a^2 < \gamma < 32a^2\right) \end{aligned} \tag{4.140}$$

where

$$\gamma = \langle n, s_1\rangle \tag{4.141}$$

is zero-mean Gaussian random variable. That is, $E\{\gamma|E_0\} = 0$*. Furthermore,*

$$E\left(\gamma^2|E_0\right) = \sigma^2 \|\mathbf{s}_1\|^2 = 64a^2\sigma^2 \tag{4.142}$$

Hence,

$$\begin{aligned} P(C|E_0) &= \frac{1}{\sqrt{128\pi\sigma^2 a^2}} \\ &\quad \times \int_{-32a^2}^{32a^2} \exp\left(-\frac{\gamma^2}{128a^2\sigma^2}\right) d\gamma \\ &= \operatorname{erf}\left(4\sqrt{SNR}\right) \end{aligned} \tag{4.143}$$

where

$$SNR = \frac{a^2/2}{\sigma^2} \tag{4.144}$$

may be viewed as the "signal-to-noise ratio" as it measures the ratio of the "power" of the signal to that of the "noise" and

$$\operatorname{erf}(x) = \frac{2}{\sqrt{\pi}}\int_0^x e^{-t^2} dt \tag{4.145}$$

denotes the error function. Next, we have

$$\begin{aligned} P(C|E_1) &= P(\{U_2 < U_1\} \cap \{U_0 < U_1\} | E_1) \\ &= P\left(\left\{\langle \mathbf{n} + \mathbf{s}_1, -\mathbf{s}_1 \rangle - 32a^2 < \langle \mathbf{n} + \mathbf{s}_1, \mathbf{s}_1 \rangle - 32a^2\right\} \cap \left\{\langle \mathbf{n} + \mathbf{s}_1, \mathbf{s}_1 \rangle - 32a^2 > 0\right\}\right) \\ &= P\left(\gamma > -32a^2\right) \end{aligned} \tag{4.146}$$

Given the description of γ,

$$\begin{aligned} P(C|E_1) &= \frac{1}{\sqrt{128\pi\sigma^2 a^2}} \int_{-32a^2}^{\infty} \exp\left(-\frac{\gamma_1^2}{128a^2\sigma^2}\right) d\gamma \\ &= \frac{1}{2} + \frac{1}{2} \operatorname{erf}\left(4\sqrt{SNR}\right) \end{aligned} \tag{4.147}$$

Finally,

$$\begin{aligned} P(C|E_2) &= P((U_1 < U_2) \cap (U_0 < U_2) | E_2) \\ &= P\left(\langle \mathbf{n} + \mathbf{s}_1, \mathbf{s}_1 \rangle - 32a^2 < \langle \mathbf{n} - \mathbf{s}_1, -\mathbf{s}_1 \rangle - 32a_2^2\right) \cap \langle \mathbf{n} - \mathbf{s}_1, -\mathbf{s}_1 \rangle - 32a^2 > 0) \\ &= P\left(\gamma < 32a^2\right) \\ &= \frac{1}{\sqrt{128\pi\sigma^2 a^2}} \int_{-\infty}^{32a^2} \exp\left(-\frac{\gamma^2}{128a^2\sigma^2}\right) d\gamma \\ &= \frac{1}{2} + \frac{1}{2} \operatorname{erf}\left(4\sqrt{SNR}\right) \end{aligned} \tag{4.148}$$

We then have (equally-likely scenario)

$$P(ER) = \frac{2}{3} - \frac{2}{3} \operatorname{erf}\left(4\sqrt{SNR}\right) \tag{4.149}$$

which can also be written as

$$P(ER) = \frac{2}{3} - \frac{2}{3} \operatorname{erf}\left(\frac{d_{\min}}{\sqrt{8\sigma^2}}\right) \tag{4.150}$$

where $d_{\min} = \sqrt{64a^2} = \|\mathbf{s}_1 - \mathbf{s}_0\| = \|\mathbf{s}_2 - \mathbf{s}_0\|$ *may be viewed as the "minimum" distance between any pair of signals in the signal set.* $P(ER)$ *is depicted in Fig. 4.21 for a large range of SNR. It is obvious that the error rate decreases exponentially with SNR. Furthermore, for SNR of as little as 0.4, one could expect an error rate of about* 1.5×10^{-4}. *Also, we noticed that at 0 SNR (when there is no signal strength), the error rate increases to 2/3 (the chance of guessing the correct case reduces to 1/3, which is the chance of randomly choosing 1 out of the 3 equally likely outcomes).*

Previous example underscored an important point; that is, the success rate of the detection mechanism in an AWGN environment depends only on the "proximity" of the signals to each other in the signal space and not on the shape of the signals,

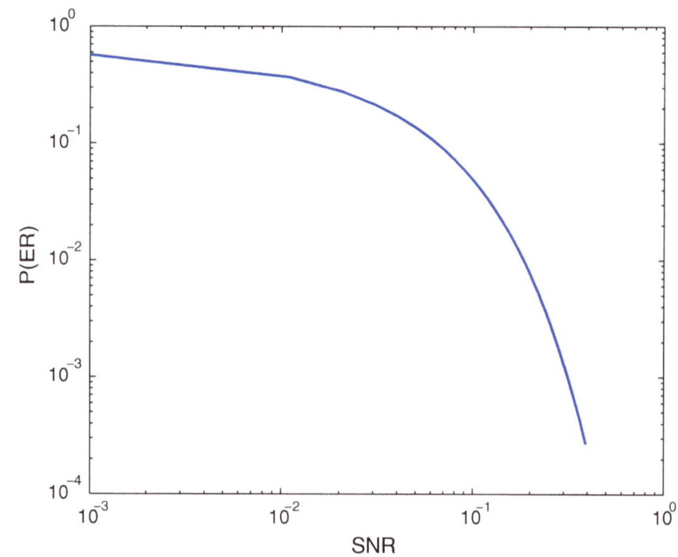

Fig. 4.21 The probability of error as a function of SNR

although as will be seen shortly, the success rate is a function of the relative positioning of the signals in the signal space. The above assertion is a direct consequence of modeling the noise as AWGN, which offers spherical symmetry in the signal space (note that the AWGN has a pdf that is a function of $\|\mathbf{n}\|^2$ where $\mathbf{n}$ is the Gaussian noise vector). In fact, the receiver metric (4.135) in the example above clearly shows that the "Euclidean distances" between the observed signal and the candidate signals (i.e., $\|\mathbf{x}-\mathbf{s}_k\|^2$ for all k) form the decision variables. It is not difficult to see that $\|\mathbf{x}-\mathbf{s}_k\|^2$ is invariant to rotation and "spatial" translation as it is a norm (length) in the Euclidean space (a vector does not change its length when it is moved or rotated). Hence, the error rate is invariant to a spatial (in the signal space) rotation and translation of the signal set. However, the relative positioning of the signals in the signal set does matter, which further implies that the error rate is solely a function of the relative geometry of the signal set (known as signal constellation) and noise variance.

To see this point, let $\left\{\mathbf{s}'_k;\, k=0,1,2\right\}$ be such that

$$\|\mathbf{x}-\mathbf{s}_k\|^2 = \left\|\mathbf{x}'-\mathbf{s}'_k\right\|^2 \text{ for all } k, \tag{4.151}$$

where $\mathbf{x}'$ is a transformed (in location and orientation, but not length) version of $\mathbf{x}$, given by

$$\mathbf{x}' = \mathbf{x}^{(r)} + b \tag{4.152}$$

with $\mathbf{x}^{(r)}$ denoting the rotated version of $\mathbf{x}$ (and hence $\|\mathbf{x}\| = \left\|\mathbf{x}^{(r)}\right\|$) and b denoting a scalar. Then, (4.151) holds if

$$\mathbf{s}_k' = \mathbf{s}_k^{(r)} + b \text{ for all } k. \tag{4.153}$$

This is due to the fact that

$$\left\|\mathbf{x}^{(r)} - \mathbf{s}_k^{(r)}\right\|^2 = \left\|(\mathbf{x} - \mathbf{s}_k)^{(r)}\right\|^2 = \|\mathbf{x} - \mathbf{s}_k\|^2.$$

Note that the same transformation does not change the variance of the noise. That is,

$$E\left\{\|\mathbf{n}\|^2\right\} - E^2\left\{\|\mathbf{n}\|\right\} = E\left\{\left\|\mathbf{n}'\right\|^2\right\} - E^2\left\{\left\|\mathbf{n}'\right\|\right\} \tag{4.154}$$

when

$$\mathbf{n}' = \mathbf{n}^{(r)} + b$$

For $\left\{\mathbf{s}_k'; k = 0, 1, 2\right\}$ and $\{\mathbf{s}_k; k = 0, 1, 2\}$ to satisfy (4.153), the two signal sets must have the same geometry, but not the same location nor orientation. Note that (4.153) also preserves the minimum "distance" (see (4.150) in the previous example) of the signal constellation, thereby keeping the error rate the same.

The above discussion focused on AWGN noise, and as we have seen in this text, the Gaussian assumption is usually invalid when one considers biological signals. However, for scenarios where a large number of samples of biological signals are taken, and the observation is perhaps a large sum of (>100) those samples, one can invoke the central limit theorem to assert Gaussian property. In those cases, then, one can use the above analysis as the basis for the hypothesis testing purposes.

Let us now consider a scenario that the two of the signals are orthogonal.

Example 11 In an experiment, we are collecting data with a biological origin. The signal is corrupted by AWGN due to electronic noise $n(t)$. We expect to either get no signal, a 30 Hz, a_1 volt sine signal (phase is important), or a 45 Hz, a_2 volt cosine wave. Each scenario corresponds to a specific condition that may be present in the subject under examination. Design a detector that minimizes the probability of making a wrong diagnosis. The decision must be made after observing a 1 s sample of the signal. Assume that the sampling rate is 128 samples/s and that the variance of the samples of the AWGN is σ^2. For the decision rule that was obtained above, give an expression for the probability of error. Assume that the three scenarios are equally-likely.

Solution: *Without the loss of generality, we consider the following observation interval*

$$\Gamma = [0, t_s, 2t_s, \ldots, 127t_s] \tag{4.155}$$

where $t_s = \frac{1}{128}$. *Since the observed interval is 1 s and the sampling rate is 128, all vectors are now 128in length. Assume that the observed signal is* $x(t)$. *We also have the corresponding signals in time domain as* $s_k(t) = \begin{cases} 0 & k=0 \\ a_1 \sin(60\pi t) & k=1 \\ a_2 \cos(90\pi t) & k=2 \end{cases}$.

Clearly, we see three scenarios as follows:

$$\begin{aligned} H_0 &\Leftrightarrow \mathbf{x} = \mathbf{n} + \mathbf{s}_0 \\ H_1 &\Leftrightarrow \mathbf{x} = \mathbf{n} + \mathbf{s}_1 \\ H_2 &\Leftrightarrow \mathbf{x} = \mathbf{n} + \mathbf{s}_2 \end{aligned} \tag{4.156}$$

where

$$\begin{aligned} \mathbf{x} &= [x(0), x(t_s), x(2t_s), \ldots, x(127t_s)]^{\mathsf{T}} \\ &= [x_0, x_1, \ldots, x_{127}]^{\mathsf{T}} \end{aligned} \tag{4.157}$$

$$\begin{aligned} \mathbf{n} &= [n(0), n(t_s), n(2t_s), \ldots, n(127t_s)]^{\mathsf{T}} \\ &= [n_0, n_1, \ldots, n_{127}]^{\mathsf{T}}, \end{aligned} \tag{4.158}$$

$$\mathbf{s}_0 = [0, 0, \ldots, 0]^{\mathsf{T}}, \tag{4.159}$$

$$\begin{aligned} \mathbf{s}_1 &= \left[0, a_1 \sin\left(2\pi \frac{30}{128}\right), a_1 \sin\left(4\pi \frac{30}{128}\right), \ldots, a_1 \sin\left(254\pi \frac{30}{128}\right)\right]^{\mathsf{T}} \\ &= \left[s_{1,0}.s_{1,1}, \ldots, s_{1,127}\right]^{\mathsf{T}}, \end{aligned} \tag{4.160}$$

and

$$\begin{aligned} \mathbf{s}_2 &= \left[0, a_2 \cos\left(2\pi \frac{45}{128}\right), a_2 \cos\left(4\pi \frac{45}{128}\right), \ldots, a_2 \cos\left(254\pi \frac{45}{128}\right)\right]^{\mathsf{T}} \\ &= \left[s_{2,0}.s_{2,1}, \ldots, s_{2,127}\right]^{\mathsf{T}}. \end{aligned} \tag{4.161}$$

In order to go any further, once again, we need $\left\{f_{\mathbf{x}}\left(\mathbf{x}|E_j\right);\ j = 0, 1, 2\right\}$. *We note that under all three hypotheses, the observed vector x is Gaussian. Hence, the mean and covariance matrices in the presence of signal are needed. Since adding a deterministic value (signals) to a random vector does not change its covariance matrix, x under all three hypotheses has the same covariance matrix. Similar to the previous example, the covariance matrix for x is* $\sigma^2 I$. *Also,* $E\{\mathbf{x}|E_0\} = 0$, $E\{\mathbf{x}|E_1\} = s_1$, *and* $E\{\mathbf{x}|E_2\} = s_2$. *Hence,*

$$
\begin{aligned}
H_0 \Leftrightarrow \quad f_{\mathbf{x}}(\mathbf{x}|E_0) &= \frac{1}{(2\pi\sigma^2)^{64}} \exp\left\{-\sum_{j=0}^{127} \frac{x_j^2}{2\sigma^2}\right\} \\
H_1 \Leftrightarrow f_{\mathbf{x}}(\mathbf{x}|E_1) &= \frac{1}{(2\pi\sigma^2)^{64}} \exp\left\{-\sum_{j=0}^{127} \frac{(x_j - s_{1,j})^2}{2\sigma^2}\right\} \\
H_2 \Leftrightarrow f_{\mathbf{x}}(\mathbf{x}|E_2) &= \frac{1}{(2\pi\sigma^2)^{64}} \exp\left\{-\sum_{j=0}^{127} \frac{(x_j - s_{2,j})^2}{2\sigma^2}\right\}
\end{aligned} \tag{4.162}
$$

This can be equivalently written as

$$
\begin{aligned}
H_0 \Leftrightarrow \quad f_{\mathbf{x}}(\mathbf{x}|E_0) &= \frac{1}{(2\pi\sigma^2)^{64}} \exp\left\{-\frac{\|\mathbf{x}-\mathbf{s}_0\|^2}{2\sigma^2}\right\} \\
H_1 \Leftrightarrow \quad f_{\mathbf{x}}(\mathbf{x}|E_1) &= \frac{1}{(2\pi\sigma^2)^{64}} \exp\left\{-\frac{\|\mathbf{x}-\mathbf{s}_1\|^2}{2\sigma^2}\right\} \\
H_{M-1} \Leftrightarrow f_{\mathbf{x}}(\mathbf{x}|E_2) &= \frac{1}{(2\pi\sigma^2)^{64}} \exp\left\{-\frac{\|\mathbf{x}-\mathbf{s}_2\|^2}{2\sigma^2}\right\}
\end{aligned} \tag{4.163}
$$

where

$$
\|\mathbf{x}\|^2 = \sum_{j=0}^{127} x_j^2 \tag{4.164}
$$

with $\|\mathbf{x}\|^2$ *denoting the norm of the vector* x. *Since* $\ln(.)$ *is a monotonically increasing function of its argument and pdfs are all non-negative, we can take the* ln *of the pdfs. That is,*

$$
\begin{aligned}
H_0 \Leftrightarrow \ln f_{\mathbf{x}}(\mathbf{x}|E_0) &= -\frac{\|\mathbf{x}-\mathbf{s}_0\|^2}{2\sigma^2} + c \\
H_1 \Leftrightarrow \ln f_{\mathbf{x}}(\mathbf{x}|E_1) &= -\frac{\|\mathbf{x}-\mathbf{s}_1\|^2}{2\sigma^2} + c \\
H_2 \Leftrightarrow \ln f_{\mathbf{x}}(\mathbf{x}|E_2) &= -\frac{\|\mathbf{x}-\mathbf{s}_2\|^2}{2\sigma^2} + c
\end{aligned} \tag{4.165}
$$

The constant c *does not enter into the decision making process, and hence can be ignored. Using* $\hat{H} = H_k$ *iff* $\Lambda_{ki}(\mathbf{x}) = \frac{f_{\mathbf{x}}(\mathbf{x}|E_k)}{f_{\mathbf{x}}(\mathbf{x}|E_i)} > \frac{P_i}{P_k}$ *for all* $i \neq k$, *the decision rule reduces to*

$$
\hat{H} = H_k \text{ iff } \ln f_{\mathbf{x}}(\mathbf{x}|E_k) > \ln f_{\mathbf{x}}(\mathbf{x}|E_i) \text{ for all } i \neq k \tag{4.166}
$$

We also have the following simplification:

$$
\hat{H} = H_k \text{ iff } \mathbf{x} \text{ is ``closest'' to } \mathbf{s}_k \tag{4.167}
$$

Note that we need to compute $\|\mathbf{x} - \mathbf{s}_k\|^2$; $k = 0, 1, 2$ *to render a decision. Similar to the previous example, the decision rule reduces to*

$$
\hat{H} = \arg\min \left\{\|x - \mathbf{s}_k\|^2\right\}. \tag{4.168}
$$

The functional block diagram for this detector was given in Fig. 4.20. Similar to the previous example,

$$\hat{H} = \arg\max \left\{ \langle \mathbf{x}, \mathbf{s}_k \rangle - \frac{1}{2} \|\mathbf{s}_k\|^2 \right\} \tag{4.169}$$

To compute the probability of error, we need to characterize the decision variables $U_k; k = 0, 1, 2$ *given as follows*

$$\begin{cases} U_0 = 0 & H_0 \\ U_1 = \langle \mathbf{x}, \mathbf{s}_1 \rangle - 32a_1^2 & H_1 \\ U_2 = \langle \mathbf{x}, \mathbf{s}_2 \rangle - 32a_2^2 & H_2 \end{cases}$$

where we have assumed that

$$\|\mathbf{s}_k\|^2 = \begin{cases} \frac{a_1^2}{2} \times 128 = 64a_1^2 & k = 1 \\ \frac{a_2^2}{2} \times 128 = 64a_2^2 & k = 2 \end{cases}$$

We then have

$$\begin{aligned} P(ER) &= \frac{1}{3} \sum_{j=0}^{2} P\left(ER|E_j\right) \\ &= 1 - \frac{1}{3} \sum_{j=0}^{2} P\left(C|E_j\right) \end{aligned} \tag{4.170}$$

where $P\left(C|E_j\right)$

$$\begin{aligned} P(C|E_0) &= P\left(\{U_1 < U_0\} \cap \{U_2 < U_0\} |E_0\right) \\ &= P\left(\left\{\langle \mathbf{x}, \mathbf{s}_1 \rangle < 32a_1^2\right\} \cap \left\{\langle \mathbf{x}, \mathbf{s}_2 \rangle < 32a_2^2\right\} |E_0\right) \\ &= P\left(\left\{\langle \mathbf{n}, \mathbf{s}_1 \rangle < 32a_1^2\right\} \cap \left\{\langle \mathbf{n}, \mathbf{s}_2 \rangle < 32a_2^2\right\}\right) \\ &= P\left(\left\{\gamma_1 < 32a_1^2\right\} \cap \left\{\gamma_2 < 32a_2^2\right\}\right) \end{aligned} \tag{4.171}$$

with $\gamma_1 = \langle n, s_1 \rangle$ *and* $\gamma_2 = \langle n, s_2 \rangle$ *denoting a pair of zero-mean, correlated Gaussian random variables. That is,* $E\{\gamma_1|E_0\} = E\{\gamma_2|E_0\} = 0$. *Furthermore,*

$$E\left(\gamma_1^2|E_0\right) = \sigma^2 \|\mathbf{s}_1\|^2 = 64a_1^2\sigma^2 \tag{4.172}$$

and

$$E\left(\gamma_2^2|E_0\right) = \sigma^2 \|\mathbf{s}_2\|^2 = 64a_2^2\sigma^2. \tag{4.173}$$

Finally,

$$E\{\gamma_1\gamma_2|E_0\} = \sigma^2\langle \mathbf{s}_1, \mathbf{s}_2\rangle = 128\sigma^2 \int_0^1 s_1(t)\, s_2(t)\, dt \\ \approx 0 \tag{4.174}$$

which renders the two Gaussian random variables uncorrelated (and, hence, independent). Hence,

$$\begin{aligned} P(C|E_0) &= \frac{1}{128\pi\sigma^2 a_1 a_2} \\ &\quad \times \int_{-\infty}^{32a_1^2} \exp\left(-\frac{\gamma_1^2}{128a_1^2\sigma^2}\right) d\gamma_1 \\ &\quad \times \int_{-\infty}^{32a_2^2} \exp\left(-\frac{\gamma_2^2}{128a_2^2\sigma^2}\right) d\gamma_2 \\ &= \frac{1}{4}\left(1+\mathrm{erf}\left(4\sqrt{SNR_1}\right)\right)\left(1+\mathrm{erf}\left(4\sqrt{SNR_2}\right)\right) \end{aligned} \tag{4.175}$$

where $SNR_i = \frac{a_i^2/2}{\sigma^2}$ *may be viewed as the signal-to-noise ratio as it measures the ratio of the power of the signal* i *to that of the noise.*

$$\begin{aligned} P(C|E_1) &= P(\{U_2 < U_1\} \cap \{U_0 < U_1\}|E_1) \\ &= P(\left\{\langle \mathbf{n}+\mathbf{s}_1, \mathbf{s}_2\rangle - 32a_2^2 < \langle \mathbf{n}+\mathbf{s}_1, \mathbf{s}_1\rangle - 32a_1^2\right\} \\ &\quad \cap \left\{\langle \mathbf{n}+\mathbf{s}_1, \mathbf{s}_1\rangle - 32a_1^2 > 0\right\}) \\ &= P\left(\left\{\gamma_2 - \gamma_1 < 32\left(a_1^2 + a_2^2\right)\right\} \cap \left\{\gamma_1 > -32a_1^2\right\}\right). \end{aligned} \tag{4.176}$$

Given the description of γ_1 *and* γ_2*, we have*

$$\begin{aligned} P(C|E_1) &= \frac{1}{128\pi\sigma^2 a_1 a_2} \\ &\quad \times \int_{-32a_1^2}^{\infty} \exp\left(-\frac{\gamma_1^2}{128a_1^2\sigma^2}\right) d\gamma_1 \\ &\quad \times \int_{-\infty}^{32(a_2^2+a_1^2)+\gamma_1} \exp\left(-\frac{\gamma_2^2}{128a_2^2\sigma^2}\right) d\gamma_2 \\ &= \frac{1}{2\sqrt{\pi}} \int_{-4\sqrt{SNR_1}}^{\infty} e^{-t^2}\left(1+\mathrm{erf}\left(\frac{4SNR_2 + 4SNR_1 + \sqrt{SNR_1}t}{\sqrt{SNR_2}}\right)\right) dt \end{aligned} \tag{4.177}$$

Finally,

$$
\begin{aligned}
P\left(C|E_2\right) &= P\left(\{U_1 < U_2\} \cap \{U_0 < U_2\} | E_2\right) \\
&= P(\left\{\langle \mathbf{n} + \mathbf{s}_2, \mathbf{s}_1\rangle - 32a_1^2 < \langle \mathbf{n} + \mathbf{s}_2, \mathbf{s}_2\rangle - 32a_2^2\right\} \\
&\quad \cap \left\{\langle \mathbf{n} + \mathbf{s}_2, \mathbf{s}_2\rangle - 32a_2^2 > 0\right\}) \qquad (4.178)\\
&= P\left(\left\{\gamma_1 - \gamma_2 < 32\left(a_2^2 + a_1^2\right)\right\} \cap \left\{\gamma_2 > -32a_2^2\right\}\right) \\
&= \frac{1}{128\pi\sigma^2 a_1 a_2} \times \int_{-32a_2^2}^{\infty} \exp\left(-\frac{\gamma_2^2}{128a_2^2\sigma^2}\right) d\gamma_2 \\
&\quad \times \int_{-\infty}^{32\left(a_2^2+a_1^2\right)+\gamma_2} \exp\left(-\frac{\gamma_1^2}{128a_1^2\sigma^2}\right) d\gamma_1 \qquad (4.179)\\
&= \frac{1}{2\sqrt{\pi}} \int_{-4\sqrt{SNR_2}}^{\infty} e^{-t^2} \left(1 + \operatorname{erf}\left(\frac{4SNR_2 + 4SNR_1 + \sqrt{SNR_2}t}{\sqrt{SNR_1}}\right)\right) dt
\end{aligned}
$$

Then,

$$
\begin{aligned}
P\left(ER\right) = 1 - \frac{1}{3}\Bigg(&\frac{1}{4} + \frac{1}{4}\operatorname{erf}\left(4\sqrt{SNR_1}\right) + \frac{1}{4}\operatorname{erf}\left(4\sqrt{SNR_2}\right) \\
&+ \frac{1}{4}\operatorname{erf}\left(4\sqrt{SNR_1}\right)\operatorname{erf}\left(4\sqrt{SNR_2}\right) \qquad (4.180)\\
&+ \frac{1}{2\sqrt{\pi}} \int_{-4\sqrt{SNR_2}}^{\infty} e^{-t^2} \left(1 + \operatorname{erf}\left(\frac{4SNR_2 + 4SNR_1 + \sqrt{SNR_2}t}{\sqrt{SNR_1}}\right)\right) dt \\
&+ \frac{1}{2\sqrt{\pi}} \int_{-4\sqrt{SNR_1}}^{\infty} e^{-t^2} \left(1 + \operatorname{erf}\left(\frac{4SNR_2 + 4SNR_1 + \sqrt{SNR_1}t}{\sqrt{SNR_2}}\right)\right) dt\Bigg) \qquad (4.181)
\end{aligned}
$$

As can be see, the probability of error remains a function of the two SNR's. $P\left(ER\right)$ is depicted in Fig. 4.22 as a function of SNR_1 and SNR_2. It becomes immediately obvious that an increase in both SNR's result in a small error rate. For $SNR_1 \to 0$ and $SNR_2 \to 0$, the probability of error is simply the probability of guessing which one of the three hypothesis is correct, which results in $P\left(ER\right) = \frac{2}{3}$. For $SNR_1 \to 0$ and large SNR_2, the probability of error is simply guessing between H_0 and H_1 (each with the probability of occurring of $\frac{1}{3}$), which results in $P\left(ER\right) = \frac{1}{3} \times \frac{1}{2} + \frac{1}{3} \times \frac{1}{2} = \frac{1}{3}$. The same is true when $SNR_2 \to 0$ and SNR_1 is large. Interestingly enough, for SNR_1 and SNR_2 greater than only 0.5, the error rate becomes smaller than 1 %. This is quite encouraging as it underscores the high success rate of the detector in the face of strong noise.

The previous example presented a more complicated case where the error rate seems to be a function of two SNR's. This is due to the fact that we have two signals with different power levels. Nonetheless, for large signal power levels (i.e., $SNR_1 = SNR_2 \gg 1$), we have the following simplification:

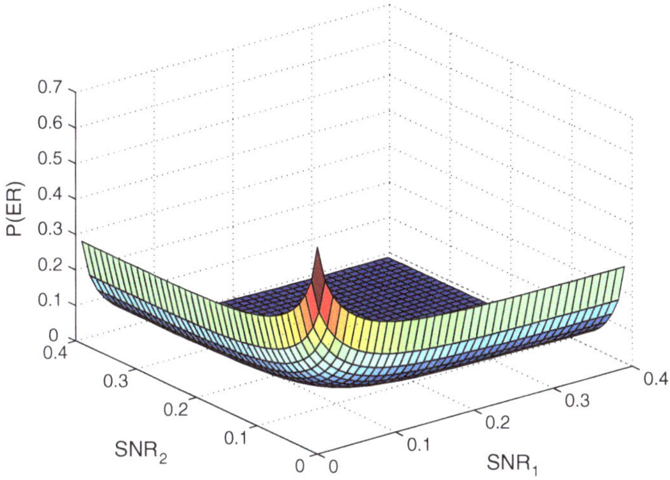

Fig. 4.22 The probability of error as the function of two SNR's

$$P\left(ER\right) \approx \frac{2}{3} - \frac{2}{3}\,\text{erf}\left(4\sqrt{SNR}\right) \tag{4.182}$$

$$= \frac{2}{3} - \frac{2}{3}\,\text{erf}\left(\frac{d_{\min}}{\sqrt{8\sigma^2}}\right) \tag{4.183}$$

where

$$d_{\min} = \|\mathbf{s}_1 - \mathbf{s}_0\| = \|\mathbf{s}_2 - \mathbf{s}_0\| = \sqrt{64a^2}$$

(note that $SNR_1 = SNR_2 = \frac{a^2/2}{\sigma^2}$ implies $a_1 = a_2$). Also, note that the distance between $\mathbf{s}_2$ and $\mathbf{s}_1$ is given by

$$d_{12} = \|\mathbf{s}_2 - \mathbf{s}_1\| = \sqrt{2}d_{\min} > d_{\min}$$

since

$$\langle \mathbf{s}_1, \mathbf{s}_2 \rangle = 0 \text{ and } \|\mathbf{s}_2 - \mathbf{s}_1\|^2 = \|\mathbf{s}_2\|^2 + \|\mathbf{s}_2\|^2$$

This implies that the minimum distance is achieved when one considers $\mathbf{s}_1$ and $\mathbf{s}_0$ or $\mathbf{s}_2$ and $\mathbf{s}_0$. Figure 4.23 depicts the exact and approximated error rates. As can be seen, the approximation is exceptionally accurate for a large range of SNRs, including small SNR values. In fact, one can say that the approximation is valid for almost all values of SNR of interest. Finally, if one assumes that SNR_2 is very large, then the signals $\mathbf{s}_1$ and $\mathbf{s}_0$ are at $d_{\min}$ distance from each other. Effectively, we now have only 2 possible cases,each having $\frac{1}{3}$ probability of occurrence. That is, H_0 and H_1 are the only ones we can make a decision in favor of (even though each has the probability of occurrence of $\frac{1}{3}$) as H_3 becomes the impossible event ($a_2 \to \infty$). In that event,

$$P\left(ER\right) \approx \frac{1}{3} - \frac{1}{3}\,\text{erf}\left(4\sqrt{SNR_1}\right) \tag{4.184}$$

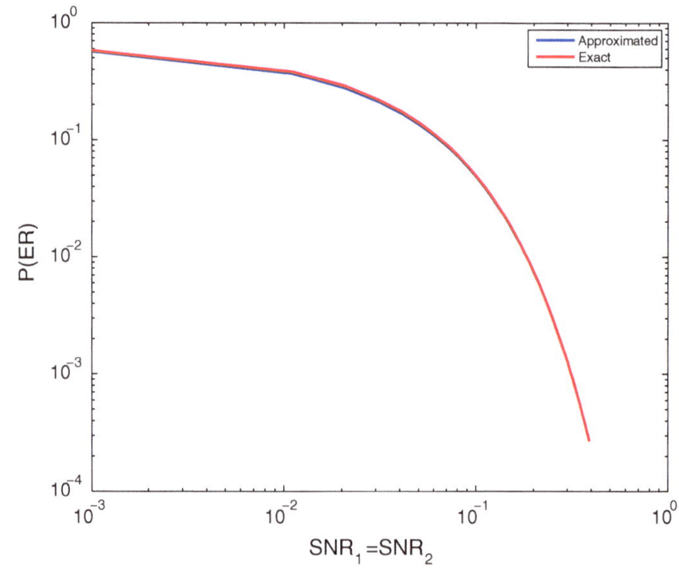

Fig. 4.23 The probability of error when the two SNR's are identical

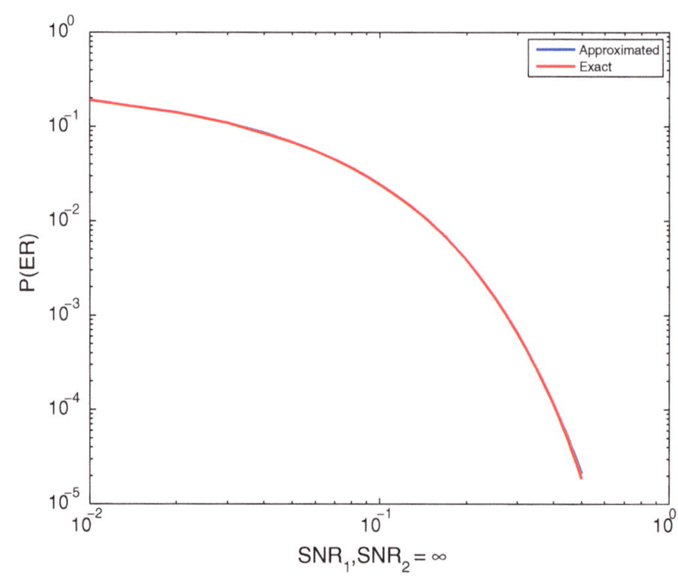

Fig. 4.24 The probability of error as a function of SNR_1 while $SNR_2 = \infty$

Figure 4.24 depicts the exact and approximated (given above) error rates. As expected, we have a perfect match. Once again, we can see that the error rate is not a function of the specific signaling scheme, but rather a function of the minimum distance of the signal constellation. Even in the case of small SNR values, one can argue that the error rate is a function of the minimum distance and not the actual shape of the signals. That is, any set of signals that yield the above SNR values will yield the same error rate provided that the signal space representations of the two signal sets are merely a rotation and/or spatial translation apart. Reader has to be cautious about using these results indiscriminately as the format of the error rate in the above example was a direct consequence of the signal space representation of the signals (two orthogonal signals with the third located in the origin).

In any scenarios where additive noise offers spherical symmetry, a similar result ensues. Let us examine this principle with yet another example where spherical symmetry is satisfied in a norm other than L_2 norm.

Example 12 In an experiment, we are to determine the presence or absence of an anomaly. The signal collected is shown to exhibit the following properties. For a healthy individual, the signal amplitude assumes a double-exponential (Laplacian) characteristics with variance σ^2. In the present of the anomaly, the signal is a double-exponential random variable (variance of σ^2) with a sinusoidal mean whose frequency and amplitude are 10 Hz and 0.1 V, respectively. The chance of the anomaly to be present is 50 %. Design a detector that observes a 1 s interval (assume 256 samples/s sampling rate) of the signal and renders an optimum decision. Assume that samples which are taken in this manner are independent. Compute the probability of error for the detector you have designed.

Solution: *Let the observed signal be* x. *Hence,*

$$\begin{aligned} H_0 &\Leftrightarrow \mathbf{x} = \mathbf{x}_0 \\ H_1 &\Leftrightarrow \mathbf{x} = \mathbf{x}_1 \end{aligned} \tag{4.185}$$

where

$$f_{\mathbf{x}_0}(\mathbf{x}) = \left(\frac{\alpha}{2}\right)^N \exp\left\{-\alpha \sum_{j=0}^{N-1} |x_k|\right\} \tag{4.186}$$

and

$$f_{\mathbf{x}_1}(\mathbf{x}) = \left(\frac{\alpha}{2}\right)^N \exp\left\{-\alpha \sum_{k=0}^{N-1} |x_k - s_k|\right\} \tag{4.187}$$

with $\alpha = \sqrt{\frac{2}{\sigma^2}} > 0$ *and* $s_k = 0.1 \sin\left(\frac{20\pi k}{256}\right)$. *Furthermore,* N *denotes the number of samples (which is 256). Note that* s_k *is the* k*th sample of the sinusoid at 256 Hz sampling rate. From the problem statement, we have* $p_0 = 0.5$ *and* $p_1 = 0.5$. *Since the MAP-HT is used here, the decision rule reduces to (note that* $f_{\mathbf{x}}(\mathbf{x}|H_0) = f_{\mathbf{x}_0}(\mathbf{x})$ *and* $f_{\mathbf{x}}(\mathbf{x}|H_1) = f_{\mathbf{x}_1}(\mathbf{x})$*)*

$$f_{\mathbf{x}_1}(\mathbf{x})\, p_1 \underset{H_0}{\overset{H_1}{\gtrless}} f_{\mathbf{x}_0}(\mathbf{x})\, p_0 \tag{4.188}$$

which results in the following log-likelihood function:

$$\Lambda(\mathbf{x}) = \ln\left(\frac{f_{\mathbf{x}_1}(\mathbf{x})}{f_{\mathbf{x}_0}(\mathbf{x})}\right) \underset{H_0}{\overset{H_1}{\gtrless}} \ln\left\{\frac{p_0}{p_1}\right\} \Rightarrow \tag{4.189}$$

$$\sum_{k=0}^{N-1} \{\alpha |x_k - s_k| - \alpha |x_k|\} \underset{H_1}{\overset{H_0}{\gtrless}} \Gamma_{th} \tag{4.190}$$

where $\Gamma_{th} = \ln\left\{\frac{p_1}{p_0}\right\} = 0$. *We, then, have*

$$\sum_{k=0}^{N-1} |x_k - s_k| \underset{H_1}{\overset{H_0}{\gtrless}} \sum_{k=0}^{N-1} \alpha |x_k| \tag{4.191}$$

Note that, if one considers the L_1 *norm, then*

$$\|\mathbf{x} - \mathbf{s}\|_1 \underset{H_1}{\overset{H_0}{\gtrless}} \|\mathbf{x}\|_1 \tag{4.192}$$

where

$$\|\mathbf{x}\|_1 = \sum_{k=0}^{N-1} |x_k| \tag{4.193}$$

denotes the L_1 *norm of* x. *It is not hard to see that the above pdf satisfies the "spherical symmetry" in* L_1 *space. The above can be further implies as (we have reintroduced a non-zero threshold so that a general result can be obtained)*

$$g(\mathbf{x}) \underset{H_1}{\overset{H_0}{\gtrless}} \Gamma_{th} \tag{4.194}$$

where

$$g(\mathbf{x}) = \sum_{k=0}^{N-1} g_k(x_k) \tag{4.195}$$

with

$$g_k(x_k) = \alpha |x_k - s_k| - \alpha |x_k|$$

One can then conclude that

$$\begin{aligned} P(ER) &\approx p_0 P(ER|H_0) + p_1 P(ER|H_1) \\ &\approx p_0 \int_{-\infty}^{\Gamma_{th}} f_g(g|H_0)\,dg + p_1 \int_{\Gamma_{th}}^{\infty} f_g(g|H_1)\,dg \end{aligned} \tag{4.196}$$

This clearly requires one to obtain the pdf of $g(\mathbf{x})$, *which is a tedious task. One, however, can take advantage of an approximation; that is,* $g(\mathbf{x})$ *is formed by summing a large number of* ($N = 256$) *i.i.d random variables (i.e.,* $g_k(x_k)$*). Hence, one can invoke the central limit theorem, which allows us to model* $g(\mathbf{x})$ *under the two hypotheses as a Gaussian random variable. Let*

$$\begin{aligned} m_0 &= E\{g(\mathbf{x})|H_0\} = \sum_{k=0}^{N-1} E\{g_k(x_k)|H_0\} \\ &= \sum_{k=0}^{N-1} m_{0,k}, \end{aligned} \tag{4.197}$$

$$\begin{aligned} m_1 &= E\{g(\mathbf{x})|H_1\} = \sum_{k=0}^{N-1} E\{g_k(x_k)|H_1\} \\ &= \sum_{k=0}^{N-1} m_{1,k}, \end{aligned} \tag{4.198}$$

$$\begin{aligned} \sigma_{g,0}^2 &= E\left\{(g(\mathbf{x}) - m_0)^2 |H_0\right\} \\ &= \sum_{k=1}^{N} E\left\{\left(g_k(x_k) - m_{0,k}\right)^2 |H_0\right\} \\ &= \sum_{k=1}^{N} \sigma_{g,0,k}^2, \end{aligned} \tag{4.199}$$

and

$$\begin{aligned} \sigma_{g,1}^2 &= E\left\{(g(\mathbf{x}) - m_1)^2 |H_1\right\} \\ &= \sum_{k=0}^{N-1} E\left\{\left(g_k(x_k) - m_{1,k}\right)^2 |H_1\right\} \end{aligned}$$

$$= \sum_{k=0}^{N-1} \sigma_{g,1,k}^2 \tag{4.200}$$

Furthermore, we have (note that $\int_{-\infty}^{\infty} |x| \exp(-\alpha |x|)\, dx = \frac{2}{\alpha^2}$ *and* $\int_{-\infty}^{\infty} |x - s_k| \exp(-\alpha |x|)\, dx = \frac{2(-as_k + e^{\alpha s_k})}{\alpha^2}$*)*

$$m_{0,k} = \frac{\alpha}{2} \int_{-\infty}^{\infty} g_k(x_k) \exp(-\alpha |x_k|)\, dx_k, \tag{4.201}$$

$$m_{1,k} = \frac{\alpha}{2} \int_{-\infty}^{\infty} g_k(x_k) \exp(-\alpha |x_k - s_k|)\, dx_k, \tag{4.202}$$

$$\sigma_{g,0,k}^2 = \frac{\alpha}{2} \int_{-\infty}^{\infty} g_k^2(x_k) \exp(-\alpha |x_k|)\, dx_k - m_{0,k}^2, \tag{4.203}$$

and

$$\sigma_{g,1,k}^2 = \frac{\alpha}{2} \int_{-\infty}^{\infty} g_k^2(x_k) \exp(-\alpha |x_k - s_k|)\, dx_k - m_{1,k}. \tag{4.204}$$

The last two equations are obtained by using the fact that the variance of a sum of N independent variables is the sum of the variances of those random variables. Using the Gaussian assumption, we then have

$$\begin{aligned} P(ER) \approx & \frac{p_0}{\sqrt{2\pi\sigma_{g,0}^2}} \int_{-\infty}^{\Gamma_{th}} \exp\left(-\frac{(x - m_0)^2}{2\sigma_{g,0}^2}\right) dx \\ & + \frac{p_1}{\sqrt{2\pi\sigma_{g,1}^2}} \int_{\Gamma_{th}}^{\infty} \exp\left(-\frac{(x - m_1)^2}{2\sigma_{g,1}^2}\right) dx \end{aligned} \tag{4.205}$$

which leads to

$$\begin{aligned} P(ER) \approx & \frac{p_0}{2}\left(1 - \operatorname{erf}\left(\frac{m_0 - \Gamma_{th}}{\sqrt{2}\sigma_{g,0}}\right)\right) \\ & + \frac{p_1}{2}\left(1 + \operatorname{erf}\left(\frac{m_1 - \Gamma_{th}}{\sqrt{2}\sigma_{g,1}}\right)\right) \end{aligned} \tag{4.206}$$

Although the error rate is the function of 4 parameters, the dependency on the SNR only is not clear at all. In fact, m_0, m_1, $\sigma_{g,0}^2$, *and* $\sigma_{g,1}^2$ *are all functions of* s_k'*s.*

$P(ER)$*'s for various values of* $SNR = \frac{0.1^2/2}{\sigma^2}$ *for two distinct signals are plotted in Fig. 4.25. Note that the error rate for a fixed noise variance also follows a path that is determined solely by the signal energy and not by its shape. This is also due*

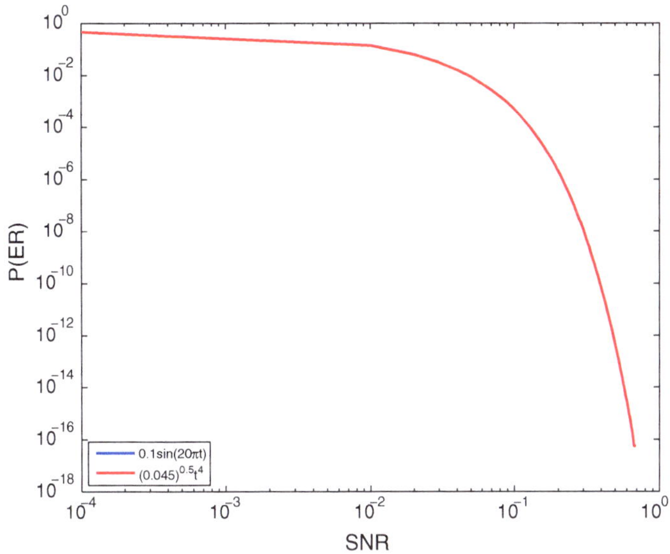

Fig. 4.25 The probability of error as a function of SNR

to the fact that the signals considered have identical energy levels and the observed vector x enjoys spherical symmetry in L_1 *space.*

In the next example, we examine a case where the spherical symmetry is violated, thereby resulting in a detection mechanism that is signal dependent.

Example 13 An experiment is designed to measure the firing of neurons for a healthy and a diseased brain using Electrocorticography (ECoG) or intracranial EEG (iEEG). The number of firings of neurons in a 1 s period is the observed signal. It is determined that the firing mechanism obeys a Poisson point process, resulting in a Poisson count statistics. For the healthy individual, this particular region of the brain produces a Poisson point process with the intensity

$$\lambda_0(t) = \lambda_s [U(t) - U(t - 0.1)] + \lambda_b \tag{4.207}$$

where $U(t)$ is a unit step function, λ_s denotes the average number of firings per second due to brain activity, and λ_b is the average number of firing per second due to background noise (device noise). When a hypothetically unhealthy brain is examined, the same Poisson point process is shown to have the following intensity

$$\lambda_1(t) = \frac{\lambda_s}{2} [U(t) - U(t - 0.1) + U(t - 0.9) - U(t - 1)] + \lambda_b \tag{4.208}$$

Design a detector that provides the minimum probability of error. The chance of someone with the above condition is 5 %. Compute the probability of error.

Solution: *Let the observed Poisson point process be* $x(t)$. *Given the time-varying nature of the "intensity" or the average event counts, we identify 3 distinct intervals of* $[0, 0.1]$, $(0.1, 0.9)$, *and* $[0.9, 1]$. *However, only two intervals exhibit distinct characteristics for the two hypotheses; that is* $[0, 0.1]$ *and* $[0.9, 1]$. *Hence, we consider the event counts over those intervals only. Let the event counts be defined as* $k_1 = \int_0^{0.1} x(\tau)\, d\tau$ *and* $k_2 = \int_{0.9}^{1} x(\tau)\, d\tau$. *Note that for the Poisson point processes the non-overlapping observation intervals yield independent event counts. Then,*

$$\begin{aligned} H_0 \Leftrightarrow\ & p(k_1 = l_1, k_2 = l_2 | H_0) \\ =\ & p(k_1 = l_1 | H_0)\, p(k_2 = l_2 | H_0) && (4.209) \\ =\ & \frac{(K_s + K_b)^{l_1}\, K_b^{l_2}}{l_1! l_2!} \exp(-K_s - 2K_b) && (4.210) \end{aligned}$$

$$\begin{aligned} H_1 \Leftrightarrow\ & p(k_1 = l_1, k_2 = l_2 | H_1) \\ =\ & p(k_1 = l_1 | H_1)\, p(k_2 = l_2 | H_1) && (4.211) \\ =\ & \frac{(K_s/2 + K_b)^{l_1}\, (K_s/2 + K_b)^{l_2}}{l_1! l_2!} \exp(-K_s - 2K_b) && (4.212) \end{aligned}$$

where $K_s = 0.1\lambda_s$ *and* $K_b = 0.1\lambda_b$. *Also,* $p_0 = 0.95$ *and* $p_1 = 0.05$. *We then have*

$$\Lambda(k) = \frac{P(k = l | H_1)}{P(k = l | H_0)} \underset{H_0}{\overset{H_1}{\gtrless}} \frac{p_0}{p_1} = 19$$

Finding the log-likelihood function, i.e., $\ln(\Lambda(k))$, *and simplifying the result, we arrive at*

$$l_2 \underset{H_0}{\overset{H_1}{\gtrless}} \Gamma + \beta l_1 \tag{4.213}$$

where $\Gamma = \frac{\ln\left(\frac{p_0}{p_1}\right)}{\ln \frac{K_s/2 + K_b}{K_b}}$ *and* $\beta = \frac{\ln \frac{K_s + K_b}{K_s/2 + K_b}}{\ln \frac{K_s/2 + K_b}{K_b}}$ *(note that, since* $p_0 > p_1$, Γ *is positive. Also, by definition,* $0 \le \beta \le 1$*). Hence, the detector makes two event (firing) counts over two disjointed time intervals and compares the scaled version of the counts to form a decision variable. Now, the probability of error for this detector is*

$$P(ER) = p_0 P(ER | H_0) + p_1 P(ER | H_1), \tag{4.214}$$

which, for the decision variable shown above, reduces to (for $p_0 > p_1$*)*

$$P(ER) = p_0 \sum_{l_1=0}^{\infty} \sum_{l_2=\lceil \Gamma+\beta l_1 \rceil}^{\infty} \frac{K_{10}^{l_1}}{l_1!} \frac{K_{20}^{l_2}}{l_2!} e^{-K_{10}-K_{20}} + p_1 \sum_{l_1=0}^{\infty} \sum_{l_2=0}^{\lfloor \Gamma+\beta l_1 \rfloor} \frac{K_{11}^{l_1}}{l_1!} \frac{K_{21}^{l_2}}{l_2!} e^{-K_{11}-K_{21}} \tag{4.215}$$

where $K_{10} = (K_s + K_b)$, $K_{20} = K_b$, *and* $K_{11} = K_{21} = \left(\frac{K_s}{2} + K_b\right)$. *In this equation,* $\lceil x \rceil$ *denotes the smallest integer greater than x and* $\lfloor x \rfloor$ *is the largest integer less than x. The analysis assumes that the scenario where* $l_2 = \Gamma + \beta l_1$ *leads to ambiguity and not to an error. That is, the detector will not render a decision when the event counts* l_1 *and* l_2 *result in* $l_2 = \Gamma + \beta l_1$. *Instead, the test is rendered "inconclusive." Note that, since spherical symmetry is absent, the performance becomes a function of both* K_s *and* K_b *and not of their ratio. To see this, the error probability is plotted as a function of* K_s *for various levels of* K_b *in Fig. 4.26. As one expect, the probability of error decreases with an increase in the signal count (which may be related to the sensitivity of ECoG device in picking up neuron firings). Furthermore, the background (noisy) count has a profound impact on the performance of the detector. Also, note that the performance is clearly not a function of* $\frac{K_s}{K_b}$ *(find and compare the performances for different values of* K_s *and* K_b *for which* $\frac{K_s}{K_b}$ *is a constant). Hence, the spherical symmetry is not present and the performance is clearly a function of the two parameters and not of the ratio of the two. This figure also points to the required sensitivity of the measurement. That is, one has to make sure that the noise contribution (*K_b*) is made small in order to have a relatively useful detector (*$P(ER) < 0.001$*).*

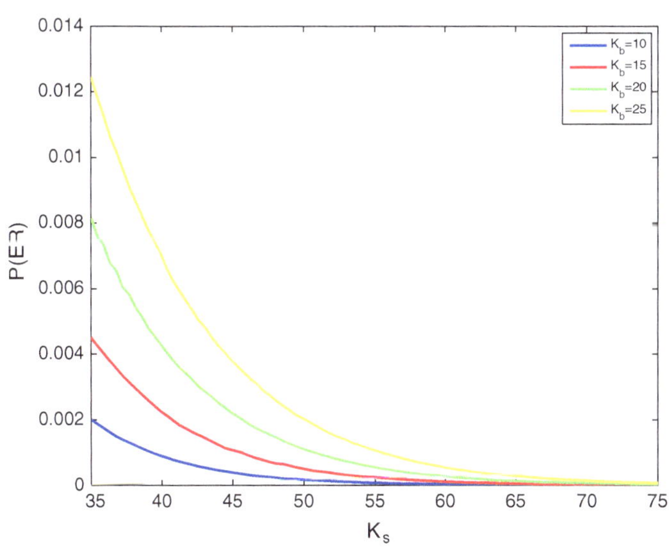

Fig. 4.26 The probability of error as a function of the average event count due to brain activity

Note that, in the above example, the average values of the event counts for the two scenarios were the same. That is,

$$E\{(k_1 + k_2)\,|H_0\} = E\{(k_1 + k_2)\,|H_1\} = K_s + K_b \tag{4.216}$$

However, it should be clear to the reader that the structure of the detector depends on the format of the intensity functions for the same set of average event counts. That is, a different set of $\lambda(t)'s$ will yield an entirely different optimum detection mechanism even though the average event counts are kept the same. Furthermore, as noted in the example, there is no spherical symmetry that one can exploit. For the AWGN and the Laplacian cases, the general shape of the structure involved computing $\|\mathbf{x} - \mathbf{s}_k\|_2$ and $\|\mathbf{x} - \mathbf{s}_k\|_1$, respectively, for any $\mathbf{s}_k$. In the case of the previous example, however, such an statement cannot be made.

4.4.3 Parameter Estimation

Similar to the signal detection scenario, we are interested in estimating the parameters of a signal for diagnosis purposes. Unlike the signal detection problem, where the decision is whether a signal is present in a noise- and/or interference-corrupted signal, the signal estimation is concerned with determining the parameters of a signal which we have determined to be present.

In that event, we need to establish criteria for the estimation "accuracy" and design estimators that achieve those criteria. In general, the problem involves observing a vector $\mathbf{x} = [x_0, x_1, \ldots, x_{l_v-1}]$, which is corrupted by noise, in order to estimate the real vector $\boldsymbol{\theta} = [\theta_1, \theta_2, \ldots, \theta_M]$. This model can also be applied to the scenario where the signal itself embodies randomness and the additive noise plays only a minor role.

The above model implies a simultaneous estimation of M parameters by observing a vector of length l_v. For most problems, $M = 1$. We make the assumption that $f_{\mathbf{x}}(\mathbf{x}|\boldsymbol{\theta})$ is known. This is a major point here, and for problem where this pdf is unknown, the pdf must be estimated! Furthermore, for the situations where $\boldsymbol{\theta}$ is a random vector, we assume that $f_{\boldsymbol{\theta}}(\boldsymbol{\theta})$ is available to us.

We therefore consider two distinct scenarios; random $\boldsymbol{\theta}$ and deterministic $\boldsymbol{\theta}$.

Before we go any further, it is useful to outline some basic properties of estimators. First, let $\hat{\boldsymbol{\theta}}$ be an estimate of $\boldsymbol{\theta}$. These properties are as follows.

- An estimator is considered to be unbiased if (the estimators are essentially random vectors)

$$E\left\{\hat{\theta}_j - \theta_j\right\} = 0 \text{ for all } j$$

  Note that this applies to random as well as deterministic random variables.

- An estimator is considered to be *consistent* if it converges to $\boldsymbol{\theta}$ in probability. That is, for an observation of length l_v

$$\lim_{l_v \to \infty} P\left(\left|\theta_j - \theta\right| > \epsilon_j\right) \to 0 \text{ for all } j \text{ and any } \epsilon_j$$

  This condition assumes that $\boldsymbol{\theta}$ is constant as $l_v \to \infty$. This is a difficult condition to satisfy for biomedical signal processing as the increase in the observation interval may lead to a varying $\boldsymbol{\theta}$. However, for the cases where a large l_v (but not large enough to violate the constant assumption for $\boldsymbol{\theta}$) results in convergence, we can use this property.
- An estimator is considered to be *efficient* if $E\left\{\left|\hat{\theta}_j - \theta_j\right|^2\right\}$ or mean-square-error (MSE), achieves the Cramer-Rao (CR) bound (this will be defined shortly). The estimate is *asymptotically efficient* if MSE approaches CR bound as $l_v \to \infty$.

4.4.3.1 Deterministic θ

In this scenario, we are given $f_{\mathbf{x}}(\mathbf{x}|\boldsymbol{\theta})$ or $f_{\mathbf{x}}(\mathbf{x}|\boldsymbol{\theta})$ is estimated using the histogram or other pdf estimation tools using the observed data. In general, the assumption is that we are provided $f_{\mathbf{x}}(\mathbf{x}|\boldsymbol{\theta})$, where $\boldsymbol{\theta}$ is an unknown. Note that the missing parameter may very well be the key parameter in the pdf; such as mean, variance, etc. The task at hand is to estimate $\boldsymbol{\theta}$.

Maximum Likelihood Estimation (MLE) The MLE technique, used for decades in signal processing, relies heavily on the availability of $f_{\mathbf{x}}(\mathbf{x}|\boldsymbol{\theta})$. The approach simply asks the following question:

Given the statistical properties of x (in terms of $f_{\mathbf{x}}(\mathbf{x}|\boldsymbol{\theta})$), what is the most likely value for θ?

The answer to this question is to find $\boldsymbol{\theta}$ that maximizes the likelihood function. That is,

$$\hat{\boldsymbol{\theta}}_{ML} = \arg\max f_{\mathbf{x}}(\mathbf{x}|\boldsymbol{\theta}) \tag{4.217}$$

or (since log function is a monotonically increasing function of its argument)

$$\hat{\boldsymbol{\theta}}_{ML} = \arg\max \ln f_{\mathbf{x}}(\mathbf{x}|`). \tag{4.218}$$

In general, finding the maximum of the likelihood function may not be a straightforward task (computationally complex). Given that acquiring MLE is mainly a function of the $f_{\mathbf{x}}(\mathbf{x}|\boldsymbol{\theta})$, which is problem dependent, a study of this principle is best done using a variety of examples.

Example 14 In an experiment, we have observed the EEG signal and have determined that the histogram supports a double-exponential pdf. See Example 6. We also assume that the samples are independent (a decimation factor of 1024 is used). Using MLE, obtain the parameter α of the double-exponential pdf when $l_v = 36,350$ samples are observed.

Solution: *In this problem, x is a vector of length l_v while the parameter to be estimated is a scalar. The double-exponential pdf is given by*

$$f(\mathbf{x}|\alpha) = \left(\frac{\alpha}{2}\right)^{l_v} \exp\left(-\alpha \|\mathbf{x}\|_1\right); \alpha > 0 \tag{4.219}$$

Then,

$$\ln f(\mathbf{x}|\alpha) = l_v \ln(\alpha/2) - \alpha \|\mathbf{x}\|_1 \tag{4.220}$$

We have

$$\begin{aligned}\frac{\partial}{\partial\alpha} \ln f(\mathbf{x}|\alpha) &= \frac{l_v}{\alpha} - \|\mathbf{x}\|_1 \\ &= 0 \Rightarrow\end{aligned} \tag{4.221}$$

$$\hat{\alpha}_{ML} = \frac{l_v}{\|\mathbf{x}\|_1} = \frac{l_v}{\sum_{i=1}^{l_v} |x_i|} \tag{4.222}$$

Note that

$$\frac{\partial^2}{\partial\alpha^2} \ln f(\mathbf{x}|\alpha) = -\frac{l_v}{\alpha^2} < 0; \alpha \neq 0 \tag{4.223}$$

implying that the above solution is the maximum of the likelihood function and not the minimum. Since the pdf is not Gaussian, the estimate does not have an intuitive interpretation (this will be examined next). In Fig. 4.27, we have depicted the histogram of data along with the PDF of double exponential pdf's when α is estimated using MLE and using $\alpha = \sqrt{\frac{2}{\sigma^2}}$ when σ^2 obtained by computing the variance of the data set. Quite clearly, MLE provides a superior performance as compared to the method used earlier for estimating α.

As noted in the previous example, the MLE does not provide an intuitive solution. For Gaussian observation, MLE provides estimates that appeal to common sense. To see this fact, let us consider the case where $\mathbf{x}$ is a Gaussian vector of uncorrelated random variables with a variance of σ^2 and the mean m. That is,

$$\begin{aligned}f_{\mathbf{x}}\left(\mathbf{x}|m,\sigma^2\right) &= \frac{1}{\left(2\pi\sigma^2\right)^{l_v/2}} \exp\left\{-\frac{\|\mathbf{x}-m\|^2}{2\sigma^2}\right\} \\ &= \frac{1}{\left(2\pi\sigma^2\right)^{l_v/2}} \exp\left\{-\sum_{i=0}^{l_v-1} \frac{(x_i-m)^2}{2\sigma^2}\right\}\end{aligned} \tag{4.224}$$

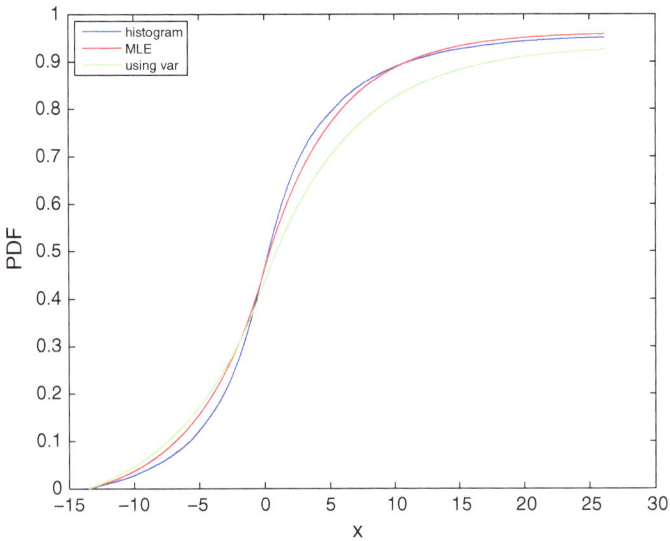

Fig. 4.27 The pdf of the observed random variable in Example 14 along with the MLE pdf and a doubly-exponential pdf with the same variance

Now, if we are interested in estimating m,

$$\frac{\partial}{\partial m} \ln f_{\mathbf{x}} \left(\mathbf{x} | m, \sigma^2 \right) = 0 \tag{4.225}$$

yields

$$\hat{m}_{ML} = \frac{\sum_{i=0}^{l_v - 1} x_i}{l_v} \tag{4.226}$$

This estimate may be viewed as the "sample" mean. Note that this provides an intuitive solution to the estimation of the mean of a random variable. That is, ensemble mean is simply the time average of the signal. Interestingly, if one attempts the estimation of the variance, a similar result emerges. That is,

$$\frac{\partial}{\partial \sigma^2} \ln f_{\mathbf{x}} \left(\mathbf{x} | m, \sigma^2 \right) = 0$$

yields

$$\hat{\sigma}_{ML}^2 = \sum_{i=0}^{l_v - 1} \frac{(x_i - m)^2}{l_v} \tag{4.227}$$

which can be viewed as the "sample" variance if mean m is known. These are attractive features of the Gaussian assumption, which as we observed in the previous sections, is not valid for a host of problems in biomedical signal processing. However, when biological signals are observed using electronic devices, the AWGN is often present, and one is compelled to estimate the parameters of the electronic noise using the aforementioned techniques.

One question that needs to be answered here is, "what is the performance of this estimator?" This requires that we establish a performance criterion for the estimation problem.

First, we note that the estimator that is found using MLE is, in principle, a random variable. As such, it has a mean and a variance. It would be ideal if the estimator has a mean that is identical to what is being estimated (unbiased). Such an estimator, as noted earlier, is known as an unbiased estimator. More precisely,

$$E\left\{\hat{\theta}_{ML}\right\} = \theta \tag{4.228}$$

If such a condition is not satisfied, the estimator is considered to be biased. For the example we have provided above, the reader can verify that the MLE estimators considered here are indeed unbiased. That is,

$$E\left\{\hat{m}_{ML}\right\} = m \tag{4.229}$$

and

$$E\left\{\hat{\sigma}^2_{ML}\right\} = \sigma^2 \tag{4.230}$$

The next step is to examine the variance of the estimators, or more importantly, the MSE for the estimator.

Performance Metric for Deterministic Parameter Estimation: Cramer-Rao (CR) Bound A logical metric to be used in assessing whether an estimator is "good" is the variance of the error (the estimator itself is a random variable) or MSE. Let Γ_j, given by,

$$\Gamma_j = E\left\{\left(\hat{\theta}_j - \theta_j\right)^2\right\};\ j = 1, 2, \ldots, M \tag{4.231}$$

be viewed as the measure of "goodness" of the estimator. Note that Γ_j for an unbiased estimator is the MSE for the estimation process, and is lower bounded by the CR bound as follows:

$$\Gamma_j \geq J^{jj} \tag{4.232}$$

where J^{jj} is the *jj*th element of J^{-1}, the inverse of the *Fisher Information Matrix* J. The *ij*th element of J is given by

$$J_{ij} = E\left\{\frac{\partial \ln f_{\mathbf{x}}(\mathbf{x}|\boldsymbol{\theta})}{\partial \theta_i}\frac{\partial \ln f_{\mathbf{x}}(\mathbf{x}|\boldsymbol{\theta})}{\partial \theta_j}\right\}$$
$$= -E\left\{\frac{\partial^2 \ln f_{\mathbf{x}}(\mathbf{x}|\boldsymbol{\theta})}{\partial \theta_i \partial \theta_j}\right\}. \qquad (4.233)$$

When the performance of an MLE is hard to compute (in some cases, the estimator is a complicated function of **x**, and hence, finding Γ_j is not a trivial task), one can resort to the CR bound, which, when it exists, establishes the lowest possible value for Γ_j for all unbiased estimators. When CR bound does not exist, unbiased estimators with error variances smaller than CR bound may exist, but a way to acquire them is generally unknown. As noted earlier, an estimator that achieves the CR bound is known to be *efficient.* MLE's are either *efficient or asymptotically efficient.* MLE's are also consistent. These fact provide a pathway to improving the estimation process; namely, letting l_v get larger until the desired accuracy has been achieved. Hence, it is best to seek an MLE solution at all times.

Example 15 For the Example 14, compute the CR bound and determine whether the MLE is an efficient estimator. Also, establish a condition under which the estimator becomes efficient.

Solution: *As we observed in the Example* 14,

$$f(\mathbf{x}|\alpha) = \left(\frac{\alpha}{2}\right)^{l_v} \exp\left(-\alpha \|\mathbf{x}\|_1\right) \qquad (4.234)$$

and

$$\ln f(\mathbf{x}|\alpha) = l_v \ln(\alpha/2) - \alpha \|\mathbf{x}\|_1 . \qquad (4.235)$$

It can readily be shown that

$$\frac{\partial^2 \ln f(\mathbf{x}|\alpha)}{\partial \alpha^2} = -\frac{l_v}{\alpha^2} \qquad (4.236)$$

Hence,

$$-E\left\{\frac{\partial^2 \ln f(\mathbf{x}|\alpha)}{\partial \alpha^2}\right\} = \frac{l_v}{\alpha^2} \qquad (4.237)$$

We then have

$$\Gamma = E\left\{(\hat{\alpha} - \alpha)^2\right\} \geq \frac{1}{-E\left\{\frac{\partial^2 \ln f(\mathbf{x}|\alpha)}{\partial \alpha^2}\right\}} = \frac{\alpha^2}{l_v} \qquad (4.238)$$

Note that as $l_v \to \infty$, CR bound$\to$ 0. This is an encouraging result. Now, let us see whether the MLE that we found earlier is unbiased. To that end, let us compute $E\left\{\hat{\alpha}_{ML}\right\}$, which is given by

$$E\left\{\hat{\alpha}_{ML}\right\} = l_v E\left\{\frac{1}{\sum_{i=0}^{l_v-1}|x_i|}\right\}$$
$$= l_v\left(\frac{\alpha}{2}\right)^{l_v}\int_{-\infty}^{\infty}dx_0\ldots\int_{-\infty}^{\infty}dx_{l_v-1}\frac{\exp\left(-\alpha\sum_{i=0}^{l_v-1}|x_i|\right)}{\sum_{i=0}^{l_v-1}|x_i|} \tag{4.239}$$

The last integration does not converge! That is, we cannot establish whether this estimator is unbiased nor if it achieves the CR bound. Note that

$$E\left\{\hat{\alpha}_{ML}^2\right\} = l_v^2 E\left\{\frac{1}{\left(\sum_{i=0}^{l_v-1}|x_i|\right)^2}\right\} \tag{4.240}$$

also does not converge. However, as we observed in the example, the estimator was successful in estimating α and producing a very good match to the experimental data.

The above example underscores an important point; that is, if the problem is ill-posed, then there is a possibility of arriving at solutions for which one may not be able to establish performance. Let us now examine the same problem in a slightly different setting. That is, let us consider a new parameter

$$\beta = \frac{1}{\alpha} \tag{4.241}$$

for the double-exponential pdf. The objective now is to estimate β instead of α. One can easily see that if β is estimated, one can obtain α. In that event,

$$f\left(\mathbf{x}|\beta\right) = \left(\frac{1}{2\beta}\right)^{l_v}\exp\left(-\frac{\|\mathbf{x}\|_1}{\beta}\right) \tag{4.242}$$

Now, an MLE for this pdf is given by

$$\ln f\left(\mathbf{x}|\beta\right) = -l_v\ln\left(2\beta\right) - \frac{\|\mathbf{x}\|_1}{\beta} \tag{4.243}$$

$$\frac{\partial}{\partial\beta}\ln f\left(\mathbf{x}|\beta\right) = -\frac{l_v}{\beta} + \frac{\|\mathbf{x}\|_1}{\beta^2} = 0 \Rightarrow \tag{4.244}$$

$$\hat{\beta}_{ML} = \frac{\|\mathbf{x}\|_1}{l_v} \tag{4.245}$$

It can be seen that

$$E\left\{\hat{\beta}_{ML}\right\} = \frac{\sum_{i=0}^{l_v-1} E\left\{|x_i|\right\}}{l_v} = \frac{l_v \beta}{l_v} = \beta \tag{4.246}$$

which implies that the MLE is unbiased. Also,

$$\begin{aligned} E\left\{\hat{\beta}_{ML}^2\right\} &= \frac{l_v E\left\{|x|^2\right\} + \left(l_v^2 - l_v\right) E^2\left\{|x|\right\}}{l_v^2} \\ &= \frac{2l_v\beta^2 + \left(l_v^2 - l_v\right)\beta^2}{l_v^2} = \beta^2 + \frac{\beta^2}{l_v} \end{aligned} \tag{4.247}$$

Now, let us consider the variance of the estimate

$$E\left\{\hat{\beta}_{ML}^2\right\} - E^2\left\{\hat{\beta}_{ML}\right\} = \frac{\beta^2}{l_v}. \tag{4.248}$$

The question that remains is whether this estimate is efficient. To that end, we need to compute the CR bound, given by

$$\begin{aligned} \frac{\partial^2}{\partial \beta^2} \ln f\left(\mathbf{x}|\beta\right) &= \frac{l_v}{\beta^2} - 2\frac{\|\mathbf{x}\|_1}{\beta^3} \\ -E\left\{\frac{\partial^2}{\partial \beta^2} \ln f\left(\mathbf{x}|\beta\right)\right\} &= -\frac{l_v}{\beta^2} + \frac{2l_v\beta}{\beta^3} \\ &= \frac{l_v}{\beta^2} \end{aligned} \tag{4.249}$$

Hence, the CR bound is

$$\text{CR bound} = \frac{\beta^2}{l_v} \tag{4.250}$$

Comparing (4.248) *and* (4.250) *leads one to the conclusion that the MLE for* β *is efficient!! Therefore, by restating the same problem, we were able to find an estimate of* β *(and, consequently,* α*) that can be established to be both unbiased and efficient. Now, let us consider the CR bound for other scenarios considered above.*

Example 16 For the Gaussian observation case where MLE was used to estimate the mean and the variance, determine the CR bounds and establish whether the estimators are efficient.

Solution: *As we observed,*

$$\hat{m}_{ML} = \frac{\sum_{i=0}^{l_v-1} x_i}{l_v} \tag{4.251}$$

which is unbiased estimator since

$$E\left\{\hat{m}_{ML}\right\} = \frac{\sum_{i=0}^{l_v-1} E\left\{x_i\right\}}{l_v} = \frac{l_v m}{l_v} = m \tag{4.252}$$

Also, note that ($x_i^{\prime}s$ are independent)

$$\begin{aligned} E\left\{\hat{m}_{ML}^2\right\} &= \frac{\sum_{i_1=0}^{l_v-1} \sum_{i_1=0}^{l_v-1} E\left\{x_i x_j\right\}}{l_v^2} \\ &= \frac{\left(l_v^2 - l_v\right) m^2 + l_v\left(\sigma^2 + m^2\right)}{l_v^2} \\ &= m^2 + \frac{\sigma^2}{l_v} \end{aligned} \tag{4.253}$$

Hence,

$$E\left\{\left(\hat{m}_{ML} - m\right)^2\right\} = \frac{\sigma^2}{l_v} \tag{4.254}$$

Note that $l_v \to \infty$, $E\left\{\left(\hat{m}_{ML} - m\right)^2\right\} \to 0$, *which implies that the estimator is consistent (mean-square sense convergence implies convergence in probability). Now, let us consider the CR bound,*

$$\ln f_{\mathbf{x}}\left(\mathbf{x}|m, \sigma^2\right) = -\frac{\sum_{i=1}^{l_v} (x_i - m)^2}{2\sigma^2} - \frac{l_v}{2} \ln\left(2\pi\sigma^2\right) \tag{4.255}$$

and

$$-\frac{\partial^2}{\partial m^2} \ln f_{\mathbf{x}}\left(\mathbf{x}|m, \sigma^2\right) = \frac{l_v}{\sigma^2} \tag{4.256}$$

Then,

$$\frac{-1}{E\left\{\frac{\partial^2}{\partial m^2} \ln f_{\mathbf{x}}\left(\mathbf{x}|m, \sigma^2\right)\right\}} = \frac{\sigma^2}{l_v} \tag{4.257}$$

Hence, the MLE is unbiased and efficient as it achieves the CR bound.

Also, when we consider the estimator of the variance,

$$\begin{aligned} E\left\{\hat{\sigma}_{ML}^2\right\} &= \sum_{i=0}^{l_v-1} \frac{E\left\{(x_i - m)^2\right\}}{l_v} \\ &= \sigma^2 \end{aligned} \tag{4.258}$$

Therefore, the MLE of the variance is unbiased. Furthermore,

$$
\begin{aligned}
E\left\{\left(\hat{\sigma}_{ML}^2\right)^2\right\} &= \sum_{i=0}^{l_v-1}\sum_{j=0}^{l_v-1}\frac{E\left\{(x_i-m)\left(x_j-m\right)\right\}}{l_v^2} \\
&= \frac{\left(l_v^2-l_v\right)\sigma^4+l_v\left(3\sigma^4\right)}{l_v^2} \\
&= \left(\sigma^2\right)^2+\frac{2\sigma^4}{l_v}
\end{aligned}
\tag{4.259}
$$

where we have used the property $E\left\{x^4\right\} = 3E^2\left\{x^2\right\}$ *when* x *is a zero-mean Gaussian random variable. Hence,*

$$
E\left\{\left(\hat{\sigma}_{ML}^2-\sigma^2\right)^2\right\} = \frac{2\sigma^4}{l_v} \tag{4.260}
$$

Once again, $l_v \to \infty$, $E\left\{\left(\hat{\sigma}_{ML}^2-\sigma^2\right)^2\right\} l_v \to \infty$, *which implies the estimator is consistent. Now, let us compute the CR bound.*

$$
\ln f_{\mathbf{x}}\left(\mathbf{x}|m,\sigma^2\right) = -\frac{\sum_{i=0}^{l_v-1}(x_i-m)^2}{2\sigma^2}-\frac{l_v}{2}\ln\left(2\pi\sigma^2\right)
$$

and

$$
-\frac{\partial^2}{\partial\left(\sigma^2\right)^2}\ln f_{\mathbf{x}}\left(\mathbf{x}|m,\sigma^2\right) = \frac{\sum_{i=0}^{l_v-1}(x_i-m)^2}{\sigma^6}-\frac{l_v}{2\sigma 4} \tag{4.261}
$$

and

$$
E\left\{\frac{\partial^2}{\partial\left(\sigma^2\right)^2}\ln f_{\mathbf{x}}\left(\mathbf{x}|m,\sigma^2\right)\right\} = \frac{l_v}{\sigma 4}-\frac{l_v}{2\sigma 4} = \frac{l_v}{2\sigma 4}
$$

Then, the CR bound is given

$$
\frac{-1}{E\left\{\frac{\partial^2}{\partial(\sigma^2)^2}\ln f_{\mathbf{x}}\left(\mathbf{x}|m,\sigma^2\right)\right\}} = \frac{2\sigma^4}{l_v} \tag{4.262}
$$

which implies that the MLE of the variance is also efficient.

Now, let us consider yet another non-Gaussian example where the efficiency of the estimator can be established.

Example 17 It has been established that the number of scatterers in a tissue exposed to optical radiation is closely approximated by a log-normal distribution. It is assumed that the parameters of the log-normal distribution are not subject to change for the organ of interest. In an experiment, a number of tests have been performed, resulting in l_v independent samples. Establish the MLE for estimating the log-mean and log-variance for this process. In each case, determine whether the estimators are unbiased and efficient.

Solution: *Let x denote the vector of the observed counts. From the description of the problem,*

$$f\left(\mathbf{x}|m,\sigma^2\right) = \frac{1}{\left(2\pi\sigma^2\right)^{l_v/2}\prod_{i=1}^{l_v} x_i}\exp\left(-\frac{1}{2\sigma^2}\left\|\ln\mathbf{x} - m\right\|_2\right); \tag{4.263}$$

Our objective is to estimate m and σ^2 *(note that* $m = E\left\{\ln x_j\right\}$ *and* $\sigma^2 = E\left\{\ln^2 x_j\right\} - m^2$ *for any j), the log-mean and log-variance, respectively. To that end,*

$$\ln f\left(\mathbf{x}|m,\sigma^2\right) = -\frac{l_v}{2}\ln\left(2\pi\sigma^2\right) - \sum_{i=0}^{l_v-1}\ln x_i - \frac{1}{2\sigma^2}\sum_{i=0}^{l_v-1}\left(\ln x_i - m\right)^2 \tag{4.264}$$

Now, let us assume σ^2 *is known, but m is not. Taking the derivative of the likelihood function with respect to m produces*

$$\frac{\partial}{\partial m}\ln f\left(\mathbf{x}|m\right) = \frac{1}{\sigma^2}\sum_{i=0}^{l_v-1}\left(\ln x_i - m\right) = 0 \rightarrow \tag{4.265}$$

$$\hat{m}_{ML} = \frac{\sum_{i=0}^{l_v-1}\ln x_i}{l_v} \tag{4.266}$$

This estimate is unbiased since

$$E\left\{\hat{m}_{ML}\right\} = \frac{E\left\{\sum_{i=0}^{l_v-1}\ln x_i\right\}}{l_v} = m \tag{4.267}$$

Also,

$$\begin{aligned} E\left\{\hat{m}_{ML}^2\right\} &= \frac{l_v E\left\{\ln^2 x\right\} + \left(l_v^2 - l_v\right)m^2}{l_v^2} \\ &= \frac{l_v\left(\sigma^2 + m^2\right) + \left(l_v^2 - l_v\right)m^2}{l_v^2} \\ &= \frac{\sigma^2}{l_v} + m^2, \end{aligned} \tag{4.268}$$

which leads to

$$E\left\{\hat{m}_{ML}^2\right\} - E^2\left\{\hat{m}_{ML}\right\} = \frac{\sigma^2}{l_v}.$$

Also,

$$\frac{\partial^2}{\partial m^2}\ln f(\mathbf{x}|m) = -\frac{l_v}{\sigma^2} \rightarrow$$
$$\text{CR bound} = \frac{-1}{E\left\{\frac{\partial^2}{\partial m^2}\ln f(\mathbf{x}|m)\right\}} = \frac{\sigma^2}{l_v}. \qquad (4.269)$$

Hence, the estimate is both unbiased and efficient. As can be seen, this is very similar to the Gaussian estimation problem. Now, if the mean is known and variance is unknown, we have

$$\frac{\partial}{\partial\sigma^2}\ln f\left(\mathbf{x}|\sigma^2\right) = -\frac{l_v}{2\sigma^2} + \frac{1}{2\sigma^4}\sum_{i=0}^{l_v-1}(\ln x_i - m)^2 = 0 \rightarrow$$
$$\hat{\sigma}_{ML}^2 = \frac{\sum_{i=0}^{l_v-1}(\ln x_i - m)^2}{l_v}. \qquad (4.270)$$

Note that

$$E\left\{\hat{\sigma}_{ML}^2\right\} = \frac{E\left\{\sum_{i=0}^{l_v-1}(\ln x_i - m)^2\right\}}{l_v} = \frac{l_v\sigma^2}{l_v} = \sigma^2, \qquad (4.271)$$

which implies that the estimate is unbiased. We also have

$$E\left\{\left(\hat{\sigma}_{ML}^2\right)^2\right\} = \frac{l_v\left(\sum_{i=0}^{l_v-1}(\ln x - m)^4\right) + (l_v^2 - l_v)\sigma^4}{l_v^2} = \frac{3l_v\sigma^4 + \left(l_v^2 - l_v\right)\sigma^4}{l_v^2} = \frac{2\sigma^4}{l_v} + \sigma^4, \qquad (4.272)$$

which leads to

$$E\left\{\left(\hat{\sigma}_{ML}^2\right)^2\right\} - E^2\left\{\hat{\sigma}_{ML}^2\right\} = \frac{2\sigma^4}{l_v}.$$

Now,

$$\frac{\partial^2}{\partial\sigma^4}\ln f\left(\mathbf{x}|\sigma^2\right) = \frac{l_v}{2\sigma^4} - \frac{1}{\sigma^6}\sum_{i=0}^{l_v-1}(\ln x_i - m)^2$$

$$-E\left\{\frac{\partial^2}{\partial\sigma^4}\ln f\left(\mathbf{x}|\sigma^2\right)\right\} = \frac{-l_v}{2\sigma^4} + \frac{l_v\sigma^2}{\sigma^6} = \frac{l_v}{2\sigma^4}.$$

Hence,

$$\text{CR bound} = \frac{2\sigma^2}{l_v}.$$

So, the MLE for the log-variance is unbiased and efficient as well.

We discussed the hypothesis testing problem for Poisson point processes case where the signal of interest is generated by brain activity. Parameter estimation is of particular interest in that arena as well. To study this subject, we resort to yet another example.

Example 18 In a study, the signal from Electrocorticograh (ECoG) is collected using an implanted probe over a period of 1 min. The probe is designed such that it counts neuron firing. It is determined that the neural firing obeys a Poisson point process and that the resulting process possesses the following intensity:

$$\lambda(t) = \lambda_s + \lambda_b \tag{4.273}$$

where λ_s and λ_b are the "signal" and "background noise" intensities, respectively, in firing/s. Given an observation of 1 min, develop an MLE for estimating λ_s. Assume that λ_b is known.

Solution: *Given that the events obey a Poisson point processes, the firing count is a Poisson random variable with PMF*

$$P\{K = k|\lambda_s\} = \frac{\left(\int_0^T \lambda(\tau)\,d\tau\right)^k}{k!}\exp\left\{-\int_0^T \lambda(\tau)\,d\tau\right\} \tag{4.274}$$

where $T = 60$. *Given that* $\lambda(t) = \lambda_s + \lambda_b$

$$P\{K = k|\lambda_s\} = \frac{(K_s + K_b)^k}{k!}\exp\{-K_s - K_b\} \tag{4.275}$$

where $K_s = \lambda_s T = 60\lambda_s$ *and* $K_b = \lambda_b T = 60\lambda_b$. *Hence, estimating* K_s *will lead to an estimate of* λ_s. *Using MLE,*

$$\begin{aligned}\Lambda(K_s) &= \ln P\{K = k|K_s\}\\ &= k\ln(K_s + K_b) - K_s - K_b - \ln(k!)\end{aligned} \tag{4.276}$$

and

$$\frac{\partial \Lambda\left(K_s\right)}{\partial K_s} = \frac{k}{K_s + K_b} - 1 = 0 \rightarrow$$
$$\hat{K}_{s,ML} = k - K_b \tag{4.277}$$

This estimate is the most logical conclusion one can arrive at; that is, the MLE suggests that we take the observed count and subtract from that the average background count. Note that due to the statistical behavior of k, the estimate may assume a negative value! However, this estimate is an MLE, and as we will observe, it is an efficient estimate (achieves the lowest error variance among all unbiased estimators). Before demonstrating this fact, let us see if this estimate is actually achieving the maximum of the likelihood function and not its minimum. To that end, we compute

$$\frac{\partial^2 \Lambda\left(K_s\right)}{\partial K_s^2} = \frac{-k}{\left(K_s + K_b\right)^2} \leq 0 \text{ for all } k. \tag{4.278}$$

Hence, $\hat{K}_{s,ML}$ produces a maximum (since $k \geq 0$) and not a minimum. Note that the estimate is unbiased. That is,

$$\begin{aligned} E\left\{\hat{K}_{s,ML}\right\} &= E\left\{k\right\} - K_b \\ &= K_s + K_b - K_b \\ &= K_s \end{aligned} \tag{4.279}$$

and

$$\begin{aligned} E\left\{\hat{K}^2_{s,ML}\right\} &= E\left\{k^2\right\} + K_b^2 - 2E\left\{k\right\} KR \\ &= \left(K_s + K_b\right) + \left(K_s + K_b\right)^2 + K_b^2 - 2\left(K_s + K_b\right) K_b \\ &= K_s + K_b + K_s^2 \end{aligned} \tag{4.280}$$

(for a Poisson count $E\left\{k^2\right\} = K_{avg}^2 + K_{avg}$ where $K_{avg} = E\left\{k\right\}$). Hence,the variance of the estimate is

$$E\left\{\hat{K}^2_{s,ML}\right\} - E^2\left\{\hat{K}_{s,ML}\right\} = \left(K_s + K_b\right) \tag{4.281}$$

Now,

$$\begin{aligned} -\frac{\partial^2 \ln P\left\{K = k|K_s\right\}}{\partial K_s^2} &= \frac{k}{\left(K_s + K_b\right)^2} \\ -E\left\{\frac{\partial^2 \ln P\left\{K = k|K_s\right\}}{\partial K_s^2}\right\} &= \frac{1}{K_s + K_b}. \end{aligned} \tag{4.282}$$

Hence,

$$\begin{aligned} \text{CR bound} &= \frac{-1}{E\left\{\frac{\partial^2 \ln P\{K=k|K_s\}}{\partial K_s^2}\right\}} \\ &= K_s + K_b. \end{aligned} \tag{4.283}$$

Equations (4.281) *and* (4.283) *suggest that the MLE in this example is an efficient estimator.*

So far, we have addressed the parameter estimation when the parameter is deterministic and the conditional pdf is known to us (or has been estimated). Since the acquisition of the likelihood function is the key step in the estimation process, next, we tackle the problem of formulating a likelihood function in the absence of the knowledge of the pdf of the signal of interest.

4.4.4 Whittle Likelihood Test (WLT)

In the previous sections, we used FFT and short-term FFT to analyze EEG and ECG time series. The key point of such analyses is the acquisition of the spectral characteristics of the signals under investigation in order to unravel the underlying (hidden) signals that are not visible to a cursory view. One can also acquire an estimate of the PSD of signals using the FFT-based techniques. Such approaches do not require the knowledge of actual pdf of the signal, although a knowledge of the correlation characteristics is required. For the MLE and other techniques discussed earlier, the statistical knowledge of the signal, such as pdf, are needed to form the estimator. One question that can be posed here is whether one can form a likelihood function for the estimation of signal parameters without having a knowledge of the pdf of the signal under investigation.

Whittle likelihood function is the answer to this question in that it provides a likelihood function using periodogram as well as the PSD of the signal. As mentioned above, PSD can be estimated using correlation function, which can be estimated for correlation-ergodic signals. To address this problem, let us first introduce the concept of periodogram.

Let $\mathbf{X} = \{\mathbf{x}_1, \mathbf{x}_2, \ldots, \mathbf{x}_N\}$ denote an $l_v \times N$ matrix of observed real, zero-mean vectors ($\mathbf{x}_n = \left[x_{0,n}, x_{1,n}, \ldots, x_{l_v-1,n}\right]^{\mathsf{T}}$), each column vector corresponding to a distinct source. The choice of l_v (which denotes time), as was seen earlier, may depend on the transient behavior of signals, etc., while N reflects the number of observed signals. Let us define the PSD of $\mathbf{X}$ (this is a slight change in the definition in order to include multiple signals) as a matrix $\mathbf{F}(\omega)$, given by

$$\mathbf{F}(\omega) = \left[f_{n,l}(\omega)\right], \tag{4.284}$$

whose elements are given by

$$f_{n,l}(\omega) = \sum_{q=-\infty}^{\infty} R_{n,l}[q]\, e^{-jq\omega}. \tag{4.285}$$

In the above, $R_{n,l}[q]$ is the (n, l)th element of the $N \times N$ matrix $\mathbf{R}_x[q]$ with

$$R_{n,l}[q] = E\left\{x_{m,n}x_{m+q,l}\right\};\; m = 0, 1, 2, \ldots, l_v - 1 \tag{4.286}$$

where a cross-W.S.S condition is assumed here ($R_{n,l}[q]$ is not a function of m for all n and l). It is obvious that $\mathbf{R}_x[q]$ may be viewed as a matrix containing the correlation and cross-correlation of the elements of $\mathbf{X}$. Note that

$$f_{n,n}(\omega) = \sum_{q=-\infty}^{\infty} R_{n,n}[q]\, e^{-jq\omega};\, q = 0, 1, \ldots, l_v - 1 \tag{4.287}$$

is, in particular, of interest as it describes the PSD of the nth source. It is quite obvious that, in order to compute $\mathbf{F}(\omega)$, only the second order statistical knowledge of $\mathbf{X}$ (namely, $\mathbf{R}_x[q]$) is needed. In some problems, one can observe PSD using experimental tools. Note that the availability of PSD is critical here.

Next, we introduce periodogram. To that end, let us also define the DFT of the columns of $\mathbf{X}$ to form yet another column vector $\mathbf{s}[q] = [s_1[q], s_2[q], \ldots, s_N[q]]^\mathsf{T}$, whose elements are given by

$$s_n[q] = \sum_{m=0}^{l_v-1} x_{m,n}e^{-jm\omega_q};\, n = 0, 1, \ldots, N - 1 \tag{4.288}$$

with $\omega_q = \frac{2\pi q}{l_v}$. We can, then, form the second-order periodogram ($N \times N$ matrix) as

$$\mathbf{P}_x[q] = \frac{1}{l_v}\mathbf{s}[q]\,\mathbf{s}^H[q]. \tag{4.289}$$

Note that the diagonal elements of $\mathbf{P}_x[q]$ are the scaled version (by $2\pi l_v$) of the absolute value squared of the signal's DFT. That is,

$$p_{n,n}[q] = \frac{1}{l_v}\left|\sum_{m=0}^{l_v-1} x_{m,n}e^{-\frac{2\pi jmq}{l_v}}\right|^2 \tag{4.290}$$

Hence, the name "periodogram" does make sense as it allows one to estimate the PSD of signals. As we discussed earlier, the PSD of a signal requires only the second order statistical knowledge of the signal, which is absent from the periodogram. As we discussed in previous chapters, the above analysis hinges upon the underlying process assuming ergodicity in power and correlation, which in turn allows one to overlook the unavailability of a complete statistical model. Otherwise, a time series analysis will not yield meaningful results. Due to asymptotic behavior of DFT, it is believed that periodogram yields an accurate estimate of the PSD. In the estimation problems posed earlier, such as MLE, the knowledge of the pdf of the signals under investigation was necessary to form the likelihood function. If such statistical information is not available, but one has access to the PSD of the signal, can we perform a likelihood test for estimating signal parameters using an alternate likelihood function? Such a likelihood test does exists and is known as the Whittle Likelihood Test (WLT).

The WLT for the problem stated above is given by

$$\Lambda_W(\mathbf{f}, \mathbf{X}) = -\sum_{n=0}^{N-1}\sum_{q=0}^{l_v-1}\left\{\frac{1}{2\pi}\frac{p_{n,n}[q]}{f_{n,n}\left(\frac{2\pi q}{l_v}\right)} + \ln\left(f_{n,n}\left(\frac{2\pi q}{l_v}\right)\right)\right\} \tag{4.291}$$

where $\mathbf{f}$ is an $N \times l_v$ matrix whose (n, q)th element is $f_{n,n}\left(\frac{2\pi q}{l_v}\right)$, the spectrum of the nth signal at the qth spectral point. Note that the likelihood function is an implicit function of $\mathbf{X}$ via $p_{n,n}[q]$ (directly) and $f_{n,n}\left(\frac{2\pi q}{l_v}\right)$ (indirectly through the correlation function of $\mathbf{X}$). WLT quite clearly tries to provide a measure of error in estimating the spectrum, given by $\frac{p_{n,n}[q]}{f_{n,n}\left(\frac{2\pi q}{l_v}\right)}$, and to use the resulting error to form an estimate of the unknown parameters in $\mathbf{X}$. In general, $\mathbf{f}$ is known while $\mathbf{X}$ bears one or more parameters that need to be estimated. Note that, unlike the MLE, the WLT is designed around the PSD of $\mathbf{X}$ and not its pdf. Now, if there are unknown parameters in $\mathbf{X}$, the above likelihood function can be used for the estimation of those parameters.

In summary, if one has access to the PSD of a signal (this can be achieved via estimating correlation or acquiring spectrum via experimental means) and a knowledge of the structure of the signal under question as it relates to the unknown parameters, WLT can be invoked to estimate those parameters.

In order to see the benefit of the WLT in dealing with estimating parameters of signals with unknown pdf, let us study this subject with the aid of an example.

Example 19 We are considering a signal that has been corrupted by a zero-mean, white additive noise with each sample having a variance of σ^2. The pdf of noise is unknown. The signal is reported to be a sinusoid with an amplitude of A (assume that A is known and that the sinusoid has a uniformly distributed phase on $[-\pi, \pi]$). However, the frequency of the sinusoid ($f = 20\,\text{Hz}$) is unknown to us. The signal is assumed to be independent of the noise. We are asked to determine the WLT for observation lengths of 1024 samples when the sampling rate is 128 samples/s. The objective is to estimate f. Use WLT to obtain an estimate of f. Also, discuss the impact of observation length on the performance of the estimator.

Solution: *Given the formulation of the problem, we recognize a number of facts. First,* $N = 1$*. Second,* $l_v = 1024$*. Also, the sampling rate is high enough (>Nyquist rate) to estimate the frequency of the sinusoid). In order to set up WLT, we need to acquire* $f_{1,1}\left(\frac{2\pi q}{l_v}\right)$*. Before doing so, let us define* $x[n]$*, which is*

$$x[n] = A\sin\left[2\pi n\frac{f}{128} + \theta\right] + w[n] \tag{4.292}$$

where $w[n]$ *denotes the additive white noise,* f *is the frequency in Hz (to be estimated), and* θ *is uniformly distributed on* $[-\pi, \pi]$*. Given that signal is independent of noise, the PSD of signal + noise is the sum of the two PSD's. That is, the PSD of the signal is given by*

$$S_x[\omega] = \sum_{n=0}^{l_v-1} R_x[n]\, e^{-jn\omega} \tag{4.293}$$

where $\omega = \frac{2\pi q}{l_v}$ *and*

$$R_x[n] = \frac{A^2}{2}\cos\left[\frac{\pi n f}{64}\right] + \sigma^2. \tag{4.294}$$

This correlation function is obtained assuming that the sinusoid has a uniformly distributed phase on $[-\pi, \pi]$ *and that noise is uncorrelated with the signal. We then have*

$$\begin{aligned} S_x\left[\frac{2\pi q}{l_v}\right] &= \frac{A^2}{2}\sum_{n=0}^{l_v-1}\cos\left[\frac{\pi n f}{64}\right]e^{-\frac{2\pi n q}{l_v}j} + \sigma^2\sum_{n=0}^{l_v-1}e^{-\frac{2\pi n q}{l_v}j} \\ &= \frac{A^2}{4}\sum_{n=0}^{l_v-1}e^{2\pi n j\left(\frac{f}{128}-\frac{q}{l_v}\right)} + \frac{A^2}{4}\sum_{n=0}^{l_v-1}e^{2\pi n j\left(-\frac{f}{128}-\frac{q}{l_v}\right)} \\ &\quad + \sigma^2 l_v \delta[q] \\ &= \frac{A^2 l_v}{4}\delta\left[q - \frac{l_v f}{128}\right] + \frac{A^2 l_v}{4}\delta\left[l_v - q - \frac{l_v f}{128}\right] \\ &\quad + \sigma^2 l_v \delta[q] \end{aligned} \tag{4.295}$$

for all $q \in [0, l_v - 1]$*. Let* $\left\lfloor \frac{l_v f}{128} \right\rfloor = q_e$ *be the quantized version of* f*. Then,*

$$S_x\left[\frac{2\pi q}{l_v}\right] = \frac{A^2 l_v}{4}\delta[q - q_e] + \frac{A^2 l_v}{4}\delta[l_v - q - q_e] + +\sigma^2 l_v \delta[q] \tag{4.296}$$

Hence,

$$f_{1,1}\left(\frac{2\pi q}{l_v}\right)=\begin{cases}\left(\frac{A^2}{4}+\sigma^2\right)l_v & q=q_e\\ \left(\frac{A^2}{4}+\sigma^2\right)l_v & q=l_v-q_e\\ \sigma^2 l_v & \text{otherwise}\end{cases} \tag{4.297}$$

Furthermore,

$$p_{1,1}[q]=\frac{1}{l_v}\left|\sum_{m=0}^{l_v-1}x[m]\,e^{-\frac{2\pi jmq}{l_v}}\right|^2 \tag{4.298}$$

where $x[m]$ *denotes the observed signal. Then, the WLT for an estimate of* f*, denoted here by* $\left\lfloor\frac{l_v\hat{f}}{128}\right\rfloor=\hat{q}_e$ *for short, is*

$$\begin{aligned}\Lambda_W(\hat{q}_e)&=-\sum_{q=0}^{l_v-1}\left\{\frac{1}{2\pi}\frac{p_{1,1}[q]}{f_{1,1}\left(\frac{2\pi q}{l_v}\right)}+\ln\left(f_{1,1}\left(\frac{2\pi q}{l_v}\right)\right)\right\}\\ &=-(l_v-2)\ln\left(\sigma^2 l_v\right)-2\ln\left(\left(\frac{A^2}{4}+\sigma^2\right)l_v\right)-\frac{p_{1,1}[\hat{q}_e]}{2\pi\left(\frac{A^2}{4}+\sigma^2\right)l_v}-\frac{p_{1,1}[l_v-\hat{q}_e]}{2\pi\left(\frac{A^2}{4}+\sigma^2\right)l_v}\\ &\quad-\sum_{\substack{q=0\\ q\neq\hat{q}_e,l_v-\hat{q}_e}}^{l_v-1}\frac{p_{1,1}[q]}{2\pi\sigma^2 l_v};\ \hat{q}_e\in[0,l_v-1]\end{aligned} \tag{4.299}$$

The above expression is computed (instead of plotting as a function of $\hat{q}_e$*, we have plotted it as a function of frequency using a sampling rate of 128) when the sinusoid assumes a 20 Hz frequency in Figs.* 4.28, 4.29 and 4.30 *for various signal level,* $\sigma^2=0.1$, *and an observation interval of* $l_v=1024$. *As can be seen, the likelihood function assumes its maximum at* $f=20$ *Hz for signal level as small as* $A=0.1$ $(\sigma^2=0.1,\ \frac{A^2}{2}=0.005\rightarrow\frac{\frac{A^2}{2}}{\sigma^2}=0.05)$, *allowing one to correctly detect the unknown frequency. As signal level is increased, the estimation technique becomes increasingly less sensitive to noise. Now, when the observation interval is reduced to only 256, a similar encouraging results emerge in Figs.* 4.31, 4.32 and 4.33, *pointing to the efficacy of this form of estimation in the absence of full statistical knowledge of the signal under investigation. It is noteworthy that reducing the observation interval (Fig.* 4.31*) made the estimation process vulnerable to the additive noise. However, with an increase in the observation interval, the estimation process yields satisfactory result.*

The above example quite clearly demonstrated the process by which one can achieve parameter estimation in the absence of a complete statistical knowledge of the signal under observation. The strength of this method is in its design, which relies on the availability of the PSD of signal (PSD may or may not be at our disposal for some problems). For many biomedical signal analysis, however, the PSD can be

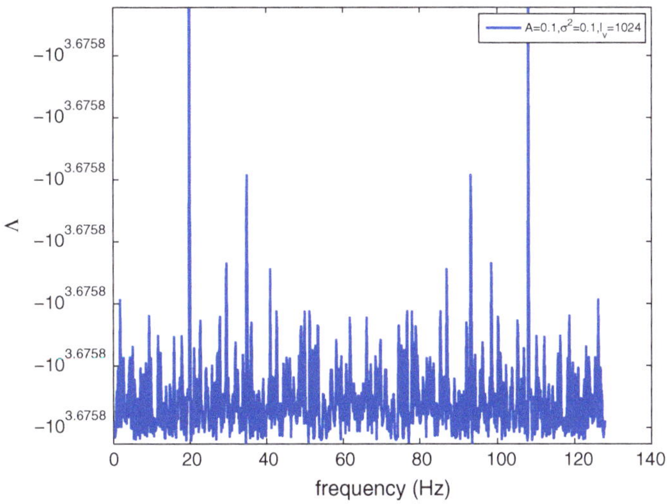

Fig. 4.28 Whittle likelihood ratio as a function of frequency for low SNR (1024 samples)

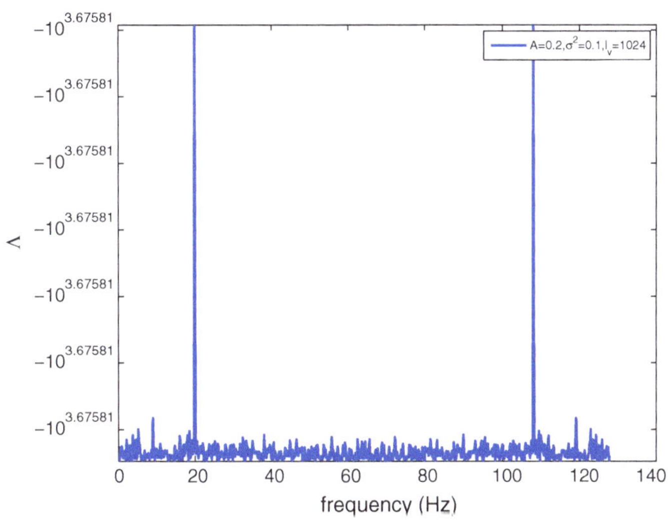

Fig. 4.29 Whittle likelihood ratio as a function of frequency for high SNR (1024 samples)

estimated by acquiring the correlation function of the signal using time-domain analysis (ergodicity in correlation) for some segments of time domain. That is, although biological signals are inherently non-stationary (as they are dependent on the behavior of the corresponding biological system, which is subject to external influences), for some segments of observation, quasi-stationay behavior can be assumed. This will allow the collection and acquisition of the correlation function, and hence, the PSD in order to invoke the WLT.

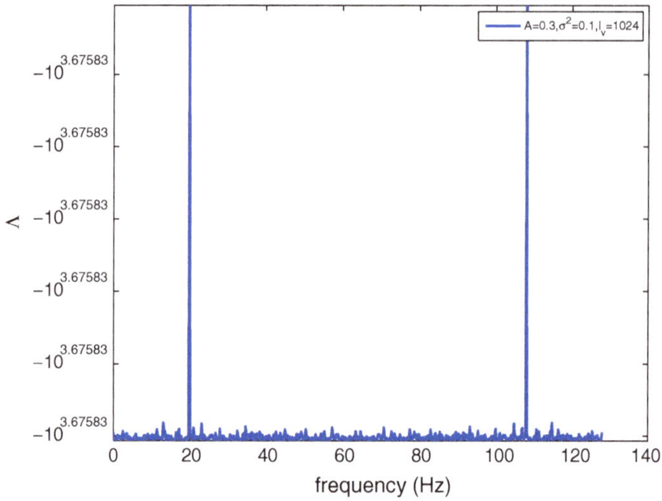

Fig. 4.30 Whittle likelihood ratio as a function of frequency for high SNR (1024 samples)

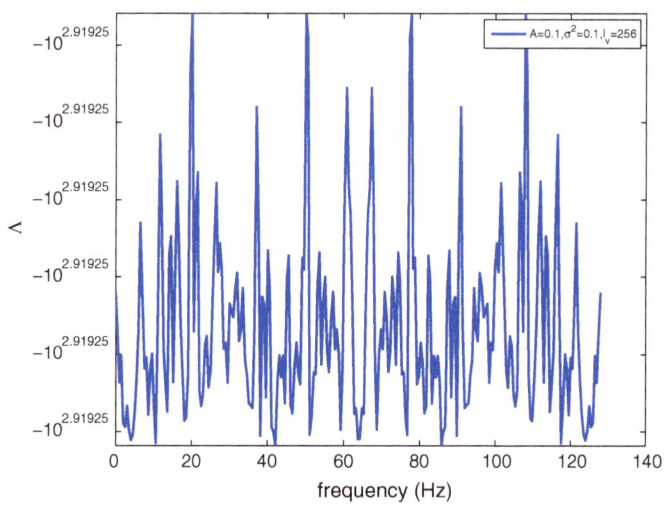

Fig. 4.31 Whittle likelihood ratio as a function of frequency for low SNR (256 samples)

4.4.5 Frequency Estimation

Frequency estimation plays a key role in some areas of biomedical signal processing. For instance, in the area of speech signal processing, one is required to estimate the major frequency components of the signal in order to classify the signal. The inherent non-stationarity of biomedical signals impose nontrivial restrictions on the standard frequency estimation methods. In the previous discussion, we addressed WLT as a

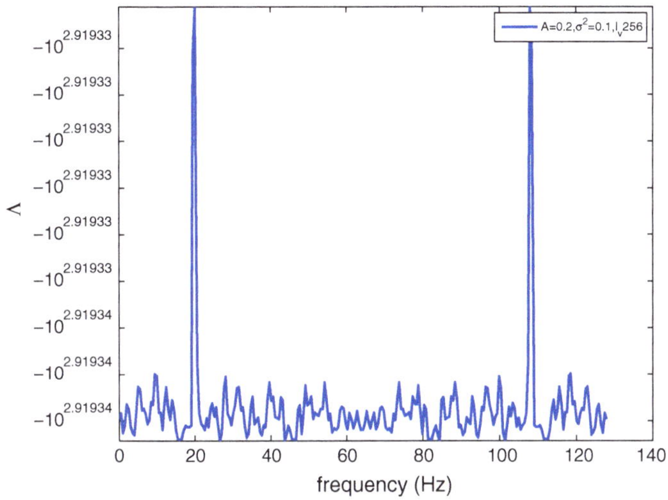

Fig. 4.32 Whittle likelihood ratio as a function of frequency for high SNR (256 samples)

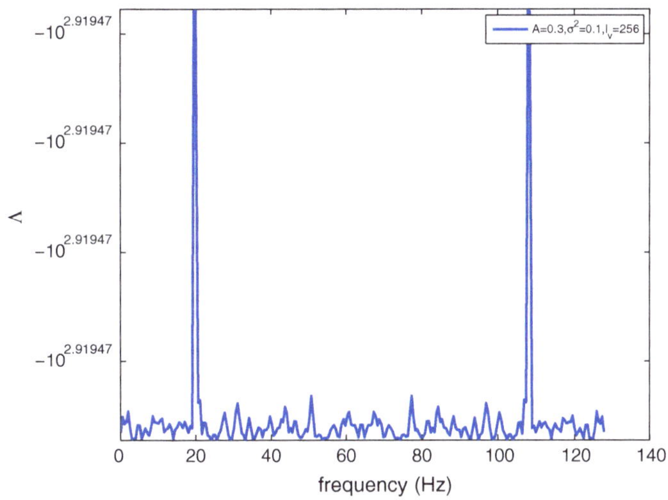

Fig. 4.33 Whittle likelihood ratio as a function of frequency for high SNR (256 samples)

means of estimation the PSD in general. However, one would be more concerned with a more computationally efficient technique to ascertain the frequency content of the signal. In this section, we address this concept when it is assumed that signal of interest is considered to be "narrowband," which implies that the spectrum of the signal is confined to a narrow slice of the frequency spectrum. That is, the PSD of the signal assumes "non-negligible" values over a small (this is relative and it may vary from problem to problem) section of the spectrum. Obviously, additive noise in the electronic devices do not fall into this category, while many biomedical signals can be classified as narrowband. For the sake of clarity, we first assume a continuos time approach and then extend the discussion to the sampled signal model. To that end, let us consider $x(t)$, which is assumed to be narrowband. In order to measure the frequency content, one approach is to obtain the "phase" of the signal and then estimate frequency by finding the derivative of the phase with respect to time. However, how do we find the "phase" of a real signal? The answer to this question can be provided in the context of a complex signal. To elaborate, for a complex signal $z(t)$, given by

$$z(t) = x(t) + jy(t), \tag{4.300}$$

phase can be defined as

$$\theta(t) = \tan^{-1}\left(\frac{y(t)}{x(t)}\right), \tag{4.301}$$

and hence, we can define the time dependent frequency as

$$\Omega(t) = \frac{d\theta(t)}{dt} \text{ in rad/s.} \tag{4.302}$$

In the case presented here, however, we only have $x(t)$. This leaves us with an ambiguity in using the above technique in order to estimate frequency. A transformation that can be used to generate the complex part of a signal, and thereby enabling us to use the above method for estimating frequency, is the Hilbert transform. To understand this concept, let us take a closer look at HT.

4.4.5.1 Hilbert Transform

HT is defined as a non-causal filtering operation with its impulse response given by

$$h_{HB}(t) = \frac{j}{\pi t}. \tag{4.303}$$

In frequency domain, the system function is defined as

$$H_{HB}(\Omega) = \begin{cases} -j & \Omega \geq 0 \\ j & \Omega < 0 \end{cases}. \tag{4.304}$$

This transform, as can be seen, cannot be implemented in the real world as it is a non-causal linear system ($h_{HB}(t) \neq 0$ for $t < 0$). However, it can always be approximated. To that end, let us consider the response of this filter to a sinusoidal signal. That is, let

$$x(t) = A\cos(\Omega_0 t + \theta) = \frac{A}{2}e^{j(\Omega_0 t + \theta)} + \frac{A}{2}e^{-j(\Omega_0 t + \theta)}. \tag{4.305}$$

Now passing this signal through the filter results in

$$y(t) = x(t) \circ h(t)$$

or

$$\begin{aligned} Y(\Omega) &= X(\Omega) H_{HB}(\Omega) = \frac{-jA}{2} e^{j(\Omega_0 t+\theta)} + \frac{jA}{2} e^{-j(\Omega_0 t+\theta)} \\ &= \frac{A}{2j} e^{j(\Omega_0 t+\theta)} - \frac{A}{2j} e^{-j(\Omega_0 t+\theta)} \Rightarrow \\ y(t) &= A\sin(\Omega_0 t+\theta) \end{aligned} \tag{4.306}$$

Similarly, for $x(t) = A\sin(\Omega_0 t+\theta) = \frac{A}{2j} e^{j(\Omega_0 t+\theta)} - \frac{A}{2j} e^{-j(\Omega_0 t+\theta)}$, we have

$$\begin{aligned} Y(\Omega) &= X(\Omega) H_{HB}(\Omega) = \frac{-A}{2} e^{j(\Omega_0 t+\theta)} - \frac{A}{2} e^{-j(\Omega_0 t+\theta)} \Rightarrow \\ y(t) &= -A\cos(\Omega_0 t+\theta). \end{aligned} \tag{4.307}$$

Hence, Hilbert filter causes a -90° phase shift for all sine and cosine components of the signal! Since this must be true for all frequency components of the signal, it is not realizable. However, if the signal is narrowband, then one can implement a unit-gain filter whose phase response remains constant over the band of interest (bandwidth of $x(t)$) at -90°. This approximation, obviously, can only be valid for a relatively small portion of the spectrum. Fortunately, many of signals which are encountered in the field of biomedical signal processing fall into this category.

Now, let us consider the following signal., which is formed by adding $x(t)$ and its Hilbert transform:

$$z(t) = x(t) + j\tilde{x}(t) \tag{4.308}$$

where

$$\tilde{x}(t) = x(t) \circ h(t). \tag{4.309}$$

Note that the combined operation may be viewed as a new filtering operation. That is, we can define

$$H_c(\Omega) = 1 + jH_{HB}(\Omega). \tag{4.310}$$

Hence,

$$z(t) = x(t) \circ h_c(t) \tag{4.311}$$

or

$$Z(\Omega) = X(\Omega) H_c(\Omega). \tag{4.312}$$

We claim that the frequency content of the signal can be estimated by measuring the phase of $z(t)$ and taking its derivative. That is,

$$\Omega(t) = \frac{d\theta(t)}{dt} \tag{4.313}$$

where

$$\theta(t) = \arg(z(t)) = \tan^{-1}\left(\frac{\tilde{x}(t)}{x(t)}\right). \tag{4.314}$$

To see this, let $x(t) = A\cos(\Omega_0 t + \theta)$. Then,

$$\begin{aligned} z(t) &= \frac{A}{2}e^{j(\Omega_0 t+\theta)} + \frac{A}{2}e^{-j(\Omega_0 t+\theta)} \\ &\quad + j\left(\frac{A}{2j}e^{j(\Omega_0 t+\theta)} - \frac{A}{2j}e^{-j(\Omega_0 t+\theta)}\right) \\ &= Ae^{j(\Omega_0 t+\theta)} \end{aligned} \tag{4.315}$$

Now,

$$\begin{aligned} \theta(t) &= \arg(z(t)) = \tan^{-1}\left(\frac{\tilde{x}(t)}{x(t)}\right) \\ &= \Omega_0 t + \theta \end{aligned} \tag{4.316}$$

and

$$\Omega(t) = \frac{d\theta(t)}{dt} = \Omega_0. \tag{4.317}$$

Hence, the Hilbert method was able to measure the frequency of $x(t) = A\cos(\Omega_0 t + \theta)$. Next, let us consider a periodic signal. That is,

$$x(t) = \sum_{n=-\infty}^{\infty} x_n e^{j\frac{2\pi nt}{T}} \tag{4.318}$$

where T denotes the period of the signal. We then have

$$\tilde{x}(t) = \sum_{n_1=-\infty}^{-1} jx_{n_1}e^{j\frac{2\pi n_1 t}{T}} - \sum_{n_2=0}^{\infty} jx_{n_2}e^{j\frac{2\pi n_2 t}{T}} \tag{4.319}$$

This in turn results in

$$z(t) = 2\sum_{n=0}^{\infty} x_n e^{j\frac{2\pi nt}{T}} \tag{4.320}$$

and

$$\theta(t) = \tan^{-1}\left\{\frac{\sum_{n=0}^{\infty} |x_n| \sin\left(\frac{2\pi nt}{T} + \phi_n\right)}{\sum_{n=0}^{\infty} |x_n| \cos\left(\frac{2\pi nt}{T} + \phi_n\right)}\right\} \tag{4.321}$$

where $\phi_n = \arg(x_n)$. This result is, obviously, disappointing as it does not allow for a true frequency discrimination. An important simplification occurs when the signal has a "significant" harmonic. In other words, when there exists a FT coefficient x_p such that

$$|x_l| \ll |x_p| \text{ for every } l \neq p. \tag{4.322}$$

In that case,

$$\begin{aligned}\theta(t) &\approx \tan^{-1}\left\{\frac{|x_p| \sin\left(\frac{2\pi pt}{T} + \phi_p\right)}{|x_p| \cos\left(\frac{2\pi pt}{T} + \phi_p\right)}\right\} \\ &\approx \frac{2\pi pt}{T} + \phi_p\end{aligned} \tag{4.323}$$

and

$$\Omega(t) \approx \frac{2\pi p}{T}. \tag{4.324}$$

Hence, we are able to extract the significant harmonic of a periodic signal using this method. This points to an important fact; that is, if the signal in question has a significant periodic component, then the Hilbert method can approximate the frequency of the signal. The accuracy of the approximation depends upon the degree by which the significant periodic component is exceeding all other components of the signal. The other impairment that is typically present in any measurement technique of this type is the impact of narrowband electronic noise. That is,

$$r(t) = x(t) + n(t) \tag{4.325}$$

where $n(t)$ is an AWGN with a 2-sided PSD of $\frac{N_0}{2}$ watts/Hz. Such a noise is wideband by definition whereas $x(t)$ is narrowband. As we learned earlier, a sub-optimum means for the detection in this case is a filtering operation. Let us assume that we can employ an ideal filter and that the uncertainty in the frequency of $x(t)$ is B Hz.

That is, $x(t) = A\cos(\Omega_0 t + \theta)$ with $0 < \Omega_0 < 2\pi B$ for some B. Hence, we can define $H(\Omega)$ as

$$H(\Omega) = \begin{cases} 1 & |\Omega| \le 2\pi B \\ 0 & \text{otherwise} \end{cases}$$

In that event,

$$\begin{aligned} r_f(t) &= x_f(t) + n_f(t) \\ &= r(t) \circ h(t) \qquad (4.326) \\ &= x(t) \circ h(t) + n(t) \circ h(t) \qquad (4.327) \end{aligned}$$

If $x(t) = A\cos(\Omega_0 t + \theta)$, then the ideal filtering operation would simply cause a phase shift. That is, $x_f(t) = A\sin(\Omega_0 t + \theta + \phi)$ whereas $n_f(t)$ is now a Gaussian noise with the PSD of

$$\begin{aligned} S_{n_f}(\Omega) &= \frac{N_0}{2}|H(\Omega)|^2 \\ &= \begin{cases} \frac{N_0}{2} & |\Omega| \le 2\pi B \\ 0 & \text{otherwise} \end{cases}. \end{aligned} \qquad (4.328)$$

Now, applying the Hilbert transform method, we have

$$\begin{aligned} z(t) &= r_f(t) + j\tilde{r}_f(t) \\ &= x(t) + j\tilde{x}(t) + n_f(t) + j\tilde{n}_f(t) \\ &= Ae^{j(\Omega_0 t + \theta)} + w(t) \end{aligned} \qquad (4.329)$$

where $w(t) = n_f(t) + j\tilde{n}_f(t)$ is a complex lowpass Gaussian noise. $z(t)$ may also be written as

$$\begin{aligned} z(t) = {} & A\cos(\Omega_0 t + \theta) + n_f(t) \\ & + j\left[A\sin(\Omega_0 t + \theta) + \tilde{n}_f(t)\right]. \end{aligned} \qquad (4.330)$$

$w(t)$ is zero-mean as $E\{w(t)\} = E\left(n_f(t)\right) + jE\left(\tilde{n}_f(t)\right) = 0$. However, in order to find its PSD, we need to examine the problem more closely. To that end, note that

$$\begin{aligned} S_w(\Omega) &= |H_c(\Omega)|^2 S_{n_f}(\Omega) \\ &= \frac{N_0}{2}|H(\Omega)|^2 |H_c(\Omega)|^2. \end{aligned} \qquad (4.331)$$

Note that

$$|H_c(\Omega)|^2 = 1 + |H_{HB}(\Omega)|^2 + jH_{HB}(\Omega) - jH^*_{HB}(\Omega). \qquad (4.332)$$

However, from the definition of the Hilbert filter,

$$H_{HB}^{*}(\Omega) = -H_{HB}(\Omega). \tag{4.333}$$

Therefore (see (4.310)),

$$|H_c(\Omega)|^2 = 1 + |H_{HB}(\Omega)|^2 + 2jH_{HB}(\Omega).$$

Given the definition of the Hilbert filter, we have

$$|H_c(\Omega)|^2 = \begin{cases} 4 & \Omega \geq 0 \\ 0 & \text{otherwise} \end{cases}. \tag{4.334}$$

This implies that

$$S_w(\Omega) = \begin{cases} 2N_0 & \Omega \geq 0 \\ 0 & \text{otherwise} \end{cases}. \tag{4.335}$$

In order to proceed with the frequency estimation, we need to obtain $\theta(t)$, given by

$$\theta(t) = \tan^{-1}\left\{\frac{A\sin(\Omega_0 t + \theta) + \tilde{n}_f(t)}{A\cos(\Omega_0 t + \theta) + n_f(t)}\right\}.$$

It is of importance for us to establish the relationship between $\tilde{n}_f(t)$ and $n_f(t)$ in order to study this expression. We note that $\tilde{n}_f(t)$ and $n_f(t)$ are a pair of real Gaussian processes (note that the linear filtering, including the Hilbert transform, will preserve the Gaussian property). We also note that the cross-power spectrum of $\tilde{n}_f(t)$ and $n_f(t)$ is

$$\begin{aligned} S_{\tilde{n}_f n_f}(\Omega) &= H_{HB}(\Omega)\, S_{n_f}(\Omega) \\ &= \begin{cases} -j\frac{N_0}{2} & 0 \leq \Omega \leq 2\pi B \\ j\frac{N_0}{2} & -2\pi B\Omega \leq \Omega < 0 \end{cases} \end{aligned} \tag{4.336}$$

This implies that

$$\begin{aligned} E\left(\tilde{n}_f(t)\, n_f(t)\right) &= R_{\tilde{n}_f n_f}(0) = \frac{1}{2\pi}\int_{-\infty}^{\infty} S_{\tilde{n}_f n_f}(\Omega)\, d\Omega \\ &= 0. \end{aligned} \tag{4.337}$$

Since $\tilde{n}_f(t)$ and $n_f(t)$ are Gaussian, the uncorrelated property, i.e., when $E\left(\tilde{n}_f(t)\right.$ $\left.n_f(t)\right) = 0$, leads to the two random process being independent.

The above expression for the instantaneous phase is a nonlinear expression of the additive noise, and hence, its analysis is fairly complicated. To see how this method is impacted by noise, we resort to an example.

Example 20 We are to estimate the frequency of a unit-amplitude, sinusoidal signal that is known to have a frequency less than or equal to 100 Hz. That is, the signal can assume any frequency in the range of [0, 100] Hz.The signal is corrupted by an AWGN. Assess the effectiveness of the Hilbert method for frequency estimation as a function the noise power when the signal assumes a frequency of 10 Hz. Assume that the observation interval is 0.5 s.

Solution: *For this problem, we are confined to an operating bandwidth. As such, it makes sense to assume that we can utilize a filter to eliminate any unwanted signal, including a portion of the wideband AGWN that is not in the bandwidth of interest. To go any further, let us formulate the problem. To that end, we can assume that the observed signal is*

$$r(t) = A\sin(\Omega_0 t + \theta) + n(t) \tag{4.338}$$

where $\Omega_0 \in [0, 100]$ *Hz. Based on the above discussion, we have*

$$\theta(t) = \tan^{-1}\left\{\frac{A\sin(\Omega_0 t + \theta) + \tilde{n}_f(t)}{A\cos(\Omega_0 t + \theta) + n_f(t)}\right\} \tag{4.339}$$

and

$$\Omega(t) = \frac{d\theta(t)}{dt} \tag{4.340}$$

In order to address this problem from a numerical standpoint, we generate the sampled version of the signal. The sampling rate is assumed to be 1000 samples/s, which should be large enough to represent any sinusoid with a frequency less than 100 Hz. Without the loss of generality we set $\theta = 0$. *Subsequently, one can generate the*

$$x_r[k] = \cos\left(20\pi 10^{-3} k\right) + \sigma^2 n[k] \tag{4.341}$$

and

$$x_i[k] = \sin\left(20\pi 10^{-3} k\right) + \sigma^2 \tilde{n}[k] \tag{4.342}$$

where $n[k]$ *and* $\tilde{n}[k]$ *are a pair of i.i.d zero-mean and unit-variance Gaussian random variables. In Fig. 4.34, a polar plot of* $x_r[k] + jx_i[k]$ *is provided for* $\sigma^2 = 10^{-2}$. *Quite clearly, one observes a circular pattern, signifying that sinusoidal nature of the real and imaginary part. Subsequently, in Fig. 4.35,* $\frac{1}{2\pi}\left|\frac{d\theta(t)}{dt}\right|$ *is depicted (the frequency in Hz). As one would expect, the observed signal points to 10 Hz as the possible frequency of the signal. However, due to* 2π *phase rotations that lead to a step function in the phase, and hence a large frequency jump, the instantaneous frequency seems to present delta-like "spikes" at regular intervals of time. These "jumps" can be ignored in view of the above observation. Hence, one can argue that*

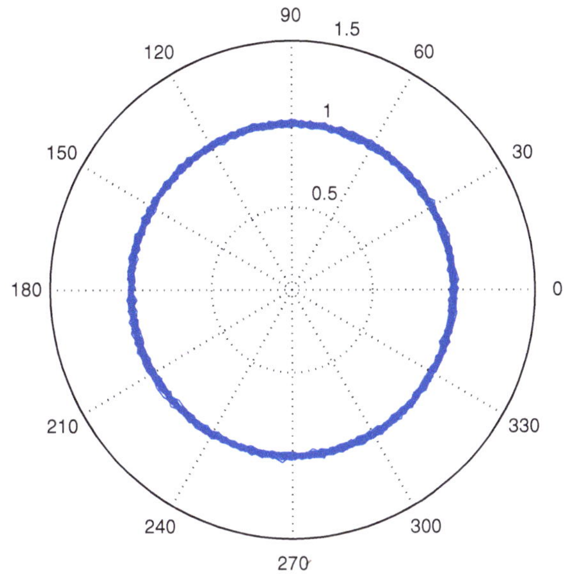

Fig. 4.34 Polar representation of the signal for $\sigma^2 = 0.01$

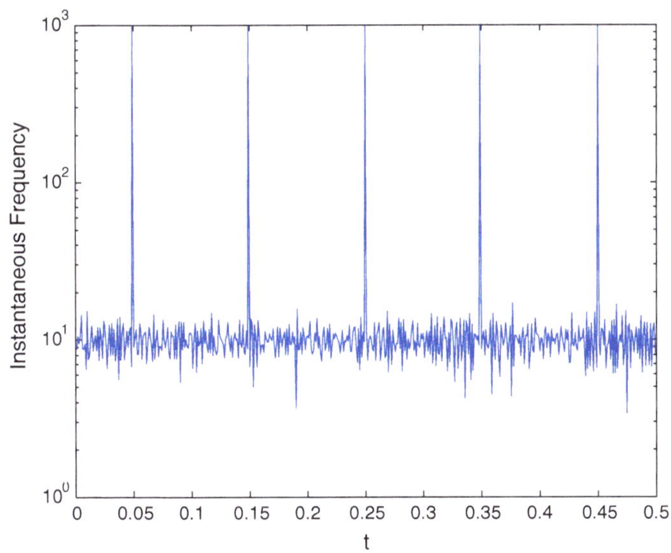

Fig. 4.35 Frequency estimation using Hilbert transform (no filtering)

the Hilbert method has successfully identified the frequency of the signal. However, the observation points to an issue regarding the noisy observation. That is, even with the small level of noise, the measured frequency has nontrivial fluctuations about its mean of 10 *Hz. Since noise offers a large bandwidth as compared to the signal, we are compelled to use a simple filtering operation to eliminate the impact of noise. To that end, an FIR low pass filter with a passband of 100 Hz and a stop frequency of 200 Hz (30 dB attenuation) is utilized. The filtering operation will have no impact on the sinusoidal component of the signal, while the noise component will be filtered out. To be specific,*

$$\begin{aligned} x_{r-f}[k] &= x_r[k] \circ h_{LP}[k] \\ &\approx \cos\left(20\pi 10^{-3}k + \theta\right) + \sigma_f^2 w[k] \end{aligned} \tag{4.343}$$

$$\begin{aligned} x_{i-f}[k] &= x_i[k] \circ h_{LP}[k] \\ &\approx \sin\left(20\pi 10^{-3}k + \theta\right) + \sigma_f^2 \tilde{w}[k] \end{aligned} \tag{4.344}$$

where now $w[k]$ and $\tilde{w}[k]$ are a pair of i.i.d zero-mean and unit-variance Gaussian random variables and σ_f^2 is the variance of the filtered signal. The filter was designed in matlab using filter design tool. The response of the filter is shown in Fig. 4.36. *Figure* 4.37 *depicts the output of the frequency estimator for the filtered signal. As can be seen, filtering has reduced the impact of wideband noise and has enhanced the accuracy of the estimation process. However, the impact of phase rotations, which result in a large variation in frequency have not been filtered out. To gain further insight, in Figs.* 4.38, 4.39 *and* 4.40, *the same problem is addressed when $\sigma^2 = 0.1$. Note that an increase in the noise level has a profound impact on the accuracy of the estimation process. In fact, the addition of the filter has limited impact as the bandwidth of the filter is wide enough to cover the entire 100 Hz band. In Figs.* 4.41, 4.42 *and* 4.43 *the same problem is addressed when $\sigma^2 = 1$. In this case, the noise is dominating the signal and resulting in a major ambiguity in the frequency estimation*

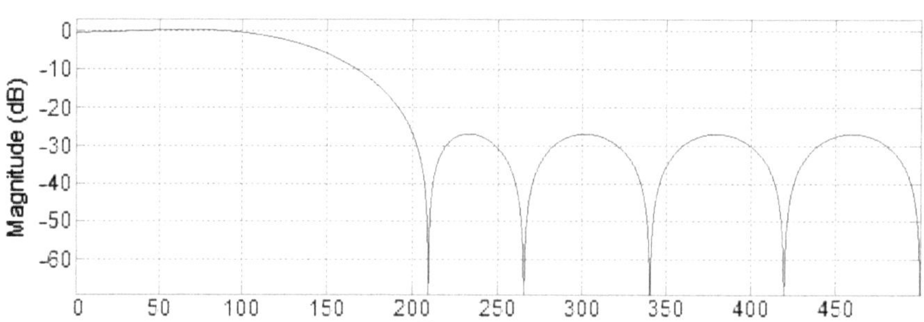

Fig. 4.36 FIR filter response. Passpand is 100 Hz with a 30 dB attenuation at 200 Hz

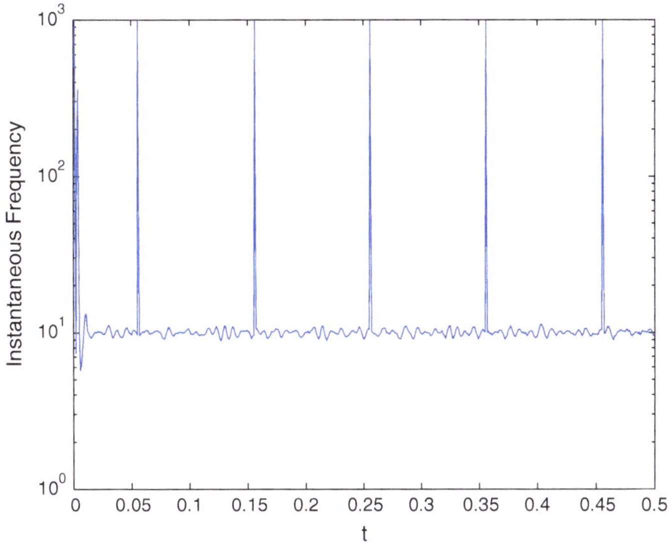

Fig. 4.37 Frequency estimation using Hilbert transform (with filtering)

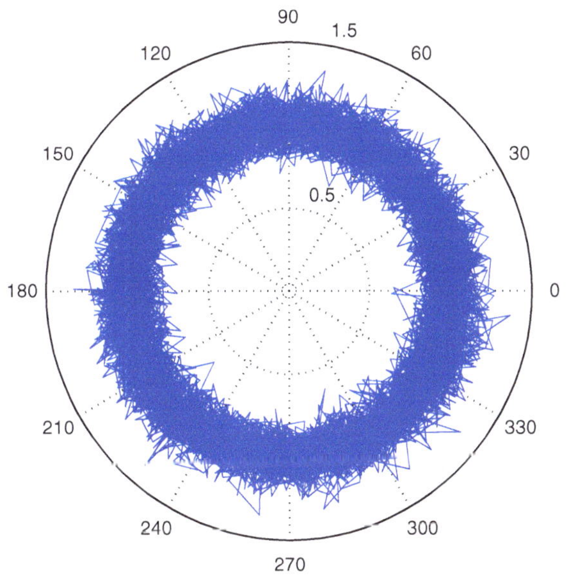

Fig. 4.38 Polar representation of the signal for $\sigma^2 = 0.1$

process, although it appears that the instantaneous frequency is centered about 10 Hz. Furthermore, the filtering operation has limited impact as the reduction of the noise variance using the filter is not significant enough to alleviate the problem. Finally,

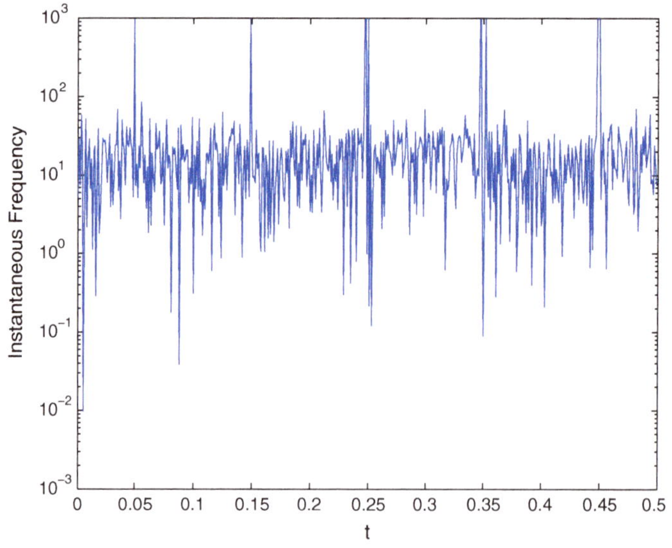

Fig. 4.39 Frequency estimation using Hilbert transform (no filtering)

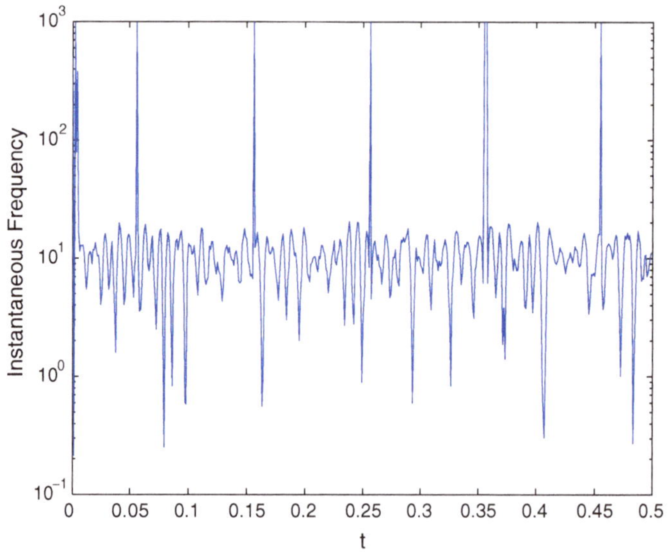

Fig. 4.40 Frequency estimation using Hilbert transform (with filtering)

one can see that the averaging of the output of the frequency estimator may not lead to a correct answer due to the large periodic spikes that are caused by the 2π phase rotation.

The above example underscores the sensitivity of the Hilbert method to noise. In fact, due to the nonlinear behavior of $\tan^{-1}$, the additive noise may result in phase

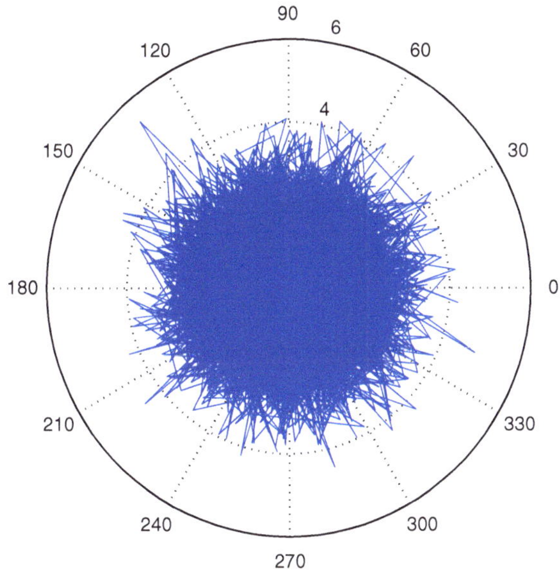

Fig. 4.41 Polar representation of the signal for $\sigma^2 = 1$

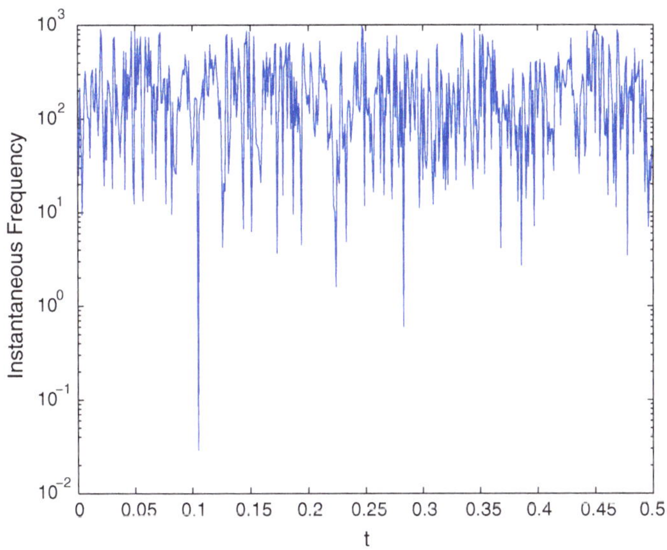

Fig. 4.42 Frequency estimation using Hilbert transform (no filtering)

rotation of 2π, which can result in large frequency errors. One way to improve the performance of the Hilbert method is the use of matched filtering discussed above. However, instead of using one single filter, a bank of filters may be used, with each filter "tuned" to a particular frequency band. An equivalent operation is correlation

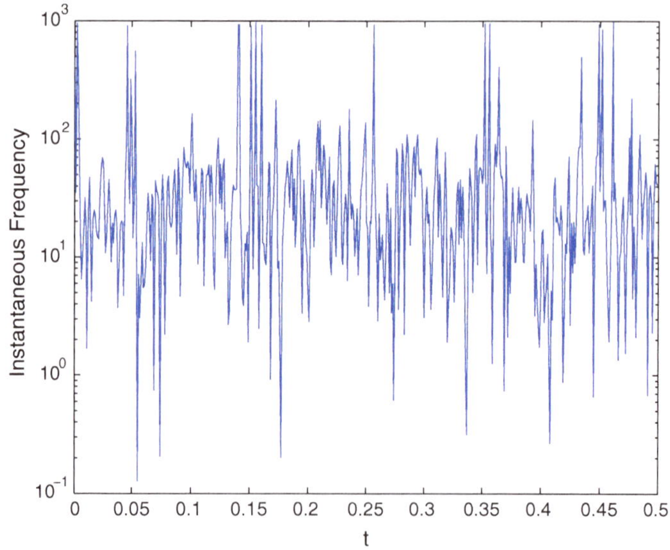

Fig. 4.43 Frequency estimation using Hilbert transform (with filtering)

with sin and cos terms at the frequencies of interest and deciding which filter produces the largest output. This operation, quite interestingly, is similar to computing the FFT of the signal and measuring the frequency at which the peak has occurred. That is,

$$\hat{f} = \frac{1}{2\pi} \arg\max_{\omega} |X[\omega]| \tag{4.345}$$

where

$$X[\omega] = \sum_{n=0}^{N-1} x[k] e^{-jk\omega} \tag{4.346}$$

where it is assumed that N samples of $x(t)$ are at our disposal. The above expression is the DFT of the signal. For practical reasons, we use FFT instead to approximate the above. That is, we compute the discrete frequency version of DFT

$$X_n = \sum_{k=0}^{N-1} r[k] e^{-j\frac{2nk\pi}{N}}; n \in [0, 1, \ldots, N-1] \tag{4.347}$$

with n corresponding to $\hat{f} = \frac{nf_s}{N}$ with f_s denoting the sampling rate in samples/s. If the observation interval consists of

$$M = NP \tag{4.348}$$

samples, we can use the following as a measure of frequency:

$$\hat{f} = \frac{1}{2\pi P} \sum_{l=1}^{P} \arg\max_{\omega} |X_l[\omega]| \tag{4.349}$$

where

$$X_l[\omega] = \sum_{k=(l-1)N}^{lN-1} x[k]\, e^{-jk\omega}. \tag{4.350}$$

The selection of N and P (hence, M) can be a design issue decided by the hardware/software capabilities of the system and the length of observation. Now, let us examine the efficacy of this technique using the above example.

Example 21 Let us consider the previous example when the frequency estimator is an FFT based mechanism. Assume that the sampling rate is 256 samples/s and that we can observe 16 s of the signal. Also, assume that the hardware/software limits us to an FFT size of 256 points. Obtain the frequency error for this method for various FFT lengths.

Solution: *Since* $N = 256$, *sampling rate is* 256 *samples/s, and observation interval is* 16 *s, we have* $M = 16 * 256$ *and* $P = 16$. *However, in order to see the impact of the observation interval, we assume values of* $P = 2, 4, 8$, *and* 16 *(i.e., observation intervals of* 2, 4 , 8, *and* 16 *s). In Fig.* 4.44, *the performance of the FFT based*

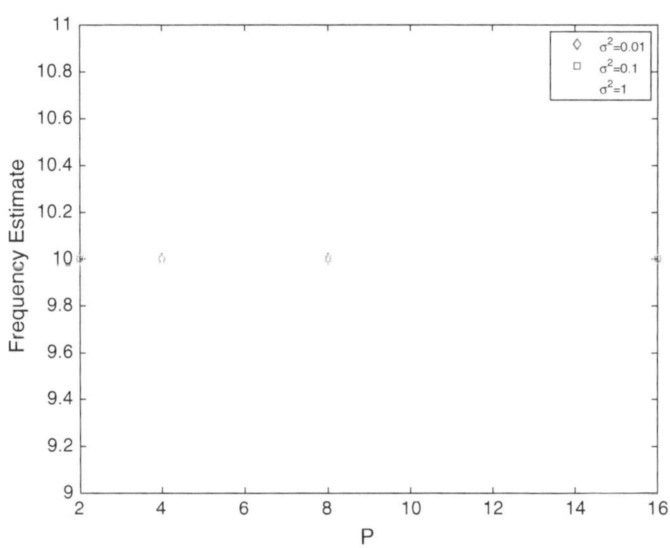

Fig. 4.44 Frequency estimation using FFT technique

technique is depicted. Interestingly enough, regardless of the power of noise and even for a small observation interval of $P = 2$ *(2 s), this algorithm yields a correct estimate of the frequency! This is due to the optimality of a correlation receiver (an FFT may be viewed as a bank of matched filters, matched to each frequency band defined by the frequency step size of the FFT) in the face of AWGN. Hence, this method is superior to the Hilbert method in the face of AWGN.*

We examined the performance of the FFT based frequency estimator in the face of AWGN and found it to be superior to its Hilbert transform counterpart. In fact, the FFT based can provide an accurate estimate of the frequency where Hilbert method clearly fails. Let us now examine the performance of these technique in the fact of biological signals. That is, some biological signals contain specific periodic terms that are masked by non-Gaussian random disturbances. We examine this problem using an example.

Example 22 Use the FFT method to identify the strongest periodic components of an EEG signal.

Solution: *We use chb02_01_edfm from MIT-BIH. This signal contains circuit noise (AWGN) along with non-Gaussian "noisy" signal component that is generated by random activities of the brain. It is questionable whether FFT can isolate periodic components of the signal in the presence of such noisy contributions, which are due to the electrical activities of the brain. To see the efficacy of the short-term FFT, we use the first row (the first probe) for* 50 *min. This window of observation is long enough to allow for the averaging of all noisy phenomena that have short correlation*

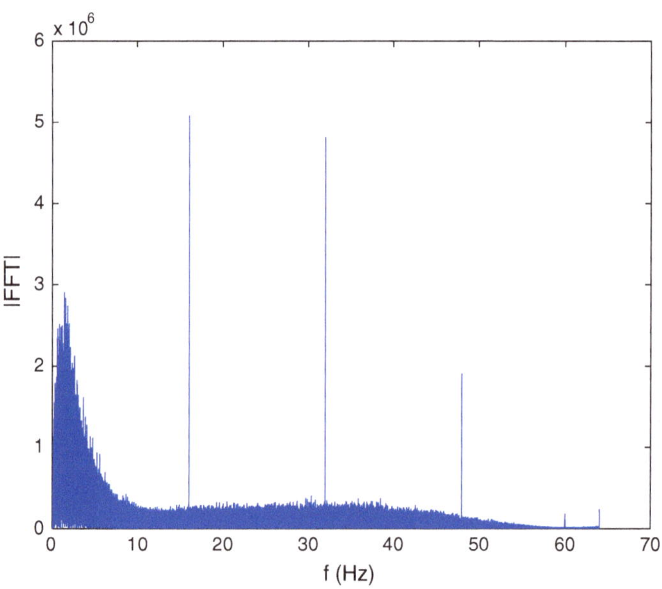

Fig. 4.45 The absolute value of the FFT of the EEG signal

times. It is assumes that the sampling rate is 256 *Hz. The mean value is subtracted out. The FFT of the signal (due to the large number of samples, the absolute value of FFT assumes a large peak) for the entire length of observation is plotted in Fig.* 4.45. *From this figure, one can clearly identify* 16 *Hz (β wave) as well as* 32 *Hz and* 48 *Hz (γ wave) periodic components. The algorithm recognizes the* 16 *Hz component as the strongest periodic component of the signal, which corresponds to the β wave present in a typical EEG.*

The above techniques focus on the detection and estimation when the signal of interest consists of simple components. A question that can be raised at this time is the following.

> If the signal of interest is known to be a mixture of multiple independent signals (including noise), how do we decompose and recover the individual components of the signal, thereby unravelling the fundamental structure of the signal?

This question will be addressed in the ensuring chapter.

Chapter 5
Signal Decomposition Methods

Signal decomposition is in large part concerned with recovering the underlying structure of the signal of interest. This subject has been studied extensively in the literature and holds an important place in the biomedical signal processing field as most signal processing problems in this arena defy the simple signal plus AWGN model. When such a model is violated, one must be concerned with the recovery of multiple signals which may be present in an observed waveform. In the previous sections and chapters of this text, we have come to learn that biological signals very often are comprised of several signals, each perhaps of clinical significance. Hence, the ability to recover the individual components of the signal is of paramount importance here.

There are two well-known techniques that one can consider; principle component analysis and independent component analysis. Wavelet analysis is yet another method of decomposing signals, but since it is not designed to isolate multiple components of the signal, we address this subject separately. We will discuss these techniques in that order in the following.

5.1 Principle Component Analysis

In many problems encountered in the field of biomedical signal processing, one usually has access to a wealth of information in time domain (time series) since signals often are collected from multitude of subjects over extended periods of time. Earlier, we demonstrated that the time-varying nature of such signals limits, if not prohibits, the use of many methods devised for the detection and estimation of signals. To elaborate, algorithms which are used routinely in the electronic communications world yield unfavorable results in this domain as they rely heavily on the stationarity (including cyclostationarity) of signals under investigation.

K. Kiasaleh, *Biological Signals Classification and Analysis*,
Lecture Notes in Bioengineering, DOI 10.1007/978-3-642-54879-6_5

Given the availability of time series, one can establish two important facts. First, one can compute the correlation function of signals under investigation. Second, equipped with this information, one can establish the correlation length (the length of time a signal remains correlated) of the signal. Such an information is valuable in that it reveals the separation of the observation one has to consider in order to insure "small" correlation among observed samples. We call this interval "window" as it reflects a period of time over which the signal remains somewhat correlated.

Once such measures are taken into account, one can form a new set of samples of the signal which reflect, perhaps more accurately, the true statistical nature of the signal. In many events, the signal is corrupted by additive electronic noise with zero mean. This calls for an averaging of the samples over the "window." In some occasions, the signal contains a non-zero mean, which needs to be extracted out. Equipped with this information, we form the new random vector

$$\mathbf{y} = \left[y_0, y_1, \ldots, y_{l_v-1}\right]^{\mathsf{T}} \tag{5.1}$$

whose elements are as follows

$$y_l = \sum_{j=lw}^{(l+1)w-1} x_j - \bar{x}; l \in \{0, 1, \ldots, l_v - 1\} \tag{5.2}$$

with w denoting the length of window and

$$\bar{x} = \lim_{p\to\infty} \frac{1}{p} \sum_{j=0}^{p-1} x_j \tag{5.3}$$

denoting the dc level (namely, average) of $x_j's$. Since an averaging for $p \to \infty$ is not possible, instead, we perform the following:

$$\bar{x} \approx \frac{1}{wl_v} \sum_{j=0}^{wl_v-1} x_j \tag{5.4}$$

which means that we average the data over the entire observation length.

In the following, we focus on classifying $\mathbf{y}$. Note that this definition assumes that

$$E\left\{x_j\right\} = \bar{x}, \tag{5.5}$$

which in turn implies mean ergodicity. In many problems, $wl_v \gg 1$, and hence, $\frac{1}{wl_v}\sum_{j=0}^{wl_v-1} x_j$ is close to $\bar{x}$, since mean ergodicity also implies that $\frac{1}{p}\sum_{j=0}^{p-1} x_j$ is a converging sequence.

One question that can be raised here is regarding the dimensionality of the signal we have observed. Since **y** is a random vector, one has to address this question from a statistical point of view, or what is known as the KL expansion discussed in Chap. 1. KL expansion allows an orthogonal expansion of the vector, thereby allowing to decompose the observed vector into a number of orthogonal vectors and their corresponding uncorrelated coefficients. Notice the assertion of "uncorrelated" and not "independent" in this discussion. That is, our focus is to decompose the observed vector into a number of components which are *uncorrelated*, but not necessarily *independent*. For Gaussian case, uncorrelated implies independence. We do not concern ourselves with Gaussian cases here, and hence, principle component decompositions does not lead to independent component analysis. In a formal way, we are interested in the expansion

$$\mathbf{y} = \sum_{q=0}^{l_v-1} b_q \boldsymbol{\phi}_q \tag{5.6}$$

where $\boldsymbol{\phi}_q = \left[\phi_{q,0}, \phi_{q,1}, \ldots, \phi_{q,l_v-1}\right]^{\mathsf{T}}; q = 0, 1, \ldots, l_v - 1$, is the qth eigenvector (column vector) with its corresponding eigenvalue λ_q, which satisfy

$$\mathbf{K}_y \boldsymbol{\phi}_q = \lambda_q \boldsymbol{\phi}_q; q \in \{0, 1, 2, \ldots, l_v - 1\} \tag{5.7}$$

with

$$\mathbf{K}_y = E\left\{\mathbf{y}\mathbf{y}^H\right\} \tag{5.8}$$

denoting the correlation matrix of **y**. Hence, the random vector **y** has been replaced with

$$\mathbf{b} = \left[b_0, b_1, \ldots, b_{l_v-1}\right]^{\mathsf{T}} \tag{5.9}$$

whose elements are uncorrelated. Note that $\mathbf{K}_y$ is an $l_v \times l_v$ matrix, and hence, one would expect l_v distinct eigenvalues. Using (5.7), we can also show that

$$\mathbf{K_y U} = \mathbf{U}\Lambda \rightarrow \mathbf{K}_y = \mathbf{U}\Lambda\mathbf{U}^{-1} \tag{5.10}$$

where

$$\mathbf{U} = \left[\boldsymbol{\phi}_0, \boldsymbol{\phi}_2, \ldots, \boldsymbol{\phi}_{l_v-1}\right] \tag{5.11}$$

and

$$\Lambda = \mathrm{diag}\left[\lambda_0, \lambda_1, \ldots, \lambda_{l_v-1}\right]. \tag{5.12}$$

As noted earlier, the objective here is to obtain the eigenvector and eigenvalues by solving (5.7). It is important to note that $\mathbf{K}_y$ enjoys the Hermitian symmetry by the virtue of its definition, which in turn leads to some interesting properties regarding the eigenvalues and eigenvectors. This problem was discussed earlier, but for the sake of providing a complete discussion here, we reiterate some key results about covariance matrices in the following discussion. To that end, we showed above that

$$\mathbf{K}_y = \mathbf{U}\Lambda\mathbf{U}^{-1}. \tag{5.13}$$

Since

$$\begin{aligned}\mathbf{K}_y &= \mathbf{K}_y^H \rightarrow \\ \mathbf{U}\Lambda\mathbf{U}^{-1} &= \left(\mathbf{U}^{-1}\right)^H \Lambda^H \mathbf{U}^H,\end{aligned} \tag{5.14}$$

then we have

$$\mathbf{U}^{-1} = \mathbf{U}^H \rightarrow \mathbf{U}^H\mathbf{U} = \mathbf{I} \text{ and } \Lambda = \Lambda^H. \tag{5.15}$$

This implies that

$$\mathbf{K}_y = \mathbf{U}\Lambda\mathbf{U}^H \tag{5.16}$$

and that the eigenvalues of the covariance matrix are real and eigenvectors are orthogonal. That is,

$$\phi_{q_1}^H \phi_{q_2} = \delta\left[q_1 - q_2\right], \tag{5.17}$$

which makes the transforming matrix **U** unitary. Then, we can obtain $b_q' s$ as follows:

$$b_q = \phi_q^H \mathbf{y};\ q \in \{0, 1, 2, \ldots, l_v - 1\} \tag{5.18}$$

or

$$\mathbf{b} = \mathbf{U}^{\mathbf{H}}\mathbf{y}. \tag{5.19}$$

Note that this is an orthogonal expansion of **y** where the basis vectors for the expansion are the columns of the unitary matrix **U**. This also can be expressed as

$$\mathbf{y} = \mathbf{U}\mathbf{b}. \tag{5.20}$$

Note that b_q's are in general random variables with their statistical characteristics dependent upon those of **y**. It can easily be shown that if **y** is zero mean (note that we made sure that this condition is satisfied), then b_q's are also zero mean with the auto-correlation

$$E\left\{b_{q_1} b_{q_2}\right\} = \lambda_q \delta\left[q_1 - q_2\right]. \tag{5.21}$$

This also implies that

$$\begin{aligned}E\left\{\mathbf{b}\mathbf{b}^H\right\} &= \mathbf{U}^H \mathbf{K}_y \mathbf{U} \\ &= \Lambda = \operatorname{diag}\left\{\lambda_0, \lambda_1, \ldots, \lambda_{l_v - 1}\right\}.\end{aligned} \tag{5.22}$$

Hence, the above operation allows expressing a correlated random vector in terms of l_v uncorrelated random variables. We cannot, however, make any further assertions regarding $b_q^{\prime}s$ as they are linear combinations of the elements of $\mathbf{y}$ with, perhaps, an unknown pdf. Unlike electronic communication problems where $\mathbf{y}$ is modeled as a Gaussian vector, resulting in $b_q^{\prime}s$ that are zero-mean, Gaussian, and uncorrelated (and hence independent) random variables, in the context of a non-Gaussian environment, we can only assume uncorrelated $b_q^{\prime}s$.

Lastly, we need to emphasize that the elements of $\mathbf{b}$, although uncorrelated, have difference variances. In some occasions, it is of interest to consider a case where we intend to make the elements of $\mathbf{b}$ not only uncorrelated, but also unit variance (white vector). Such a vector can be obtained using the transformation

$$\begin{aligned} \mathbf{b}_n &= \Lambda^{-1/2}\mathbf{b} \\ &= \Lambda^{-1/2}\mathbf{U}^H\mathbf{y} \end{aligned} \tag{5.23}$$

Note that

$$\begin{aligned} E\left\{\mathbf{b}_n\mathbf{b}_n^H\right\} &= \Lambda^{-1/2}\mathbf{U}^H\mathbf{K}_y\mathbf{U}\Lambda^{-1/2} \\ &= \Lambda^{-1/2}\Lambda\Lambda^{-1/2} \\ &= \mathbf{I}. \end{aligned}$$

In this equation, $\mathbf{I}$ denotes an $l_v \times l_v$ identity matrix. As can be seen, a key assumption ($\Lambda^{-1/2}$) for this transformation is that $\lambda_i > 0$ for all i. Note that $\lambda_i \geq 0$ for all i is insured by $\mathbf{K}_y$ being a covariance matrix (and hence at least positive semi-definite). $\lambda_i > 0$ for all i (positive definite), however, may not be satisfied for some biomedical signal processing problems, which result in a covariance matrix that is only positive semi-definite and not positive definite.

We need to add that in many problems in the field of biomedical signal processing $\mathbf{K}_y$ is not full ranked, resulting in less than l_v distinct, non-zero eigenvalues. That is, the covariance matrix cannot be guaranteed to be positive definite. Instead, one has to consider a positive semi-definite matrix in general when dealing with covariance matrices observed in the field of biomedical signal processing. Furthermore, based on the above definition, the eigenvector with the largest eigenvalue is of critical importance, as it reflects the most significant component of the expansion. It would be then of interest to consider a method where the most significant eigenvalues are obtained. That is, we are interested in a method that yields

$$\lambda_0 \geq \lambda_1 \geq \cdots \geq \lambda_{l_v-1} > 0.$$

A method that allows for achieving this goal is the singular value decomposition (SVD). Before we present the concept of SVD, it is important to note that $\mathbf{K}_y$ may not be available due to the unavailability of the statistical properties of $\mathbf{y}$. Therefore, one has to relying on an estimate of $\mathbf{K}_y$, which is often obtained via observing the desired signal over long periods of time.

Singular Value Decomposition In general, the covariance matrix is a square, full-rank matrix, leading to l_v non-zero eigenvalues. In some cases, the covariance matrix is not full ranked (i.e., some of the l_v eigenvalues will be zero). Such a covariance matrix is positive semi-definite, and the standard eigenvector decomposition discussed above cannot be used. It is also of interest to insure that the eigenvectors that yield the significant eigenvalues are identified first for a large l_v.

The concept of SVD is concerned with these problems. Furthermore, SDV can be applied to matrices that are indeed non-square. That is, if $\mathbf{A}$ denotes an $m \times n$ $(n \leq m)$ matrix (not necessariy full rank), one can find the matrices $\mathbf{U}$ $(m \times m)$, $\mathbf{V}$ $(n \times n)$, and Σ $(m \times n$ and diagonal) such that

$$\mathbf{A} = \mathbf{U}\Sigma\mathbf{V}^H \tag{5.24}$$

where Σ is a matrix of all 0's except for $\Sigma(1,1) = \lambda_0$, $\Sigma(2,2) = \lambda_1$,..., $\Sigma(r-1,r-1) = \lambda_{r-1}$ with $\lambda_0 \geq \lambda_1 \geq \cdots \geq \lambda_{r-1} > 0$ (called singular values of $\mathbf{A}$) and r $(\leq n)$ denoting the rank of the matrix $\mathbf{A}$ (hence, r non-zero eigenvalues). $\mathbf{A}$ will be full column rank when $r = n \leq m$. The proof for the above assertion is quite instructive. Hence, a brief discussion to that end is provided below.

First, let us consider $\mathbf{A}^H\mathbf{A}$. This matrix is an $n \times n$ matrix that has, obviously, Hermitian symmetry. Hence, it has up to n non-negative, real eigenvalues. Let $\lambda_0 \geq \lambda_1 \geq \cdots \geq \lambda_{r-1} > 0$ be the r non-zero eigenvalues. Also, let $\mathbf{V}_1 = \left[\mathbf{v}_0, \mathbf{v}_1, \ldots, \mathbf{v}_{r-1}\right]$ be the corresponding eigenvectors ($\mathbf{v}_j$ is an $n \times 1$ vector). Note than $\mathbf{V}_1$ is an $n \times r$ matrix. Since $\mathbf{A}^H\mathbf{A}$ has Hermitian symmetry,

$$\mathbf{V}_1\mathbf{V}_1^H = \mathbf{I} \tag{5.25}$$

Furthermore, for the remaining $n - r$ zero eigenvalues, let $\mathbf{V}_2 = \left[\mathbf{v}_r, \mathbf{v}_{r+1}, \ldots, \mathbf{v}_n\right]$, an $n \times (n-r)$ matrix, denote a matrix of the corresponding eigenvectors. Also, let

$$\mathbf{S} = \mathrm{diag}\left[\sqrt{\lambda_0}, \sqrt{\lambda_1}, \ldots, \sqrt{\lambda_{r-1}}\right]. \tag{5.26}$$

Then, from eigenvector-eigenvalue relationship,

$$\left(\mathbf{A}^H\mathbf{A}\right)\mathbf{V}_1 = \mathbf{V}_1\mathbf{S}^2 \tag{5.27}$$

or

$$\mathbf{S}^{-1}\mathbf{V}_1^H\left(\mathbf{A}^H\mathbf{A}\right)\mathbf{V}_1\mathbf{S}^{-1} = \mathbf{I}. \tag{5.28}$$

Now, let

$$\mathbf{U}_1 = \mathbf{A}\mathbf{V}_1\mathbf{S}^{-1}. \tag{5.29}$$

Note that $\mathbf{U}_1$ is an $m \times r$ matrix. Clearly, $\mathbf{U}_1$, as shown by (5.28), is a unitary matrix. That is,

$$\mathbf{U}_1^H \mathbf{U}_1 = \mathbf{I}. \tag{5.30}$$

Also,

$$\left(\mathbf{A}^H \mathbf{A}\right) \mathbf{V}_2 = \mathbf{V}_2 \mathbf{0} = \mathbf{0} \rightarrow \mathbf{A}\mathbf{V}_2 = \mathbf{0}. \tag{5.31}$$

Let us define $\mathbf{U}_2$ $(m \times (m - r))$ such that

$$\mathbf{U}_2^H \mathbf{U}_1 = \mathbf{0}. \tag{5.32}$$

Let us also define

$$\mathbf{U} = [\mathbf{U}_1, \mathbf{U}_2] \tag{5.33}$$

and

$$\mathbf{V} = [\mathbf{V}_1, \mathbf{V}_2]. \tag{5.34}$$

Then,

$$\mathbf{U}^H \mathbf{A}\mathbf{V} = \begin{bmatrix} \mathbf{U}_1^H \mathbf{A}\mathbf{V}_1 & \mathbf{U}_1^H \mathbf{A}\mathbf{V}_2 \\ \mathbf{U}_2^H \mathbf{A}\mathbf{V}_1 & \mathbf{U}_2^H \mathbf{A}\mathbf{V}_2 \end{bmatrix} \tag{5.35}$$

However,

$$\mathbf{U}_1 = \mathbf{A}\mathbf{V}_1\mathbf{S}^{-1} \rightarrow \mathbf{A}\mathbf{V}_1 = \mathbf{U}_1\mathbf{S} \tag{5.36}$$

which leads to

$$\begin{aligned} \mathbf{U}^H \mathbf{A}\mathbf{V} &= \begin{bmatrix} \mathbf{U}_1^H \mathbf{A}\mathbf{V}_1 & \mathbf{U}_1^H \mathbf{A}\mathbf{V}_2 \\ \mathbf{U}_2^H \mathbf{U}_1\mathbf{S} & \mathbf{U}_2^H \mathbf{A}\mathbf{V}_2 \end{bmatrix} \\ &= \begin{bmatrix} \mathbf{S} & \mathbf{0} \\ \mathbf{0} & \mathbf{0} \end{bmatrix} = \Sigma \end{aligned} \tag{5.37}$$

Since $\mathbf{U}$ and $\mathbf{V}$ are both unitary, (5.24) follows.

Now, if $\mathbf{A}$ has Hermitian symmetry ($\mathbf{A} = \mathbf{A}^H$) (as is the case for all square covariance matrices by definition), then it becomes clear that

$$\mathbf{U} = \mathbf{V}. \tag{5.38}$$

Before we present an example, it is interesting to note that the pseudo-inverse of $\mathbf{A}$ may be defined as

$$\mathbf{A}^{\dagger} = \mathbf{V} \begin{bmatrix} \mathbf{S}^{-1} & \mathbf{0} \\ \mathbf{0} & \mathbf{0} \end{bmatrix} \mathbf{U}^H \tag{5.39}$$

Note that

$$
\begin{aligned}
\mathbf{A}^{\dagger}\mathbf{A} &= \mathbf{V}\begin{bmatrix} \mathbf{S}^{-1} & \mathbf{0} \\ \mathbf{0} & \mathbf{0} \end{bmatrix}\mathbf{U}^{H}\mathbf{U}\begin{bmatrix} \mathbf{S} & \mathbf{0} \\ \mathbf{0} & \mathbf{0} \end{bmatrix}\mathbf{V}^{H} \\
&= \mathbf{V}\begin{bmatrix} \mathbf{S}^{-1} & \mathbf{0} \\ \mathbf{0} & \mathbf{0} \end{bmatrix}\begin{bmatrix} \mathbf{I} & \mathbf{0} \\ \mathbf{0} & \mathbf{U}_2^{H}\mathbf{U}_2 \end{bmatrix}\begin{bmatrix} \mathbf{S} & \mathbf{0} \\ \mathbf{0} & \mathbf{0} \end{bmatrix}\mathbf{V}^{H} \\
&= \mathbf{V}\begin{bmatrix} \mathbf{S}^{-1} & \mathbf{0} \\ \mathbf{0} & \mathbf{0} \end{bmatrix}\begin{bmatrix} \mathbf{S} & \mathbf{0} \\ \mathbf{0} & \mathbf{0} \end{bmatrix}\mathbf{V}^{H} = \mathbf{V}\begin{bmatrix} \mathbf{I} & \mathbf{0} \\ \mathbf{0} & \mathbf{0} \end{bmatrix}\mathbf{V}^{H} \\
&= \mathbf{V}_1\mathbf{V}_1^{H} = \mathbf{I}
\end{aligned}
\tag{5.40}
$$

The pseudo-inverse may also be found if $\mathbf{A}$ $(m \times n)$ is full column rank (i.e., rank of $\mathbf{A}$ is $n \leq m$ which implies $\mathbf{A}^{H}\mathbf{A}$ is non-singular). Then,

$$\mathbf{A}^{\dagger} = \left(\mathbf{A}^{H}\mathbf{A}\right)^{-1}\mathbf{A}^{H}. \tag{5.41}$$

If $\mathbf{A}$ it is full row rank (rank of $\mathbf{A}$ is $m \leq n$) (that is $\mathbf{A}\mathbf{A}^{H}$ is non-singular), then

$$\mathbf{A}^{\dagger} = \left(\mathbf{A}\mathbf{A}^{H}\right)^{-1}\mathbf{A}. \tag{5.42}$$

Example 1 A vector $\mathbf{y} = [y_0, y_1, y_2]^{\top}$ has been observed, which satisfies the following

$$y_0 = 2x_0 + 3x_1 \tag{5.43}$$

$$y_1 = x_0 - x_1 \tag{5.44}$$

$$y_2 = x_0 + 4x_1 \tag{5.45}$$

with $[x_0, x_1]^{\top}$ denoting a zero-mean random vector with the covariance matrix $\mathbf{K}_x = \begin{bmatrix} 1 & 0 \\ 0 & 3 \end{bmatrix}$. First, find the covariance matrix of $\mathbf{y}$ and then use eigenvector or SVD methods to express $\mathbf{y} = [y_0, y_1, y_2]^{\top}$ in terms of uncorrelated orthogonal vectors. Compare your answer with the definition of $\mathbf{y}$ given above.

Solution: *For this problem, $l_v = 3$, so one would expect 3 distinct eigenvalues. First, we recognize that*

$$\mathbf{y} = \begin{bmatrix} 2 & 1 & 1 \\ 3 & -1 & 4 \end{bmatrix}^{\mathsf{T}}\mathbf{x} \tag{5.46}$$

$$= \mathbf{B}^{\mathsf{T}}\mathbf{x} \tag{5.47}$$

where $B^{\mathsf{T}} = \begin{bmatrix} 2 & 1 & 1 \\ 3 & -1 & 4 \end{bmatrix}$. This implies that we do not need to expand $\mathbf{y}$ in terms of 3 uncorrelated random variables. In fact, 2 such random variables will suffice. This *will become evident when eigenvalues of the covariance matrix of* $\mathbf{y}$ *are computed. The covariance matrix of* $\mathbf{y}$ *is*

$$\begin{aligned}\mathbf{K}_y &= \mathbf{B}^\mathsf{T}\mathbf{K}_x\mathbf{B} \\ &= \begin{bmatrix} 31 & -7 & 38 \\ -7 & 4 & -11 \\ 38 & -11 & 49 \end{bmatrix}.\end{aligned} \tag{5.48}$$

To find the eigenvalues, we need to solve the following equation

$$\det\left\{\mathbf{K}_y - \lambda\mathbf{I}\right\} = 0. \tag{5.49}$$

Using matlab, one can quickly arrive at $\Lambda = diag\{0, 2.77, 81.23\}$. *This implies that* $\mathbf{K}_y$, *as expected, has a rank of 2 instead of 3. This also means that* $\mathbf{K}_y^{-1}$ *does not exist. The function "eig" in matlab does not yield a high-to-low ordering of the eigenvalues. Instead, using the SVD function of the matlab, we arrive at*

$$\mathbf{U} = \mathbf{V} = \begin{bmatrix} -0.6 & -0.54 & -0.57 \\ 0.16 & -0.79 & 0.57 \\ -0.77 & 0.25 & 0.57 \end{bmatrix} \tag{5.50}$$

and

$$\Sigma = \begin{bmatrix} 81.23 & 0 & 0 \\ 0 & 2.77 & 0 \\ 0 & 0 & 0 \end{bmatrix}. \tag{5.51}$$

Note that the most significant eigenvalue is 81.23, *which may be considered as the "principle" value and its corresponding eigenvector as the "principle" component of* $\mathbf{y}$. *Hence,* $\mathbf{y}$ *can be describes as*

$$\begin{aligned}\mathbf{y} &= \mathbf{U}^\mathsf{T}\left[b_0, b_1, b_2\right]^\mathsf{T} \\ &= \begin{bmatrix} -0.6 & 0.16 & -0.77 \\ -0.54 & -0.79 & 0.25 \\ -0.57 & 0.57 & 0.57 \end{bmatrix}\begin{bmatrix} b_0 \\ b_1 \\ b_2 \end{bmatrix}\end{aligned} \tag{5.52}$$

where

$$\begin{aligned}b_0 &= \phi_0^H\mathbf{y} \\ &= -0.6y_0 + 0.16y_1 - 0.77y_2,\end{aligned} \tag{5.53}$$

$$\begin{aligned}b_1 &= \phi_1^H\mathbf{y} \\ &= -0.544y_0 - 0.79y_1 + 0.25y_2,\end{aligned} \tag{5.54}$$

and

$$\begin{aligned} b_2 &= \phi_2^H \mathbf{y} \\ &= -0.57 y_0 + 0.57 y_1 + 0.57 y_2. \end{aligned} \tag{5.55}$$

Based on what we have observed, b_0 is the principle component of $\mathbf{y}$. *That is, we can make the following approximation with little consequence:*

$$\hat{y}_0 \approx b_0 \phi_0 = b_0 \left[-0.6, 0.16, -0.77\right] \tag{5.56}$$

$$\hat{y}_1 \approx 0 \tag{5.57}$$

$$\hat{y}_2 \approx 0. \tag{5.58}$$

So, upon the receipt of $\mathbf{y}$, *we compute b_0 and then form $\hat{y}_0$, which bears the most significant component of the signal.*

It is tempting to compute x using the definition $\mathbf{y}$ *using the concept of pseudo-inverse. That is*

$$\begin{aligned} \mathbf{B}^{\mathsf{T}} \mathbf{x} &= \mathbf{y} \\ \mathbf{B}\mathbf{B}^{\mathsf{T}} \mathbf{x} &= \mathbf{B}\mathbf{y} \\ \mathbf{x} &= \left(\mathbf{B}\mathbf{B}^{\mathsf{T}}\right)^{-1} \mathbf{B}\mathbf{y}. \end{aligned} \tag{5.59}$$

Note that, although B is not a square matrix, BB^{T} is a square matrix, which can be inverted if it is full rank. In this case, BB^{T} is full rank, and hence

$$\begin{aligned} \mathbf{x} &= \mathbf{C}\mathbf{y} \\ &= \begin{bmatrix} 0.33 & 0 \\ 0.467 & -0.2 \\ 0.133 & 0.2 \end{bmatrix}^{\mathsf{T}} \mathbf{y} \end{aligned} \tag{5.60}$$

or

$$x_0 = 0.333 y_0 + 0.467 y_1 - 0.133 y_2 \tag{5.61}$$

$$x_1 = -0.2 y_0 + 0.2 y_2. \tag{5.62}$$

Notice that the matrix $\mathbf{C}$ *generates a pair of uncorrelated random variables from* $\mathbf{y}$ *whose elements are correlated.*

PCA, then, is the process of decomposing a random vector into its uncorrelated components with its most significant "components" identified by the process. This by no means insures that the uncorrelated components are independent of one another, which would be of significant interest in the problems where the observed signal is a linear combination of a number of independent processes. In such problems, identifying the "independent" components may lead to the discovery of an underlying process that is the identifiers for a disease or an anomalous physiological condition.

Such decomposition may then be used for diagnosis and tracking of diseases and/or physiological conditions. Therefore, it is of significant interest to consider a process that decomposes a signal into its independent components. Such a process is known as the Independent Component Analysis (ICA), which will be introduced in the next section. Before doing so, let us consider yet another example of PCA as it deals with biological signals.

Example 2 We have observed an EEG signal for a period of 5 h (use the data from Example 4.33). We like to examine the patient's condition on an hourly basis. We consider a window of length 5 s to average noise out and provide uncorrelated samples. A vector of size 4 is formed from the 4 consecutive windowed samples. That is, 180 vectors of length 4 are observed over an hour period. Form a covariance matrix for the observed vectors using matlab for each hour of observation. Perform the PCA decomposition of the resulting vector. Show how the principle eigenvalues change with time over 5 h. The sampling rate of the raw EEG data is 256 samples/s.

Solution: *The observation interval for the above signal is limited to the first 5 h ($5 \times 3600 \times 256$ samples). The signal is first filtered by a constant window of length* 5×256. *Then, the signal is decimated by* 5×256 *to give the effect of window and hold. The resulting 3600 samples* ($5 \times 3600 \times 256/5/256$) *are divided into 5 blocks, each of length 720, representing 1 h of observation. Finally, 720/4 = 180 vectors of length 4 are generated for each 1 h block. The 180 vectors are fed into corr.m function as an* 180×4 *matrix. The above is repeated for each of the five 1-h observation intervals. The five covariance matrices (for 5 h) that are formed by using corr.m function of matlab are as follows:*

$$\mathbf{K}_y = \begin{bmatrix} 1 & -0.6237 & 0.0945 & -0.0626 \\ -0.6237 & 1 & -0.6134 & 0.2622 \\ 0.0945 & -0.6134 & 1 & -0.7030 \\ -0.0626 & 0.2622 & -0.7030 & 1 \end{bmatrix}, \tag{5.63}$$

$$\mathbf{K}_y = \begin{bmatrix} 1 & -0.7059 & -0.0166 & 0.2289 \\ -0.7059 & 1 & -0.4753 & -0.0406 \\ -0.0166 & -0.4753 & 1 & -0.6032 \\ 0.2289 & -0.0406 & -0.6032 & 1 \end{bmatrix}, \tag{5.64}$$

$$\mathbf{K}_y = \begin{bmatrix} 1 & 0.3601 & -0.0386 & -0.0525 \\ -0.3601 & 1 & -0.4940 & -0.0473 \\ -0.0386 & -0.4940 & 1 & -0.2602 \\ -0.0525 & -0.0473 & -0.2602 & 1 \end{bmatrix}, \tag{5.65}$$

$$\mathbf{K}_y = \begin{bmatrix} 1 & -0.5647 & 0.0653 & 0.028 \\ -0.5647 & 1 & -0.4986 & 0.0154 \\ 0.0653 & -0.4986 & 1 & -0.6993 \\ 0.028 & 0.0154 & -0.6993 & 1 \end{bmatrix}, \tag{5.66}$$

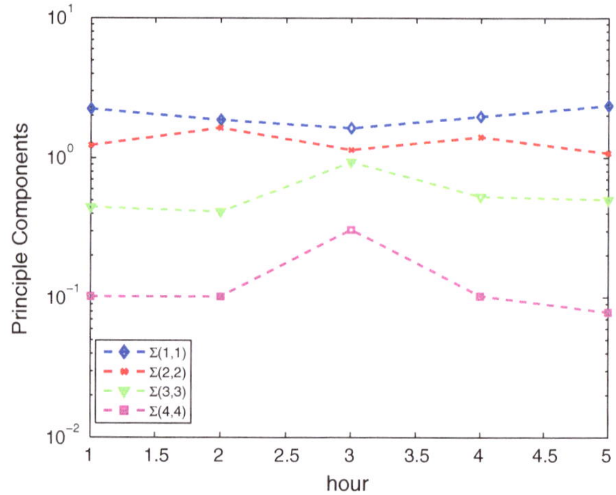

Fig. 5.1 The principle components of the EEG signal as a function of the observation time

$$\mathbf{K}_y = \begin{bmatrix} 1 & -0.6404 & 0.2505 & -0.0944 \\ -0.6404 & 1 & -0.7523 & 0.2388 \\ 0.2505 & -0.7523 & 1 & -0.5951 \\ -0.0944 & 0.2388 & -0.5951 & 1 \end{bmatrix}. \tag{5.67}$$

Using the svd.m function of matlab, the SVD decomposition is performed. Given that the above matrices satisfy the Hermitian symmetry, $U = V$ for all matrices. The diagonal elements of the corresponding Σ for each matrix are plotted in Fig. 5.1. *One can clearly identify the principle component of the vector, with its eigenvalue consistently more than an order of magnitude larger than that of the one with the smallest eigenvalue. Also, one can see that the principle components do not vary significantly with time. Now, if an event occurs, the principle component tracking may be used to determine the onset of a physiological response. We can also conclude that only the first three eigenvalues are of significant importance, and hence one has to be concerned with those components in analyzing this signal.*

5.2 Independent Component Analysis

In the previous section, we introduced the concept of PCA where the focus was to decompose an observed vector into its "uncorrelated" components. Although such an approach is of significant interest in many biomedical signal processing problems, in some instances, it is far more desirable to decompose a biologically originated signal into its "independent" components. The word "independent" is used here from a statistical perspective. As noted in the previous sections, in order to establish independence, one has to have access to the joint and marginal pdfs of the observed vectors. This is a far more difficult challenge than that posed by PCA, where the second order statistics of the observed vectors are only required to invoke SVD algorithm. It is also important to note that ICA draws from the rich history of Blind Source Separation (BSS) in the signal processing world. BSS, as the name implies, makes an attempt to extract signals from a (or multiple) source(s) by observing samples of noise-corrupted and linearly manipulated versions of the desired signals. Although the linear manipulation was asserted here, in principle, that is not a requirement for BSS. For biomedical signal processing purposes, however, the general assumption is that the process is a linear one. The word "Blind" is used here to refer to the fact that the nature of the linear mixture operation of the source signals is supposed to be unknown. Hence, the main idea is to make an attempt to estimate it using the observed signal statistics and the known statistics of the signal. For the sake of consistency, we refer to this operation as ICA as it is a special case of BSS.

Let us formulate this problem next. To that end, let

$$\mathbf{s}\left(t\right) = \left[s_0\left(t\right), s_1\left(t\right), \ldots, s_q\left(t\right)\right]^{\mathsf{T}} \tag{5.68}$$

denote a vector of q source signals. That is, the elements of this vector denote time-varying signals which we are interested in extracting. It is assumed that the source signals have known statistics and that they are independent of one another (mutually independent). More specifically,

$$f_{\mathbf{s}}\left(\mathbf{s}\right) = \prod_{j=0}^{q} f_{s_j}\left(s_j\right) \tag{5.69}$$

where $f_{s_j}\left(s_j\right)' s$ are known to us. In this case, q does not denote time (as was done in previous sections), but rather the number of independent sources we intend to identify. Now, let

$$\mathbf{x}\left(t\right) = \mathbf{A}\mathbf{s}\left(t\right) \tag{5.70}$$

be the observed vector ($l_v \times 1$) with $\mathbf{A}$ denoting an $l_v \times q$ "mixing" matrix with $l_v \geq q$. As noted above, this is a linear mixing model, which will be used throughout this section. The objective is to recover $\mathbf{s}\left(t\right)$ by observing $\mathbf{x}\left(t\right)$. Obviously, if $l_v = q$ and $\mathbf{A}$ is known and non-singular (or with linearly independent columns), then

$$\hat{\mathbf{s}}\left(t\right) = \mathbf{A}^{-1}\mathbf{x}\left(t\right). \tag{5.71}$$

Even though $l_v \neq q$, but $\mathbf{A}$ is known and is full column rank (namely, the rank of $\mathbf{A} = q \leq l_v$) (hence, $\mathbf{A}^T\mathbf{A}$ is non-singular), one may use a pseudo-inverse to find $\hat{\mathbf{s}}\left(t\right)$ as follows:

$$\hat{\mathbf{s}}\left(t\right) = \left(\mathbf{A}^T\mathbf{A}\right)^{-1}\mathbf{A}^T\mathbf{x}\left(t\right). \tag{5.72}$$

Unfortunately, that is not the case as $\mathbf{A}$ is unknown to us. In fact, we are in search of $\mathbf{A}$. However, what is available to us is $\mathbf{x}\left(t\right)$ and the statistics of $s\left(t\right)$. We note that a square matrix is know to be "non-mixing" if it has only one non-zero element in each row and each column. In that case, $\mathbf{x}(t)$ will be a copy of $\mathbf{s}(t)$, and observing $\mathbf{x}(t)$ will result in recovering $\mathbf{s}(t)$ up to perhaps permutation and possibly sign and scale changes. In what follows, however, we assume that $\mathbf{A}$ is a "mixing" matrix.

Now, let us define the matrix $\mathbf{W}$ $(q \times l_v)$ that is obtained for un-mixing the vector $\mathbf{x}(t)$. This implies

$$\mathbf{u}(t) = \mathbf{W}\mathbf{x}(t) + \mathbf{b}(t) \tag{5.73}$$

where $\mathbf{u}(t)$ is considered to be a vector with independent elements and $\mathbf{b}(t)$ is a vector that centers the data to the "origin." This is known as an *affine* transformation.

We also diagonalize the covariance matrix of the data $\mathbf{x}$ so that

$$\mathbf{K}_{\mathbf{x}} = 4\mathbf{I}. \tag{5.74}$$

This is accomplished by a process similar to eigenvector decomposition described in the previous section.

At this stage, we suppress the element of time with an understanding that the above procedure is performed over a block of observed signal and, hence, is subject to change as time evolves. This implies

$$\mathbf{u} = \mathbf{W}\mathbf{x} + \mathbf{b} \tag{5.75}$$

where now $\mathbf{u}$ denotes a $q \times 1$ vector, whose elements must be independent. Moreover, we considered $\mathbf{u}$ to be an estimate of $\mathbf{s}$ up to perhaps some permutation and scaling. Mathematically,

$$f_{\mathbf{u}}(\mathbf{u}) = \prod_{j=0}^{q-1} f_{u_j}\left(u_j\right). \tag{5.76}$$

In order to form an algorithm that yields a solution for $\mathbf{W}$ (which must be non-singular) to achieve (5.76), let us use the generalization of the mutual information function, given by

$$\begin{aligned} I(\mathbf{u}) &= \int f_{\mathbf{u}}(\mathbf{u}) \ln \frac{f_{\mathbf{u}}(\mathbf{u})}{\prod_{j=0}^{q-1} f_{u_j}\left(u_j\right)} d\mathbf{u} \\ &= -H(\mathbf{u}) + \sum_{j=0}^{q-1} H\left(u_j\right). \end{aligned} \tag{5.77}$$

Note that if $\mathbf{u}$ becomes a vector with independent elements, $I(\mathbf{u}) = 0$. Also, note that $H(\mathbf{u})$, the entropy of $\mathbf{u}$, will become identical to the sum of the entropies of the individual elements of $\mathbf{u}$ (i.e., sum of $H\left(u_j\right)$) when $\mathbf{u}$ consists of independent elements. Hence, one can establish an algorithm to compute $\mathbf{W}$ using the above function as a "metric." An algorithm that benefits from this approach is known as the *Infomax* technique, which is discussed next.

5.2.1 Infomax

First, let us define the (logistic) sigmoidal function, given by

$$h(u) = \frac{1}{1 + \exp(-u)}. \tag{5.78}$$

This function is plotted, along with its derivative, in Fig. 5.2. Note that the derivative of this function, given by

$$h'(u) = \frac{e^{-u}}{\left(1 + e^{-u}\right)^2} \tag{5.79}$$

reaches its maximum at $u = 0$, signifying that the maximum rate of change occurs at $u = 0$. Also, h' is an even function which can be considered as a pdf; note that $h'(u) \geq 0$ for all u and

$$\int_{-\infty}^{\infty} h'(u)\, du = 1. \tag{5.80}$$

That is, if we consider $h'(u)$ as a pdf of a random variable v, then v is a zero-mean random variable. Note that $h'(u)$ resembles a Gaussian pdf. This fact is critical as we exploit this fact in setting up the infomax algorithm.

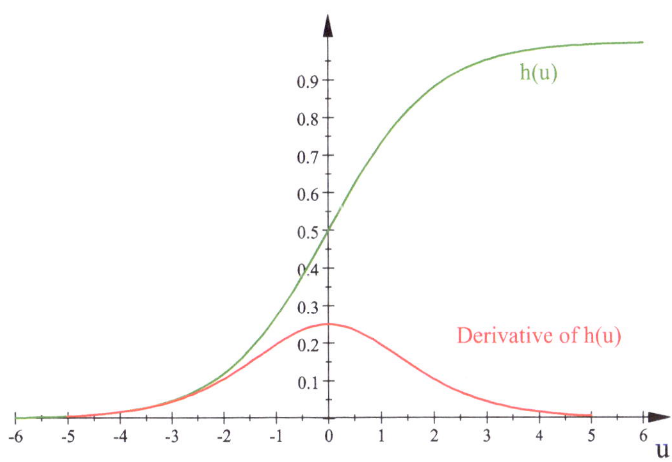

Fig. 5.2 The sigmoidal function and its derivative

At this stage, let us consider a new random vector $\mathbf{y}$ whose ith element is

$$y_i = h_i(u_i) \tag{5.81}$$

where $h_i s$ are "matching" functions. The selection of $h_i s$ is one of the steps in the ICA algorithm. Then, the pdf of $\mathbf{y}$ is given by

$$f_{\mathbf{y}}(\mathbf{y}) = \frac{f_{\mathbf{u}}(\mathbf{u})}{\left|\det(\mathbf{J}(\mathbf{u}))\right|_{\mathbf{u}=\mathbf{h}^{-1}(\mathbf{y})}} \tag{5.82}$$

where $\mathbf{J}_u(\mathbf{u})$ is the Jacobian matrix whose (i, j) element, J_{ij}^u, is

$$J_{ij}^u = \frac{\partial h_i(u_i)}{\partial u_j} \tag{5.83}$$

and $\mathbf{h}^{-1}(\mathbf{y})$ is an inverse mapping of $\mathbf{y}$ to $\mathbf{u}$ (we assume $h_i' s$ are 1-to-1 mappings). Furthermore,

$$\mathbf{h}(\mathbf{u}) = \left[h_0(u_0), h_1(u_1), \ldots, h_{q-1}(u_{q-1})\right]^{\mathsf{T}}.$$

We can see that the Jacobian is a diagonal matrix, resulting in

$$f_{\mathbf{y}}(\mathbf{y}) = \frac{f_{\mathbf{u}}(\mathbf{u})}{\prod_{j=0}^{q-1}\left|h_i'(u_i)\right|_{\mathbf{u}=\mathbf{h}^{-1}(\mathbf{y})}}.$$

Note that by the definition of the pdf function,

$$\Delta p = f_{\mathbf{y}}(\mathbf{y})\, d\mathbf{y} = f_{\mathbf{u}}(\mathbf{u})\, d\mathbf{u} \tag{5.84}$$

where Δp is the probability of $\mathbf{y}$ (or $\mathbf{u}$) falling in a small region defined by $d\mathbf{y}$ (or $d\mathbf{u}$) about the operating point. Now, let us re-examine $I(\mathbf{u})$. To that end,

$$\begin{aligned} I(\mathbf{u}) &= \int f_{\mathbf{u}}(\mathbf{u}) \ln f_{\mathbf{u}}(\mathbf{u})\, d\mathbf{u} \\ &\quad - \int f_{\mathbf{u}}(\mathbf{u}) \sum_{j=0}^{q-1} \ln f_{u_j}(u_j)\, d\mathbf{u} \\ &= \int f_{\mathbf{y}}(\mathbf{y}) \ln f_{\mathbf{u}}\left(\mathbf{h}^{-1}(\mathbf{y})\right) d\mathbf{y} \\ &\quad - \int f_{\mathbf{u}}(\mathbf{u}) \sum_{j=0}^{q-1} \ln f_{u_j}(u_j)\, d\mathbf{u} \end{aligned} \tag{5.85}$$

where we have assumed that $h_i(u_i)$ functions are 1-to-1 (as is the case for the sigmoidal function). In arriving at the above, it is assumed that, for

$$\mathbf{y} = \mathbf{h}(\mathbf{u}), \tag{5.86}$$

we have

$$f_{\mathbf{y}}(\mathbf{y})\, d\mathbf{y} \ln f_{\mathbf{u}}(\mathbf{u})\,|_{\mathbf{u}=\mathbf{h}^{-1}(\mathbf{y})} = f_{\mathbf{u}}(\mathbf{u})\, d\mathbf{u} \ln f_{\mathbf{u}}(\mathbf{u}). \tag{5.87}$$

Next, we divide and multiply $f_{\mathbf{u}}\left(\mathbf{h}^{-1}(\mathbf{y})\right)$ by $\prod_{j=0}^{q-1} \left|h'_j\left(u_j\right)\right| |_{\mathbf{u}=\mathbf{h}^{-1}(\mathbf{y})}$ (assuming that the pdfs are well behaving and that they do not assume a value of 0 anywhere in their domain), resulting in

$$\begin{aligned} I(\mathbf{u}) &= \int f_{\mathbf{y}}(\mathbf{y}) \ln \left\{ \frac{f_{\mathbf{u}}\left(\mathbf{h}^{-1}(\mathbf{y})\right) \prod_{j=0}^{q-1} \left|h'_j\left(u_j\right)\right| |_{\mathbf{u}=\mathbf{h}^{-1}(\mathbf{y})}}{\prod_{j=0}^{q-1} \left|h'_j\left(u_j\right)\right| |_{\mathbf{u}=\mathbf{h}^{-1}(\mathbf{y})}} \right\} d\mathbf{y} \\ &\quad - \int f_{\mathbf{u}}(\mathbf{u}) \sum_{j=0}^{q-1} \ln f_{u_j}\left(u_j\right) d\mathbf{u} \\ &= \int f_{\mathbf{y}}(\mathbf{y}) \ln \frac{f_{\mathbf{u}}\left(\mathbf{h}^{-1}(\mathbf{y})\right)}{\prod_{j=0}^{q-1} \left|h'_j\left(u_j\right)\right| |_{\mathbf{u}=\mathbf{h}^{-1}(\mathbf{y})}} d\mathbf{y} \\ &\quad + \int f_{\mathbf{u}}(\mathbf{u}) \sum_{j=0}^{q-1} \ln \frac{\left|h'_j\left(u_j\right)\right|}{f_{u_j}\left(u_j\right)} d\mathbf{u} \end{aligned} \tag{5.88}$$

where we also have used

$$f_{\mathbf{y}}(\mathbf{y})\, d\mathbf{y} \ln \left|h'_j\left(u_j\right)\right| |_{\mathbf{u}=\mathbf{h}^{-1}(\mathbf{y})} = f_{\mathbf{u}}(\mathbf{u})\, d\mathbf{u} \ln \left|h'_j\left(u_j\right)\right|. \tag{5.89}$$

Now, recognizing that $\frac{f_{\mathbf{u}}\left(\mathbf{h}^{-1}(\mathbf{y})\right)}{\prod_{j=0}^{q-1}\left|h'_j\left(u_j\right)\right| |_{\mathbf{u}=\mathbf{h}^{-1}(\mathbf{y})}} = f_{\mathbf{y}}(\mathbf{y})$, we have

$$\begin{aligned} I(\mathbf{u}) &= \int f_{\mathbf{y}}(\mathbf{y}) \ln f_{\mathbf{y}}(\mathbf{y})\, d\mathbf{y} + \int f_{\mathbf{u}}(\mathbf{u}) \sum_{j=0}^{q-1} \ln \frac{\left|h'_j\left(u_j\right)\right|}{f_{u_j}\left(u_j\right)} d\mathbf{u} \\ &= -H(\mathbf{y}) + E\left\{ \sum_{j=0}^{q-1} \ln \frac{\left|h'_j\left(u_j\right)\right|}{f_{u_j}\left(u_j\right)} \right\} \end{aligned} \tag{5.90}$$

Next, if we match $\left|h'_j\left(u_j\right)\right|$ to $f_{u_j}\left(u_j\right)$ for all j (resulting in $\sum_{j=0}^{q-1} \ln \frac{\left|h'_i(u_i)\right|}{f_{u_j}(u_j)} = 0$), then minimizing $I(\mathbf{u})$ is identical to maximizing the entropy of $\mathbf{y}$ (i.e., $H(\mathbf{y})$).

That is,

$$\left|h'_j\left(u_j\right)\right| = f_{u_j}\left(u_j\right) \text{ for } \forall\, j \Rightarrow I\left(\mathbf{u}\right) = -H\left(\mathbf{y}\right). \tag{5.91}$$

It is well-known that the entropy of a random vector is maximized when the random vector elements are uniformly-distributed (which in turn insures independence of the elements).

Hence, the problem of *infomax* reduces to the following:

Given that we are observing $\mathbf{x}$, *find a non-singular transformation* $\mathbf{W}$ *such that* $H\left(\mathbf{y}\right)$ *is maximized*

That is, *infomax*, which maximizes $H\left(\mathbf{y}\right)$, minimizes the redundancy in $\mathbf{u}$, thereby achieving ICA. But, how $H\left(\mathbf{y}\right)$ is related to $\mathbf{W}$? Note that

$$\mathbf{y} = \mathbf{h}\left(\mathbf{W}\mathbf{x}\right) \tag{5.92}$$

and

$$\begin{aligned} H\left(\mathbf{y}\right) &= E\left\{\ln f_{\mathbf{y}}\left(\mathbf{y}\right)\right\} \\ &= \int f_{\mathbf{y}}\left(\mathbf{y}\right) \ln f_{\mathbf{y}}\left(\mathbf{y}\right) d\mathbf{y} \\ &= \int f_{\mathbf{x}}\left(\mathbf{x}\right) \ln f_{\mathbf{y}}\left(\mathbf{y}\right) |_{\mathbf{y}=\mathbf{h}(\mathbf{W}\mathbf{x})} d\mathbf{x}. \end{aligned} \tag{5.93}$$

However,

$$f_{\mathbf{y}}\left(\mathbf{y}\right) |_{\mathbf{y}=\mathbf{h}(\mathbf{W}\mathbf{x})} = f_{\mathbf{x}}\left(\mathbf{x}\right) \frac{1}{\left|\det\left(\mathbf{J}\right)\right|}$$

where $\mathbf{J}$ is now the Jacobian matrix whose $(i,\ j)$ element is given by

$$J_{ij} = \frac{\partial h_i\left(u_i\right)}{\partial x_j} = \frac{\partial h_i\left(u_i\right)}{\partial u_j}\frac{\partial u_j}{\partial x_j} = h'_i\left(u_i\right) W_{ij}. \tag{5.94}$$

Finally,

$$H\left(\mathbf{y}\right) = H\left(\mathbf{x}\right) - E\left\{\ln\left|\det\left(\mathbf{J}\right)\right|\right\}. \tag{5.95}$$

Since we have no control over $H\left(\mathbf{x}\right)$ (as it is defined by the input processes and the mixing operation), in order to minimize $I\left(\mathbf{u}\right)$, we need to maximize $H\left(\mathbf{y}\right)$, which is identical to minimizing $E\left\{\ln\left|\det\left(\mathbf{J}\right)\right|\right\}$. This suggest an iterative algorithm. Before discussing the algorithm, let us introduce some basic identities. For a non-singular matrix $\mathbf{J}$, we have

$$\frac{\partial \det\left(\mathbf{J}\right)}{\partial \mathbf{J}} = \det\left(\mathbf{J}\right)\mathbf{J}^{-\mathsf{T}} \tag{5.96}$$

$$\frac{\partial \ln |\det(\mathbf{J})|}{\partial \mathbf{J}} = \mathbf{J}^{-\mathsf{T}}. \tag{5.97}$$

In these identities $-\mathsf{T}$ denotes inverse transpose.

Now, note that, from (5.94), we have

$$\mathbf{J} = \mathbf{F}\mathbf{W} \tag{5.98}$$

where

$$\mathbf{F} = \begin{bmatrix} h_0'(u_0) & 0 & \cdots & 0 \\ 0 & h_1'(u_1) & 0 & \vdots \\ \vdots & 0 & \ddots & 0 \\ 0 & \vdots & 0 & h_{q-1}'(u_{q-1}) \end{bmatrix}. \tag{5.99}$$

Note that

$$|\det(\mathbf{F})| = \prod_{i=0}^{q-1} \left|h_i'(u_i)\right|, \tag{5.100}$$

which leads to

$$\ln|\det(\mathbf{F})| = \sum_{i=0}^{q-1} \ln\left|h_i'(u_i)\right|. \tag{5.101}$$

Now, we have

$$\begin{aligned}\frac{\partial \ln|\det(\mathbf{J})|}{\partial \mathbf{W}} = \frac{\partial \ln|\det(\mathbf{F}\mathbf{W})|}{\partial \mathbf{W}} &= \frac{\partial \ln|\det(\mathbf{W})|}{\partial \mathbf{W}} + \frac{\partial \ln|\det(\mathbf{F})|}{\partial \mathbf{W}} \\ &= \mathbf{W}^{-\mathsf{T}} + \sum_{i=0}^{q-1} \frac{\partial \ln\left|h_i'(u_i)\right|}{\partial \mathbf{W}} \\ &= \mathbf{W}^{-\mathsf{T}} + \sum_{i=0}^{q-1} \frac{\partial \ln\left|h_i'(u_i)\right|}{\partial u_i} \frac{\partial u_i}{\partial \mathbf{W}}. \end{aligned} \tag{5.102}$$

Since

$$\mathbf{u} = \mathbf{W}\mathbf{x} + \mathbf{b}, \tag{5.103}$$

$\frac{\partial u_i}{\partial \mathbf{W}}$ is a matrix with all zero rows except for the ith row, which is $\mathbf{x}^\mathsf{T}$. We then have

$$\frac{\partial \ln|\det(\mathbf{J})|}{\partial \mathbf{W}} = \mathbf{W}^{-\mathsf{T}} + \boldsymbol{\kappa}(\mathbf{u})\,\mathbf{x}^\mathsf{T} \tag{5.104}$$

where $\boldsymbol{\kappa}(\mathbf{u})$ is a vector-valued function, given by

$$\boldsymbol{\kappa}(\mathbf{u}) = \left[\kappa_0(u_0), \kappa_1(u_1), \ldots, \kappa_{q-1}(u_{q-1})\right]^{\mathsf{T}} \tag{5.105}$$

with

$$\kappa_i(u_i) = -\frac{\partial \ln \left|h_i'(u_i)\right|}{\partial u_i}. \tag{5.106}$$

Note that we are interested in finding a $\hat{\mathbf{W}}$ such that $E\left\{\frac{\partial \ln|\det(\mathbf{J})|}{\partial \mathbf{W}}\right\}|_{\mathbf{W}=\hat{\mathbf{W}}} = 0$. It is rather obvious that, if one achieves $\frac{\partial \ln|\det(\mathbf{J})|}{\partial \mathbf{W}}|_{\mathbf{W}=\hat{\mathbf{W}}} = 0$, one has

$$E\left\{\frac{\partial \ln|\det(\mathbf{J})|}{\partial \mathbf{W}}\right\}|_{\mathbf{W}=\hat{\mathbf{W}}} = 0. \tag{5.107}$$

Thereby, $\hat{\mathbf{W}}$ minimizes $E\{\ln|\det(\mathbf{J})|\}$ and maximizes $H(\mathbf{y})$.

Hence, we may set up a search (iterative) mechanism (steepest descent) that uses the above as a means of estimating $\mathbf{W}$. That is, if $\mathbf{W}_n$ denotes $\mathbf{W}$ at the nth stage of iteration, then

$$\mathbf{W}_{n+1} = \mathbf{W}_n + \zeta_n\left[\mathbf{W}_n^{-\mathsf{T}} - \boldsymbol{\kappa}(\mathbf{W}_n\mathbf{x}_n)\mathbf{x}_n^{\mathsf{T}}\right] \tag{5.108}$$

where $\mathbf{x}_n^{\mathsf{T}}$ denotes that nth block of data, which is observed and is then fed into the algorithm. ζ_n is introduced to control the convergence behavior of the algorithm. Note that this algorithm requires that $\mathbf{W}_n$ be non-singular at every step of iteration.

The last equation may be viewed as the *infomax* algorithm, which becomes identical the maximum-likelihood estimation of the mixing operator if

$$\left|h_j'(u_j)\right| = f_{u_j}(u_j) \text{ for } \forall\, j. \tag{5.109}$$

One issue with this algorithm is the requirement of inverting $\mathbf{W}_n$ at each iteration. It has been shown that a modified algorithm achieves the same result by multiplying the incremental correction $\zeta_n\left(\mathbf{W}_n^{-\mathsf{T}} - \boldsymbol{\kappa}(\mathbf{W}_n\mathbf{x}_n)\mathbf{x}_n^{\mathsf{T}}\right)$ by $\mathbf{W}_n^{\mathsf{T}}\mathbf{W}_n$ from the left, resulting in the following simplified (computationally) algorithm:

$$\mathbf{W}_{n+1} = \mathbf{W}_n + \zeta_n\left[\mathbf{I} - \boldsymbol{\kappa}(\mathbf{W}_n\mathbf{x}_n)\mathbf{x}_n^{\mathsf{T}}\mathbf{W}_n^{\mathsf{T}}\right]\mathbf{W}_n. \tag{5.110}$$

Since $\mathbf{x}_n^{\mathsf{T}}\mathbf{W}_n^{\mathsf{T}} = \mathbf{u}_n^{\mathsf{T}}$, we finally have

$$\mathbf{W}_{n+1} = \mathbf{W}_n + \zeta_n\left[\mathbf{I} - \kappa(\mathbf{u}_n)\mathbf{u}_n^{\mathsf{T}}\right]\mathbf{W}_n \tag{5.111}$$

where $\mathbf{u}_n$ is the observed $\mathbf{u}$ in the nth iteration, computed using

$$\mathbf{u}_n = \mathbf{W}_n\mathbf{x}_n. \tag{5.112}$$

The resulting algorithm, as can be seen, does not require inverting a matrix, which can be significant for some problems.

***A*-priori pdfs** As we observed above, the remaining task for completing the *infomax* algorithm is identifying the pdf **u**, namely $f(\mathbf{u})$, in order to obtain $\kappa()$. These may be viewed as the *a*-priori pdf. Since many of the problems in the biomedical signal processing world assume either super-Gaussian or sub-Gaussian characteristics, the sigmoidal function described above provides a good universal fit for the *infomax* problem. Before examining the goodness of the sigmoidal function, let us formally introduce the concept of super- and sub-Gaussian pdfs. For the case of super- and sub-Gaussian, the pdfs, when the mean has been subtracted out (the data is assumed to be adjusted by its mean), are defined as

$$f_u(u) = \frac{1}{C_{\alpha,\beta}} \exp\left\{-\frac{|u|^\alpha}{2\beta}\right\}; \begin{cases} \alpha > 2 \text{ super-Gaussian} \\ \alpha < 2 \text{ sub-Gaussian} \end{cases} \tag{5.113}$$

where $C_{\alpha,\beta} = \int_{-\infty}^{\infty} \exp\left\{-\frac{|x|^\alpha}{2\beta}\right\} dx$. Since α and β are unknown, it is instructive to seek approximations for these pdfs in general. To learn about the shape of these pdfs, $\exp\left\{-\frac{|u|^\alpha}{2}\right\}$ is plotted in Fig. 5.3 for various values of α. It is interesting to note that, for super-Gaussian, the pdf is confined to a limited range about $u = 0$, which implies that the random variable u is not likely to assume large values. To the contrary, for

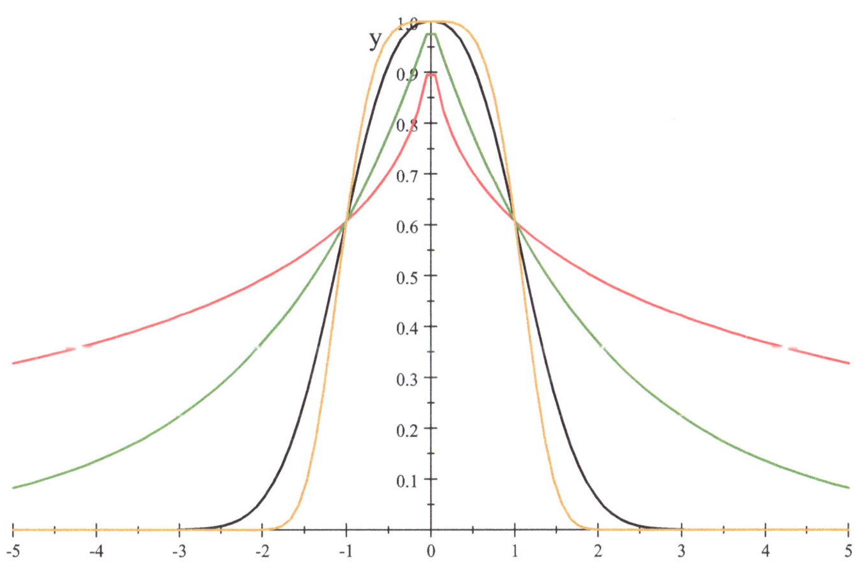

Fig. 5.3 Super- and sub-Gaussian pdfs. $\alpha = 4$ (*sienna*) and $\alpha = 2.5$ (*black*) for super-Gaussian. $\alpha = 1$ (*green*, also known as Laplacian) and $\alpha = 0.5$ (*red*) for sub-Gaussian

sub-Gaussian case, the random variable u is likely to assume large values. Note that the case of $\alpha = 1$ is the case of Laplacian pdf, which was studied earlier.

Next, we examine $\kappa(u)$ when $h(u)$ is a sigmodial function. Note that

$$\kappa_{\text{sig}}(u) = -\frac{\partial \ln \left| h'(u) \right|}{\partial u} = \frac{1 - e^{-u}}{1 + e^{-u}} = \tanh\left(\frac{u}{2}\right) = 2h(u) - 1. \tag{5.114}$$

We also have

$$\kappa_{\text{sup}}(u) = -\frac{\partial \ln f(u)}{\partial u} = \frac{\alpha |u|^{\alpha-1}}{2\beta} \text{sgn}(u); \alpha > 2 \tag{5.115}$$

$$\kappa_{\text{sub}}(u) = -\frac{\partial \ln f(u)}{\partial u} = \frac{\alpha |u|^{\alpha-1}}{2\beta} \text{sgn}(u); \alpha < 2. \tag{5.116}$$

In Figs. 5.4 and 5.5, $\kappa_{\text{sup}}(u)$ and $\kappa_{\text{sub}}(u)$, respectively, are depicted for appropriate ranges of u (note that super-Gaussian random variable is limited to a small range of u while sub-Gaussian has a much wider range of u) along with the approximate sigmoidal function, given by $\kappa_{\text{sig}}(u)$. As can be seen, $\kappa_{\text{sig}}(u)$ offers a reasonably good approximation of the two scenarios.

For this reason, we use $\kappa_{\text{sig}}(u)$ in problems related to the field of biomedical signal processing.

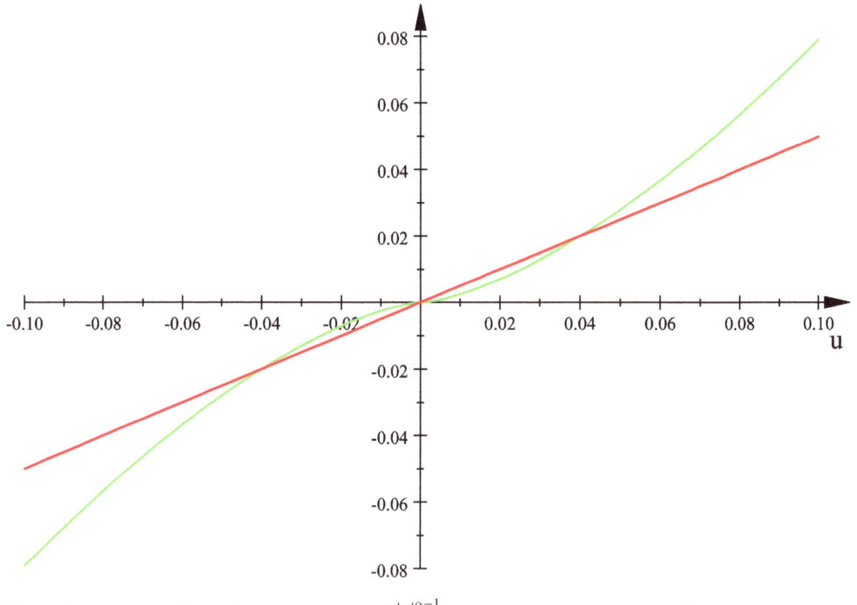

Fig. 5.4 The super-Gaussian case ($\kappa_{\text{sup}} = \frac{\alpha|u|^{\alpha-1}}{2\beta}\text{sgn}(u)$) with $\alpha = 2.5$ and $\beta = 0.5$ (*green*) and $\tanh(u/2)$ (*red*) for small range of u

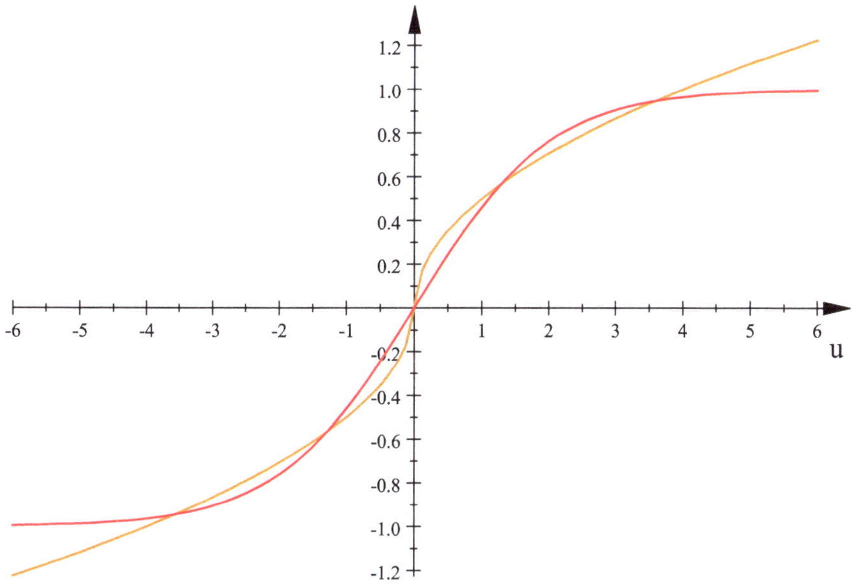

Fig. 5.5 The sub-Gaussian case ($\kappa_{\text{sub}}(u) = \frac{\alpha|u|^{\alpha-1}}{2\beta}\text{sgn}(u)$) with $\alpha = 1.5$ and $\beta = 1.5$ (*sienna*) and $\tanh(u/2)$ (*red*) for large range of u

Example 3 In this problem, we are observing three signals. The signals are known to be linear combinations of 3 independent ECG signals belonging to three different patients, denoted by $s_1(t)$, $s_2(t)$, and $s_3(t)$. Assume that the signals are generated as follows:

$$x_1(t) = s_1(t) + 2s_2(t), \tag{5.117}$$

$$x_2(t) = s_1(t) + s_2(t) - s_3(t), \text{ and} \tag{5.118}$$

$$x_3(t) = 2s_1(t) + 3s_2(t) + s_3(t). \tag{5.119}$$

This structure, however, is unknown to us. Given that we have observed 1 min of the signal $x(t) = [x_1(t), x_2(t), x_3(t)]^\mathsf{T}$, perform the ICA algorithm and demonstrate the success of the algorithm by comparing the decomposed signal with the original signals.

Solution: *For this problem, $l_v = q = 3$ and*

$$\mathbf{A} = \begin{bmatrix} 1 & 2 & 1 \\ 1 & 1 & -1 \\ 2 & 3 & 1 \end{bmatrix}.$$

However, this matrix is not known to us. The ECG signals are depicted in Fig. 5.6

These signals, for all intends and purposes, are independent. Specifically, $x_1(t)$, $x_2(t)$, and $x_3(t)$ are 60 s of the first probe (there are 23 probes indicated by 23 row

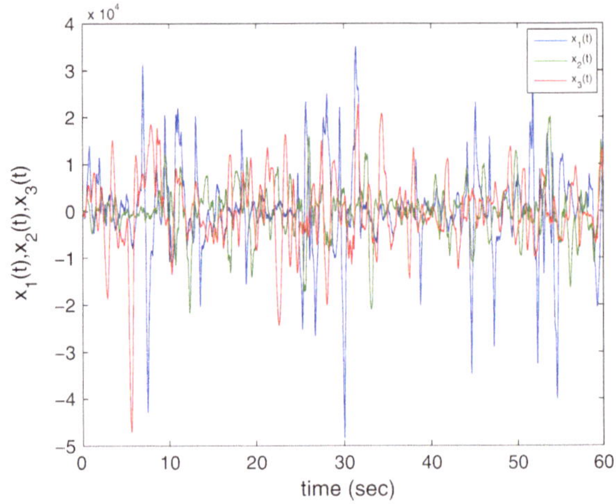

Fig. 5.6 Three independent EEG signals to test the ICA algorithm

in each record) chb03_01.edfm, chb02_01.edfm, and chb01_01.edfm, respectively. Note that the three signals denote 60 s of the first hour of the recorded EEG signals of 3 different patients. To invoke the ICA algorithm, we first filter these signals using a 128-tap FIR filter with all 1 taps (an averaging filter). Second, we use $\zeta_n = -2$. *The choice of* ζ_n *is quite critical to the convergence of the ICA algorithm. For this example, we used trial and error to arrive at the above number. Using the entire length of the available data (256 * 60 = 15360), the algorithm iterates 15,360 times. The outcome of the iteration is depicted (the signals are normalized to their maximums and we have depicted* $-u_1(t)$, *as the algorithm estimates the signal within a multiplicative constant and sign change) in Fig. 5.7. We have intentionally used only 33 s of the*

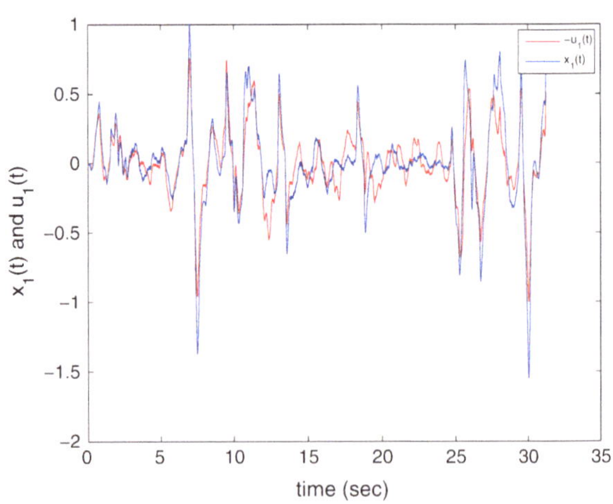

Fig. 5.7 The outcome of ICA

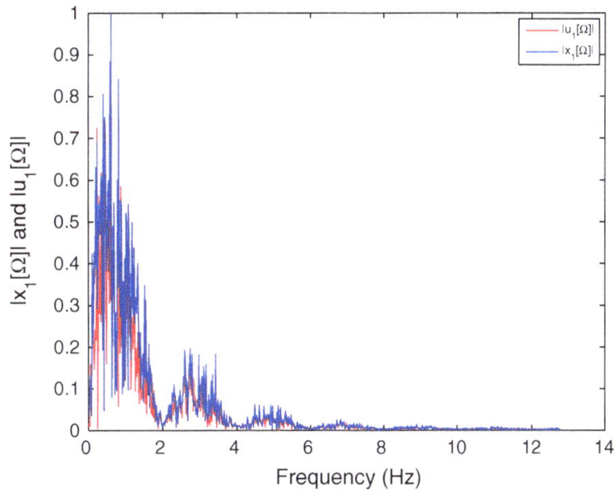

Fig. 5.8 The absolute values of the FFTs of the outcome of ICA and the original signal

two signals to reveal any discrepancy between the two signals. As can be seen, the extracted signal tracks the original signal. To gain further insight, the magnitudes of the FFTs of the two signals are depicted in Fig. 5.8. This figure also confirms the similarity of the two signals from the spectral content point of view.

5.3 Wavelet Decomposition (WD)

Wavelet decomposition (WD) or transform (WT) is yet another important tool for analyzing non-stationary signals, such as biomedical signals. In this section, we present some key results and examples related to WT in both continuous and discrete domain. Since we consider continuous time first, we can refer to WT as continuous WT (CWT). Naturally, a discrete version of WT will be discussed next.

It is important to note that our intention is not to be exhaustive here, but rather to demonstrate the potential of WT in studying signals with inherent statistical nonstationarity. Before doing so, we remind the reader of some key points. Throughout this text, we have used, rather seamlessly, Fourier Transform as a tool for investigating harmonic properties of signals. Although FT has a much broader use in the non-biological signal processing world, its use in biomedical signal processing is rather hampered by the non-stationarity of such signals. However, we have shown that the sort term FT (STFT) can provide valuable information about biological signals. One question that can be raised at this junction is the following. "*What should be the length of the window for STFT?*" A rather simple answer to this question is that the observation interval must be short enough to allow stationarity (in terms of signal properties) for the signal of interest. Further, the observation interval must be long enough to provide meaningful statistical information about the signal. That is, *does the observation interval allow for achieving convergence (for ergodic signals) as established in Chap.* 1? Note that if the signal properties change "too fast," one may not be able to glean meaningful information about the signal by examining its FT. We can also argue that the signal ergodicity plays a key role in allowing us to use time series of a biological signal to extract meaningful statistical information. We investigated the problem of ergodicity earlier, and the reader is encouraged to review Chap. 1 for further information. Hence, it becomes apparent that one needs to pay a close attention to the selection of the observation length of a biological signal. In particular, we recognize that one needs to be aware of the short-term as well as long-term properties of the signal. A sound approach will be to use a short enough time span for which the properties of the signal are not changing. Then, examine whether the length of the observation interval is long enough to allow convergence for the parameters that we are interest in estimating. Since STFT takes the time-varying nature of non-stationary signals into account, it becomes clear that we need to study STFT as a prelude to WD.

Hence, before WT is discussed in some detail, we proceed to introduce the concept of STFT in a formal way.

5.3.1 Short Term Fourier Transform (STFT)

Let $x(t)$ denote a time series of a biological signal. We can, then, define the STFT as

$$X(\tau, \Omega) = \int_{\infty}^{\infty} x(t)\, g(t-\tau)\, e^{-j\Omega t} dt \tag{5.120}$$

where $g(t)$ is a window function. Note that $g(t)$ is often time-limited, although that is not a restrict requirement, leading to the concept of "short-term," even though the limits of the integration in the above equation are $+\infty$ and $-\infty$. It is critical to note that we are interested in finding a mechanism to isolate signals to time and frequency components. This implies searching for a $g(t)$ that allows isolation in time and frequency. Similar to the uncertainty principle in quantum physics, we have a lower bound here as well. From the signal theory, we understand that time-bandwidth product for all signals is lower bounded by $\frac{1}{2}$. That is, if we choose $g(t)$ that occupies the root-mean-square (RMS) bandwidth (known as the Gabor bandwidth) of W_{rms} Hz and an RMS duration of T_{rms} seconds, then

$$T_{rms} W_{rms} \geq \frac{1}{2}$$

where

$$W_{rms} = \sqrt{\frac{\int_{-\infty}^{\infty} \Omega^2 |G(\Omega)|^2 d\Omega}{\int_{-\infty}^{\infty} |G(\Omega)|^2 d\Omega}} \tag{5.121}$$

and

$$T_{rms} = \sqrt{\frac{\int_{-\infty}^{\infty} t^2 |g(t)|^2 dt}{\int_{-\infty}^{\infty} |g(t)|^2 dt}} \tag{5.122}$$

with $G(\Omega)$ denoting the FT of $g(t)$. Note that if $\frac{|G(\Omega)|^2}{\int_{-\infty}^{\infty}|G(\Omega)|^2 d\Omega}$ is treated as a pdf (this expression is always positive and integrates to 1), W_{rms} is the "standard deviation" or RMS value in frequency while T_{rms} can be viewed as the RMS duration in time. The signal that does achieve the lower bound is

$$g(t) = \frac{1}{\sqrt{2\pi\sigma^2}} e^{-\frac{t^2}{2\sigma^2}} \tag{5.123}$$

where

$$T_{rms} = \frac{\sigma}{\sqrt{2}}. \tag{5.124}$$

Also,

$$G(\Omega) = e^{-\frac{\sigma^2\Omega^2}{2}} \tag{5.125}$$

leading to

$$W_{rms} = \frac{1}{\sqrt{2}\sigma}. \tag{5.126}$$

The above window function is somewhat impractical as it extends to infinity in time and frequency. One obvious and practical choice for $g(t)$ is an NRZ pulse shape. That is,

$$g(t) = \frac{1}{T}\{U(t) - U(t-T)\} \tag{5.127}$$

with $U(t)$ denoting the unit step function. This pulse is depicted in Fig. 5.9. Now,

$$X(nT, \Omega) = \frac{1}{T}\int_{nT}^{(n+1)T} x(t)\, e^{-j\Omega t} dt \tag{5.128}$$

may be viewed as the FT of the nth segment of the signal. This is, perhaps, the most intuitive definition of STFT as it implies acquiring the FT of the T seconds, non-overlapping segments of the signal for all time. This is particularly of significance when we have access to a priori information about the behavior of the signal; that is, the effectiveness of this approach hinges upon knowing that the signal is "stationary" (in its properties) over T seconds long segments of the signal. Now, if the same can be said about the properties of the signal in the frequency domain, one can choose to use an alternate window function, given by

$$G(\Omega) = \frac{2\pi}{\Omega_0}\{U(\Omega) - U(\Omega - \Omega_0)\}, \tag{5.129}$$

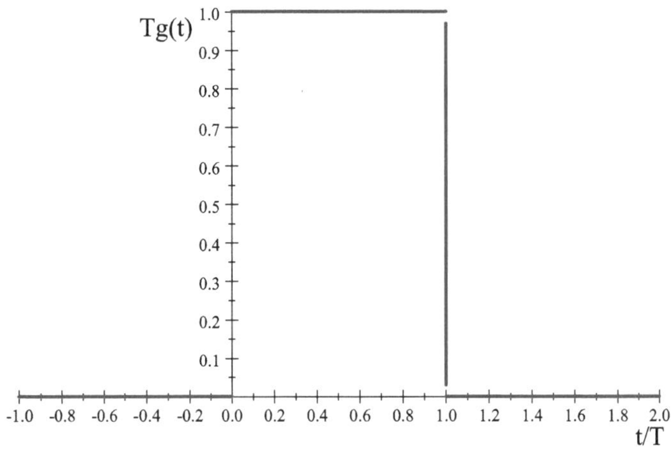

Fig. 5.9 $g(t) = \frac{1}{T}\{U(t) - U(t-T)\}$

which leads to

$$\begin{aligned} g(t) &= -\frac{\sin\left(\frac{\Omega_0 t}{2}\right)}{\frac{\Omega_0 t}{2}} e^{j\frac{\Omega_0 t}{2}} \\ &= -\sin c\left(\frac{\Omega_0 t}{2}\right) e^{j\frac{\Omega_0 t}{2}}. \end{aligned} \tag{5.130}$$

$G(\Omega)$ is depicted in Fig. 5.10. $G(\Omega)$ is depicted in Fig. 5.10. Note that $g(t)$ is an infinite-length window function. This window function is known as the "Shannon Window" and has its roots in the sampling theorem discussed in Chap. 1. Note that although $g(t)$ is infinite in length, it has a finite "significant" length of $\frac{4\pi}{\Omega_0}$. The absolute value of this window function for $\Omega_0 = 2\pi$ (1 Hz frequency window) is depicted in the next page. Note that the significant interval for this function is 2 s (between -1 and 1). In fact, one can argue that this pulse is significant for the interval of $\left[-\frac{2\pi}{\Omega_0}, \frac{2\pi}{\Omega_0}\right]$ for any Ω_0. It becomes immediately obvious that this form of window creates a non-trivial amount of overlap between integration intervals for $X(\tau, \Omega)$ for different values of τ. In order to gain appreciation for each method, let us consider the following example.

Example 4 Let us consider two scenarios. In one case, we are interested in detecting the presence of a 1-s burst of a 10 Hz, unit amplitude sinusoid in AWGN with a sample variance of 0.5. Assume a sampling rate of 1 kHz. Assume that the burst occurs between [0, 10] seconds. In the other case, the sinusoidal signal is present for [0.5, 11], but the exact frequency is unknown. We, however, know that the frequency of the signal is limited to the range [0, 10] Hz. Develop detection strategies for the two scenarios using STFT.

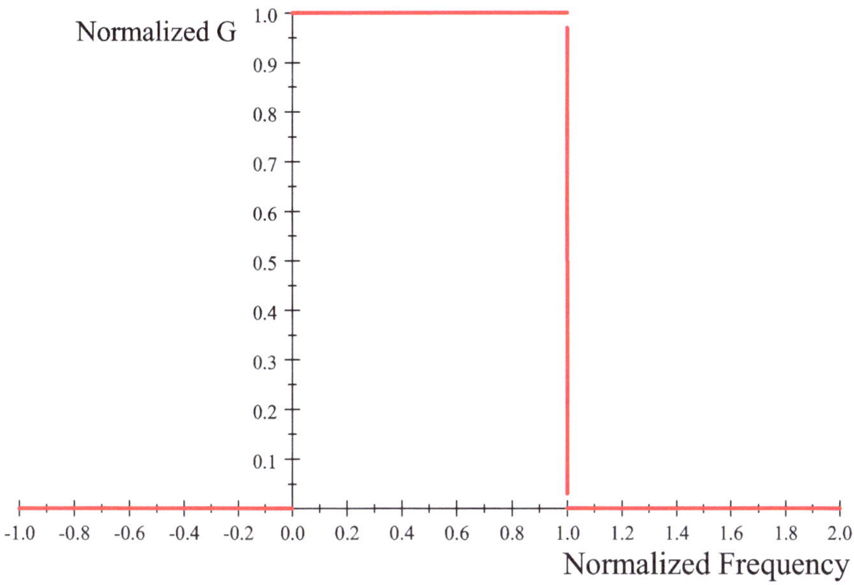

Fig. 5.10 $G(\Omega) = \frac{2\pi}{\Omega_0}\{U(\Omega) - U(\Omega - \Omega_0)\}$

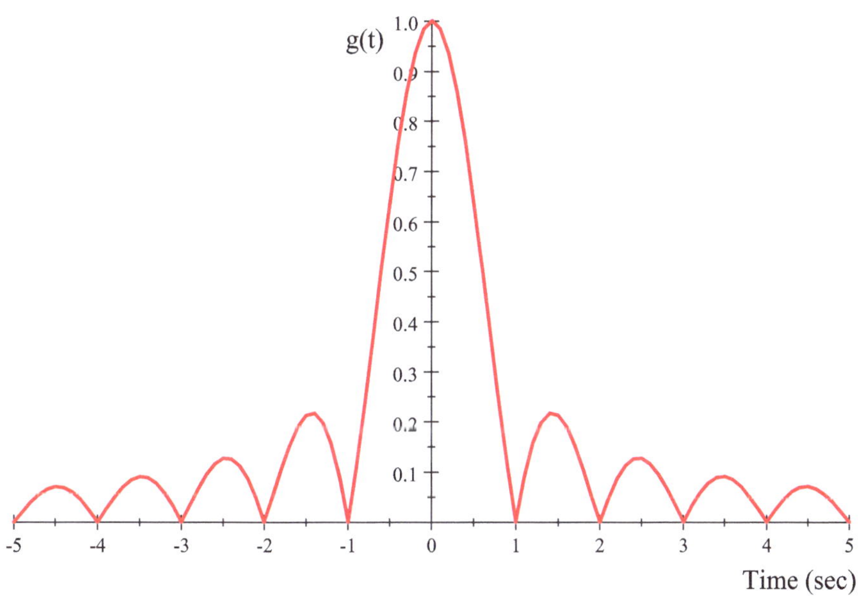

Fig. 5.11 Shannon window $|g(t)|$ for $\Omega_0 = 2\pi$ rad/s (1 Hz)

Solution: *For the first scenario, it is clear that we must search 1-s intervals in* $[0, 10]$. *Hence, we use* $g(t) = \frac{1}{T}\{U(t) - U(t-T)\}$. *This leads to*

$$X(n, \Omega) = \int_{n}^{n+1} x(t)\, e^{-j\Omega t} dt;\ n = 0, 2, \ldots, 9 \tag{5.131}$$

We assume that the signal is present over the interval $[4, 5]$, *corresponding to* $n = 4$. *Now, in order for us to develop a detection strategy, we compute the following* (Fig. 5.11):

$$F_n = \arg\max_{\Omega} \{|X(n, \Omega)|\}. \tag{5.132}$$

Now,

$$\hat{n} = \arg\max_{n} F_n.$$

For the second scenario, however, we us

$$X(\tau, \Omega) = \int_{0}^{10} x(t)\, g(t-\tau)\, e^{-j\Omega t} dt \tag{5.133}$$

where

$$\begin{aligned} g(t) &= \Im^{-1}\left\{\frac{2\pi}{\Delta}\{U(\Omega) - U(\Omega - \Delta)\}\right\} \\ &= -\sin\text{c}\left(\frac{\Delta t}{2}\right) e^{j\frac{\Delta t}{2}} \end{aligned} \tag{5.134}$$

where $\Delta = 20\pi$. *Hence,*

$$g(t) = -\sin\text{c}(10\pi t)\, e^{j10\pi t}. \tag{5.135}$$

In the second case, since we are interested in finding the frequency of the signal, we divide the STFT into N segments. That is, we compute

$$F_n(\tau) = X\left(\tau, \frac{(n-1)\Delta}{N}\right);\ n = 1, 2, \ldots, N.$$

Now,

$$F_n = \arg\max_{\tau} \{|F_n(\tau)|\} \tag{5.136}$$

and

$$\hat{n} = \arg\max_{n} F_n. \tag{5.137}$$

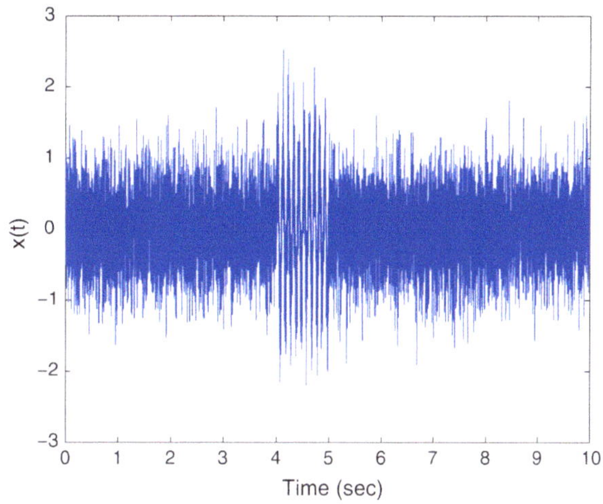

Fig. 5.12 A 10 Hz sinusoidal burst confined to [4, 5] interval and corrupted by AWGN

Note that $\hat{n}$, *in the first scenario, leads to finding the interval in which the signal is present, whereas* $\hat{n}$ *identifies the frequency slot to which this signal belongs for the second scenario. For the first scenario, as noted earlier, we assume that the sinusoid is only present in the interval* [4, 5]. *The resulting signal is depicted in* Fig. 5.12.

The signal is well masked by the additive noise. In Fig. 5.13, $|X(n, \Omega)|$ *is plotted for various n (denoting the 10 time slots). As can be seen, the STFT clearly detects the presence of the 10 Hz sinusoidal burst. Subsequently,* F_n *is depicted in Fig.* 5.14, *where one can identify the 5th time slot (namely,*[4, 5]) *as the slot that the signal is present. Clearly, STFT has been able to successfully identify the presence of the signal in the correct time slot when the signal is exposed to a substantial level of additive noise.*

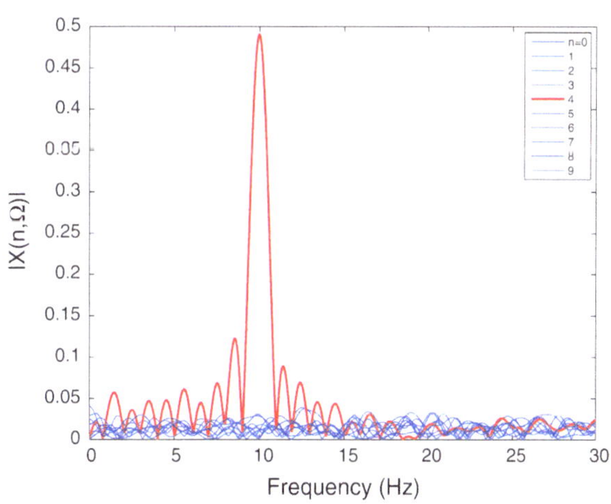

Fig. 5.13 $|X(n, \Omega)|$ as a function of Ω for the 10 Hz sinusoidal burst in [4, 5] interval

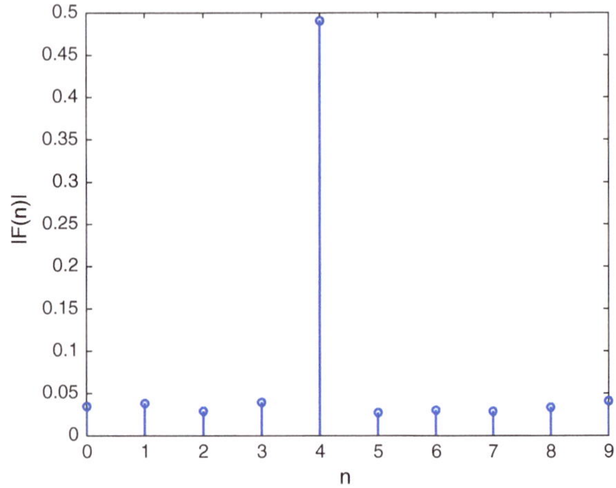

Fig. 5.14 F_n for the 10 Hz sinusoidal burst in [4, 5] interval

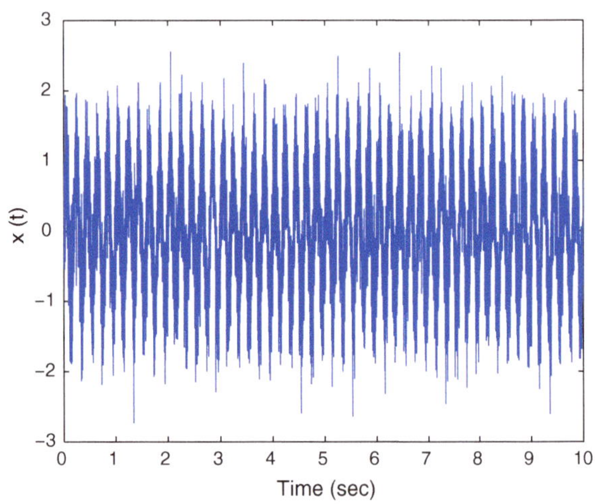

Fig. 5.15 A 5 Hz sinusoidal signal confined to [0, 10] interval and corrupted by AWGN

Now, let us consider the second scenario. In this case, we assume that the signal assumes a frequency of 5 Hz, which is confined to $[0, 10]$ *Hz interval. The resulting signal, which is now entirely corrupted by noise, is depicted in Fig.* 5.15. *For quantization, we assume* $N = 10$, *which implies that we have a frequency resolution of 1 Hz. In Fig.* 5.16, $\left|X\left(\tau, \frac{n-1}{N}\Delta\right)\right|$ *is depicted. As can be seen, this function is largest for* $n = 6$ *(corresponding to a frequency of 5 Hz, as* $n = 1$ *corresponds to 0 Hz). Furthermore,* F_n, *shown in Fig.* 5.16, *also points to 6 Hz as the candidate frequency. Hence, this algorithm was able to correctly estimate the frequency of the sinusoid buried in AWGN* (Fig. 5.17).

The above example clearly underscores the usefulness of STFT when we study signals with either uncertainty in time, frequency, or both. Note that this method may be used for signal detection in noise. In particular, when signal possesses

Fig. 5.16 $X\left(\tau, \frac{(n-1)\Delta}{N}\right)$ as a function of τ for $N = 10$ and $\Delta = 20\pi$

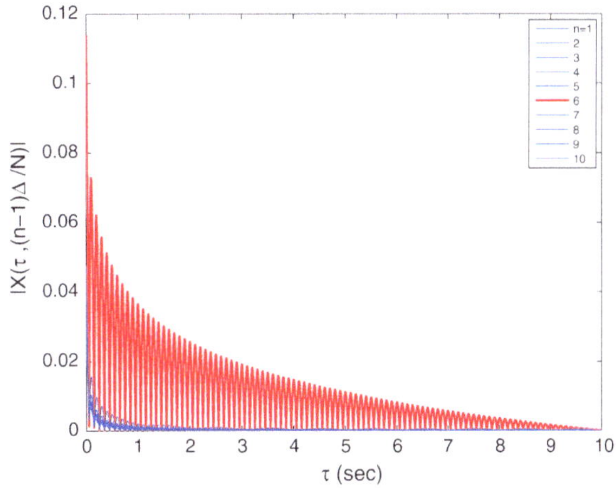

Fig. 5.17 F_n for the 5 Hz sinusoidal signal corrupted by AWGN for [0, 10] interval

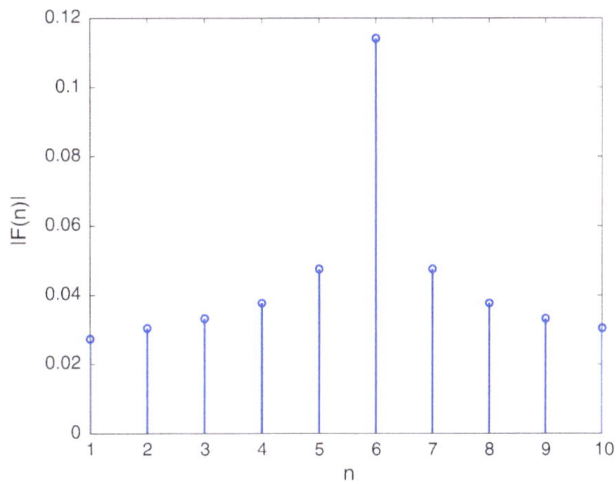

time-varying behavior, STFT becomes quite useful. However, one has to be cognizant of the fact that we were provided information about the transient nature of the signal for the above example. In particular, the fact that sinusoidal burst is 1 s long is privileged information when dealing with biological signals. Very often, the properties of the transient signal is unknown or of interest. Before we discuss the case of biological signals, let us discuss the concept of STFT when time and frequency are quantized. This may be viewed as a form of discrete STFT. As we observed in the previous example, we had to resort to discretization the time and frequency domain to perform the analysis. Note that STFT was defined above as

$$X(\tau, \Omega) = \int_{\infty}^{\infty} x(t)\, g(t-\tau)\, e^{-j\Omega t} dt. \qquad (5.138)$$

Clearly, this must be computed for all τ and all Ω. From a practical standpoint, however, this is impractical as this approach results in infinite values for $X(\tau, \Omega)$. In practice, we may quantize the time and frequency into an $N \times M$ matrix. That is, we compute

$$X_{n,m} = X((n-1)T, m\Delta)| = \int_{(n-1)T}^{nT} x(t)\, g(t-(n-1)T)\, e^{-jm\Delta t} dt; \\ n \in [1, N], m [\in 1, M] \qquad (5.139)$$

where

$$g(t) = \sin c\left(\frac{\Delta t}{2}\right) e^{j\left(\frac{\Delta t}{2}+\pi\right)} \qquad (5.140)$$

and T and Δ are step sizes in time and frequency, respectively. In this case, it is assume that $g(t)$ is confined to $[0, T]$. This form of STFT, in effect, searches N windows of length T seconds each, resulting in a total observation length of NT seconds. Furthermore, the search also considers M frequency slots, each Δ rad/s wide, resulting in a total search bandwidth of $M\Delta$ rad/s. From the "detection" point of view, $X_{n,m}$ may be viewed as the output of the mth correlator (of a bank of M correlators) in the nth signaling period. The correlator is mixing the incoming signal with $g(t)\, e^{-jm\Delta t}$, which may be viewed as a sinusoidal (complex) pulse, in order to decide which one of the M possible frequencies is present in the nth time slot. The real part of this function is depicted in Fig. 5.18. Note that the decay due to sinc function

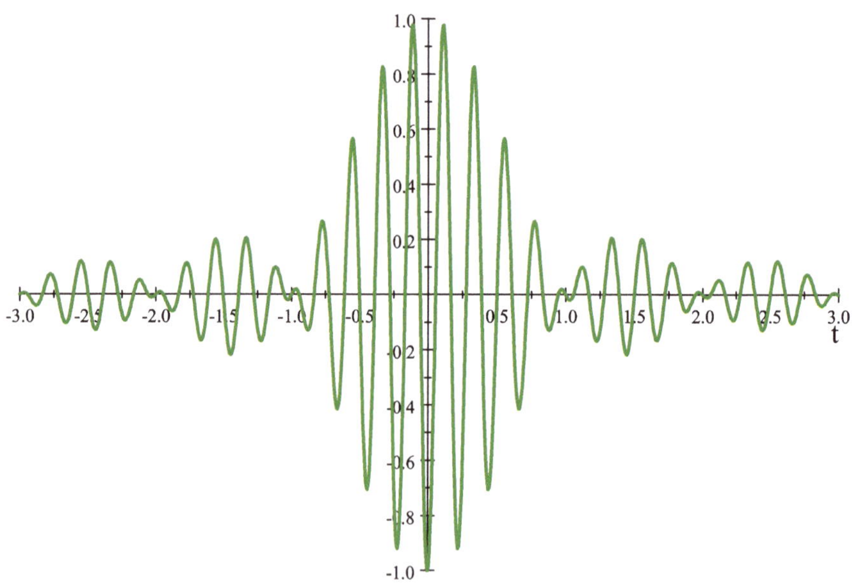

Fig. 5.18 Re $\{g(t)^{-jm\Delta t}\} = \text{sinc}(\pi t)\cos(\pi t + \pi - 10\pi t)$ $(\Delta = 2\pi, m = 5)$

results in a finite length of the window operation, although the window continues to exists for all t. In the world of communications, this is known as M-ary frequency shift keying (MFSK), a popular signaling scheme in digital communications.

Graphically, we can perceive the search as a "time" and "frequency" filtering using a finite length filter in both time and frequency, as shown in Fig. 5.19 where it is assumed that the search is performed for all time and all frequency slots, whereas signal has assumed specific time and frequency assignment in accordance to the colored boxes.

Caution must be exercised here; note that the pulse shape $g(t)$ is only bandlimited if observed over an infinite length in time, whereas the above requires the search over T seconds and intervals. However, if

$$\Delta T \gg 1, \tag{5.141}$$

then the above operation results in a non-negligible error in providing an accurate frequency search or time isolation using $g(t)$. The above quantity is known as the time-frequency (TF) product. This condition, however, leads to a sobering reality of STFT; that is, in order to achieve high frequency resolution, time resolution must be compromised and vice versa. As we learned earlier, TF product is lower bounded and the lowest value can be achieved using a Gaussian pulse. However, if we choose to use the rectangular window, the TF number may now be used to form a search strategy; for instance, if a frequency resolution of Δ is desired, then T must be selected such that

$$T \gg \Delta^{-1}. \tag{5.142}$$

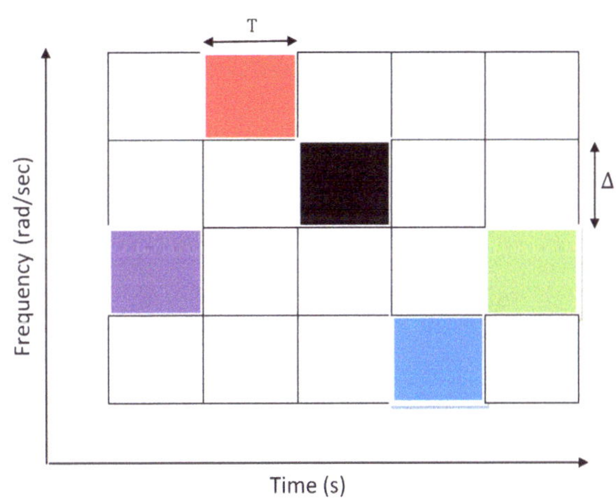

Fig. 5.19 Time and Frequency representation of signals using STFT

Similarly, if T is known and fix,

$$\Delta \gg \frac{1}{T}. \tag{5.143}$$

Note that, in each case, the above conditions provide limiting boundaries on the achievable resolution. That is, if one is interested in achieving a high resolution in frequency ($\Delta \ll 1$) using the above window option, then $T \gg 1$, and vice versa. These conditions may or may not be satisfied, depending upon the problem under investigation. For instance, if the frequency of the signal is known to change faster than T described above, then one must reduce T, thereby increasing Δ. This in turn results in low resolution in the frequency domain and a "blurring" effect in the frequency domain. Hence, the above conditions provide a means of assessing whether STFT will yield an accurate isolation of the signal in frequency and time. We can now examine STFT when biological signals are considered. To that end, let us consider the following example.

Example 5 Use STFT for analyzing EEG and ECG signals. In the case of ECG, the objective is extract the heart beat rate, which is confined to 30–180 bpm with a resolution of 1 bpm. In the case of EEG, we are interested in identifying the presence of a 16 Hz tone in the EEG signal. The search may consider 10–20 Hz range, and the resolution should be at least 1 Hz. Assume for the two cases that we have access to MIT-BIH with data extending over 10 min.

Solution: *For the first scenario, it is clear that* $\Delta = 2\pi/60$ *Hz and* $M = 150$. $T = 120$ *is then selected* ($\Delta T = 4\pi$). *We use Normal Sinus Rhythm data base (nsrdb) of the MIT-BIH repository. In particular, the record 16,265 is used. The record is obtained with a sampling rate of 128 Hz. Since the observation interval is 600 s,* $N = 5$. *Next,* $|X_{n,m}|$ *is computed and is depicted in Fig. 5.20. As can be*

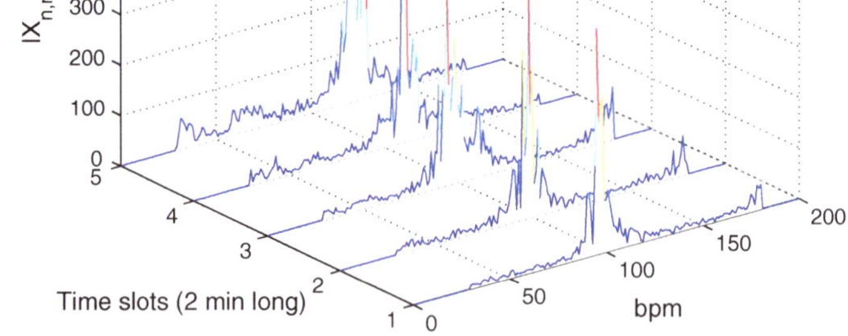

Fig. 5.20 STFT for ECG signal

seen, the signal exhibits strong peaks around 100 bpm. Furthermore, the location of the peaks changes from time slot to time slot. A color plot in Fig. 5.21 *reaffirms the presence of periodic components around 100 bpm, which alludes to the presence of a periodic QRS complex (the strength is proportional to the wavelength of the color). Note that the heart rate, as expected, seems to vary slightly from one time slot to another* (Fig. 5.21).

For the case of EEG, we use chbmit record from the MIT-BIH repository. In particular, 10 min of chb01_01_edfm (first of the 23 signals) with $\Delta = 2\pi$ *and* $M = 10$ *is used. In this case, we use a 10 min interval with a window of length 2 min (*$T = 120$, $N = 5$, $\Delta T = 240\pi$*). Figure* 5.22 *and its corresponding color plot in Fig.* 5.23 *depict the presence of a number of periodic components at several frequencies in the range of* $[10, 20]$ *Hz. However, at about 16 Hz, one can identify a strong periodic component that remains somewhat consistently strong across the 10 min interval (see the beige path at 16 Hz in Fig.* 5.23 *across all time slots).*

As we saw above, the STFT exploits temporal and spectral properties of the signal to extract useful information. Along the same line, we can introduce a different set of functions to extract important information about our signals. This tool is the wavelet decomposition (WD), also known as the wavelet transform (WT). We next describe WT is some detail.

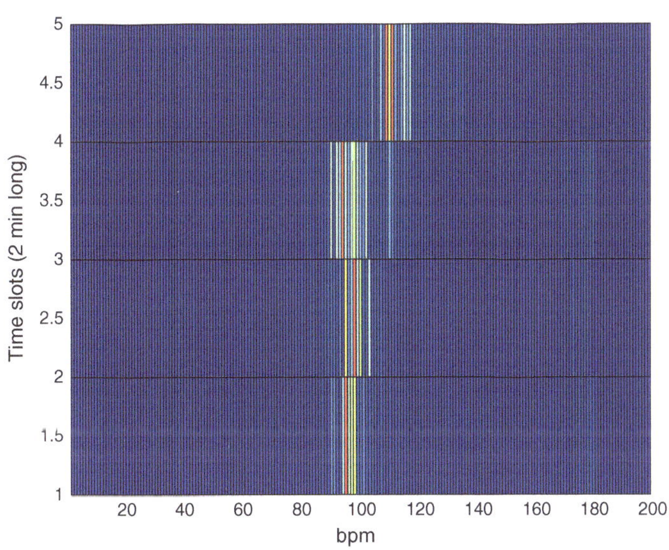

Fig. 5.21 The strength (*red* strongest, *blue* weakest) of the SHTF as a function of time

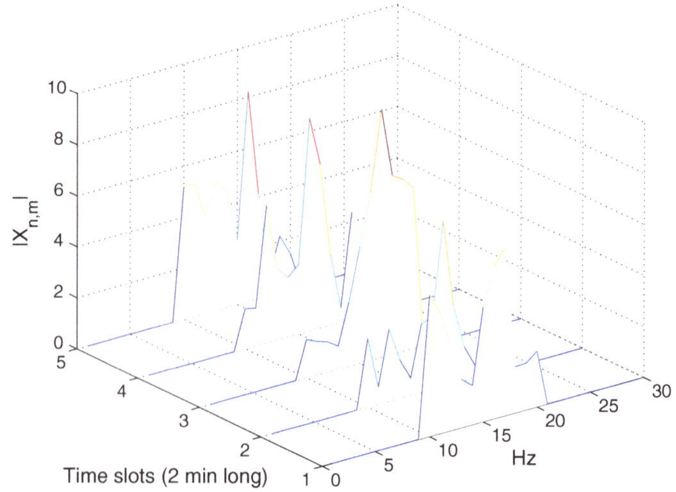

Fig. 5.22 STFT for EEG signal

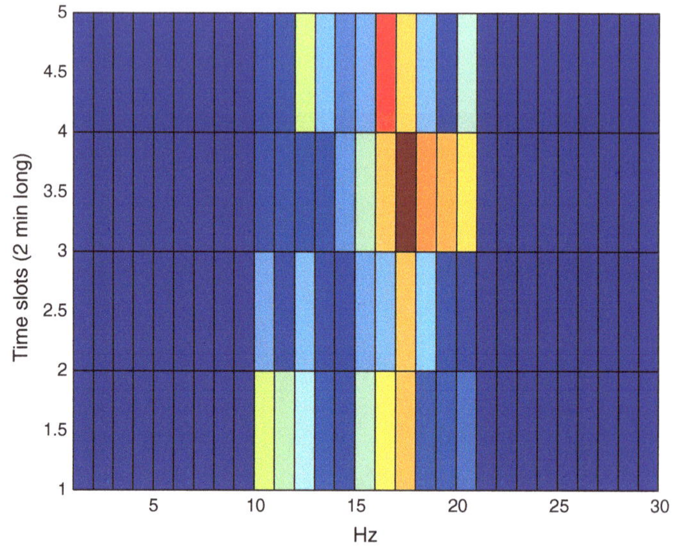

Fig. 5.23 The strength (*red* strongest, *blue* weakest) of the STFT as a function of time

5.3.2 *Continuous WT (CWT)*

Similar to STFT, we can define CWT of a signal as

$$w(\tau, s) = \frac{1}{\sqrt{|s|}} \int_{-\infty}^{\infty} x(t') \psi\left(\frac{t' - \tau}{s}\right) dt' \tag{5.144}$$

where $\psi(t)$ is known as the "mother" wavelet. Note that s may be interpreted as a "dilation" or "contraction" parameter in time (spectral isolation), whereas τ denotes a shift in time (temporal isolation) for a time-bandwidth limited $\psi(t)$. The above expression seems to be the correlation of $x(t)$ with delayed and dilated/contracted versions of $\psi(t)$. This may be viewed as a generalization of STFT. To elaborate, similar to STFT, τ denotes a delay (or translation) in time, but Ω has been replaced with a temporal dilation/contraction. This is not surprising as $e^{-j\Omega t}$ in STFT may be replaced with $e^{-j\frac{t}{s}}$ for $s = \frac{1}{\Omega}$. Now, $e^{-j\Omega t}$ may now be viewed as dilation or contraction of e^{-jt} depending upon the value of Ω. If we replace e^{-jt} with $\psi(t)$, a generalization ensues, allowing one to use a family of functions and not just e^{-jt} to analyze $x(t)$. It is important to note that the standard FT cannot provide time and frequency localization, whereas STFT (and CWT) is designed to achieve localization in both time and frequency. One can argue that STFT, however, defined above using the sinc pulse is a form of CWT since it is capable of providing frequency-time localization.

Equivalently, the above can be represented as

$$w(\tau, s) = \int_{-\infty}^{\infty} x(t')\, \psi_{\tau,s}(t')\, dt' \tag{5.145}$$

where

$$\psi_{\tau,s}(t) = \frac{1}{\sqrt{|s|}}\psi\left(\frac{t-\tau}{s}\right). \tag{5.146}$$

The key question at this stage is how to choose $\psi(t)$. There are some well-known CWT wavelets which will be discussed in some detail next. Before we proceed, it is critical for us to recognize that not all functions can be used as a wavelet. The key factor here is that this transformation must provide a means of recovering the original signal from its WT. In the case of FT, this was insured by the duality of the FT and its inverse. In general, we can define an *admissibility* condition for $\psi(t)$ as

$$0 < C_\psi = \frac{1}{2}\int_{-\infty}^{\infty} \frac{|\Psi(\Omega)|^2}{\Omega^2} d\Omega < \infty \tag{5.147}$$

where $\Psi(\Omega)$ is the FT of $\psi(t)$. The above condition is a sufficient, but not necessary condition. A sufficient condition for *admissibility,* which is easier to verify, is to show that $\psi(t)$ vanishes with time in such a manner that

$$\int_{-\infty}^{\infty} \psi(t)\, dt = 0. \tag{5.148}$$

Furthermore, we must be able to recover $x(t)$ using

$$x(t) = \frac{1}{C_\psi}\int_{0}^{\infty}\int_{-\infty}^{\infty} w(s,\tau)\, \frac{1}{\sqrt{|s|}}\psi\left(\frac{t-\tau}{s}\right) d\tau \frac{ds}{s^2}, \tag{5.149}$$

which implies that

$$\int_{0}^{\infty} \int_{-\infty}^{\infty} \frac{1}{|s|^3} \psi\left(\frac{t-\tau}{s}\right) \psi\left(\frac{t'-\tau}{s}\right) d\tau ds = C_{\psi} \delta\left(t-t'\right). \tag{5.150}$$

This confirms that the inverse operation can be performed and that the original function can be recovered from its WT.

5.3.2.1 The Fourier Transform Perspective

If $x(t)$ adheres to the requirements of a signal which has a FT, then one can show that

$$\begin{aligned} w(\tau, s) &= \frac{1}{\sqrt{|s|}} \int_{-\infty}^{\infty} x\left(t'\right) \psi\left(\frac{t'-\tau}{s}\right) dt' \\ &= \int_{-\infty}^{\infty} X(\Omega)\, \Psi(s\Omega)\, e^{j\Omega\tau} d\Omega. \end{aligned} \tag{5.151}$$

This appears to be the output of a linear system whose system function is

$$H(\Omega) = \Psi(s\Omega) \tag{5.152}$$

at time τ. That is, WT, in this case, is identical to the output of a filter whose system function is a dilated/contracted (spectrally) version of the basic system $\Psi(\Omega)$ at time τ when the input is $x(t)$. This interpretation provides a better understanding of the mechanism by which WT isolates various properties of a signal. The spectral selectivity of $\Psi(\Omega)$ also comes to question when one is in the process of selecting an appropriate wavelet transform.

For instance, if the signal $x(t)$ can be described as a band-pass (BP) signal, then an appropriate wavelet will have a mother wavelet whose FT exhibits a BP characteristics. In yet another scenario, the signal of interest exhibits temporal selectivity while its spectral behavior may be classified as that of a base-band (BB) signal. In that event, a mother wavelet that is base-band in the Fourier domain is, perhaps, more appropriate.

There are a number of CWT that can be found in the literature. In general, as noted above, the selection of the proper mother wavelet is problem dependent. In the ensuing discussion, we focus on two well-known CWT. In particular, we focus on Morlet (Gabor) wavelet, denoting a BP wavelet, and Marr's wavelet, representing a BB wavelet.

5.3.2.2 Morlet (Gabor) Wavelet (GW)

First, let us consider the Gaussian function

$$g(t) = \frac{1}{\sqrt{2\pi}} e^{-\frac{t^2}{2}}. \tag{5.153}$$

Note that

$$G(\Omega) = e^{-\frac{\Omega^2}{2}}. \tag{5.154}$$

This implies that the above signal offers isolation in temporal and frequency domains. Hence, it should be used as a wavelet. However, the *admissibility* condition cannot be satisfied for this function. That is, $\int_{-\infty}^{\infty} \psi(t)\, dt \neq 0$. An alternative is the Morlet wavelet (MW) where the above function is combined with a symmetrical, zero-mean complex exponential. That is, let us consider

$$g(t) = \frac{e^{2\pi j f_0 t} - K}{\sqrt{2\pi}} e^{-\frac{t^2}{2}} \tag{5.155}$$

for some K (K is added to insure $\int_{-\infty}^{\infty} \psi_{GW}(t)\, dt = 0$). This implies that the mother GW for some f_0 is given by

$$\psi_{GW}(t) = \frac{1}{\sqrt{2\pi}} \left[e^{2\pi j f_0 t} - e^{-2\pi^2 f_0^2} \right] e^{-\frac{t^2}{2}}, \tag{5.156}$$

which leads to

$$w_{GW}(\tau, s) = \frac{1}{\sqrt{2\pi |s|}} \int_{-\infty}^{\infty} x(t') \times \left[e^{-2\pi j f_0 \frac{t'-\tau}{s}} - e^{-2\pi^2 f_0^2} \right] e^{-\frac{(t'-\tau)^2}{2s^2}} dt'. \tag{5.157}$$

Also, note that

$$W_{GW}(\Omega) = e^{-(\Omega - 2\pi f_0)^2/2} - e^{-\Omega^2/2 - 2\pi^2 f_0^2} \tag{5.158}$$

with $W_{GW}(0) = \int_{-\infty}^{\infty} \psi_{GW}(t)\, dt = 0$. Furthermore, $\frac{|\Psi_{GW}(\Omega)|^2}{\Omega^2}$ is finite at $\Omega = 0$, which is a necessary condition for C_ψ to be finite. The real and imaginary parts of $\psi_{GW}(t)$ are depicted in Figs. 5.24 and 5.25, respectively.

The temporal behavior of these signals points to the time-limited attribute, which is a trademark of all wavelet transforms. Note that GW provides a more selective temporal selectivity as compared to STFT. Interestingly enough, the spectral behavior of these waveforms is also points to the spectral isolating feature of the wavelet (see Fig. 5.26 where $f_0 = 2$). Note that the FT of wavelet is centered at $\Omega_0 = 2\pi f_0 = 4\pi$. This is due to the Gaussian nature of the wavelet, which leads to a Gaussian FT shape in the frequency domain.

As noted earlier, the type of mother wavelet used in a problem is often linked to the properties of the signal under investigation. In the case of GW, we can clearly see a BP attribute for the mother wavelet (see 5.26), and hence, this wavelet is more suitable for studying BP signals. It would be of interest to examine the GW when dilation and shift is time are imposed. In Figs. 5.27 and 5.28, the real and imaginary parts GW for a shift of $\tau = 2$ s and a contraction factor $s = 0.5$ are depicted, respectively.

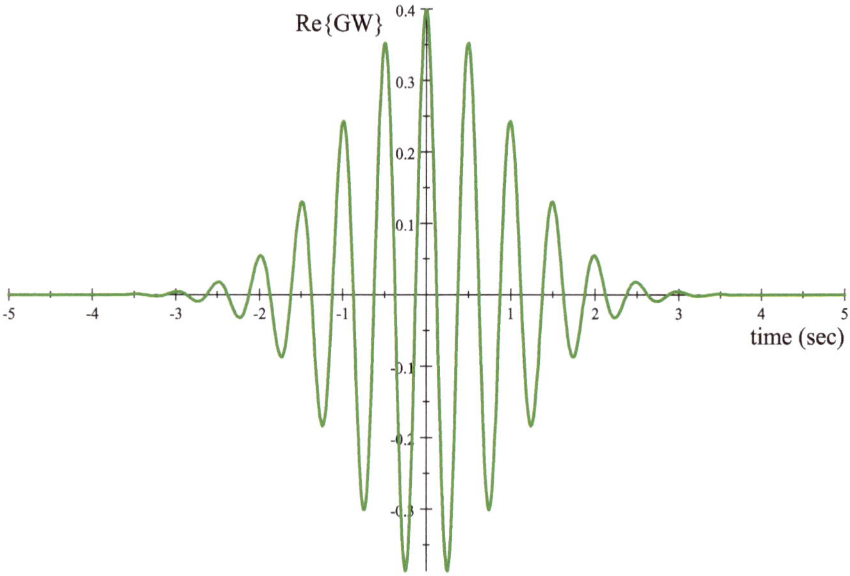

Fig. 5.24 $\frac{1}{\sqrt{2\pi}}\left(\cos(4\pi t) - e^{-8\pi^2}\right)\exp\left\{-\frac{t^2}{2}\right\}$ $(f_0 = 2)$

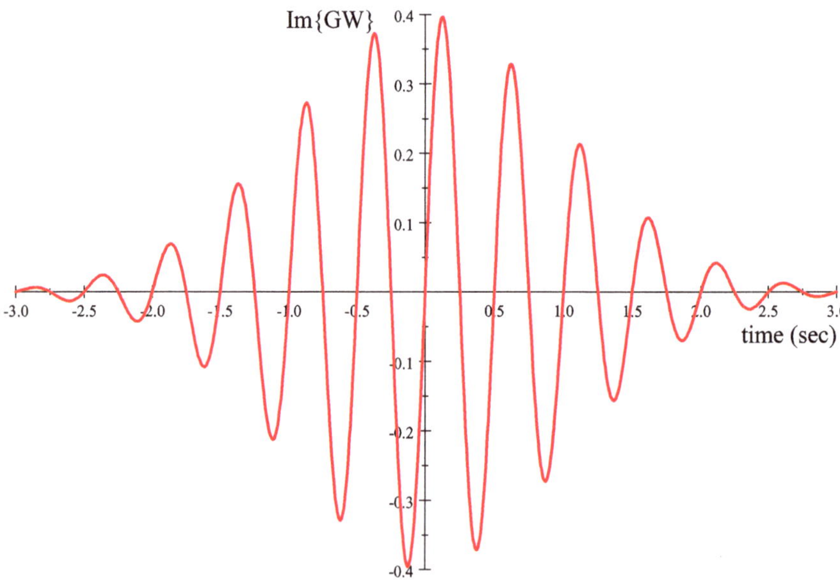

Fig. 5.25 $\frac{1}{\sqrt{2\pi}}\left(\sin(4\pi t) - e^{-8\pi^2}\right)\exp\left\{-\frac{t^2}{2}\right\}$ $(f_0 = 2)$

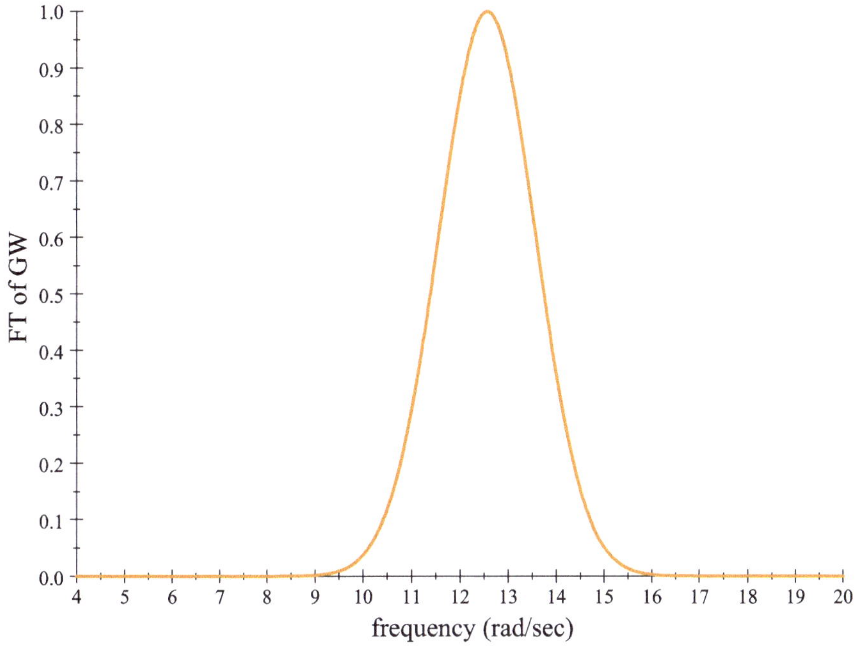

Fig. 5.26 $\Psi_{GW}(\Omega) = e^{-(\Omega-4\pi)^2/2} - e^{-\Omega^2/2-8\pi^2}$ for $f_0 = 2$

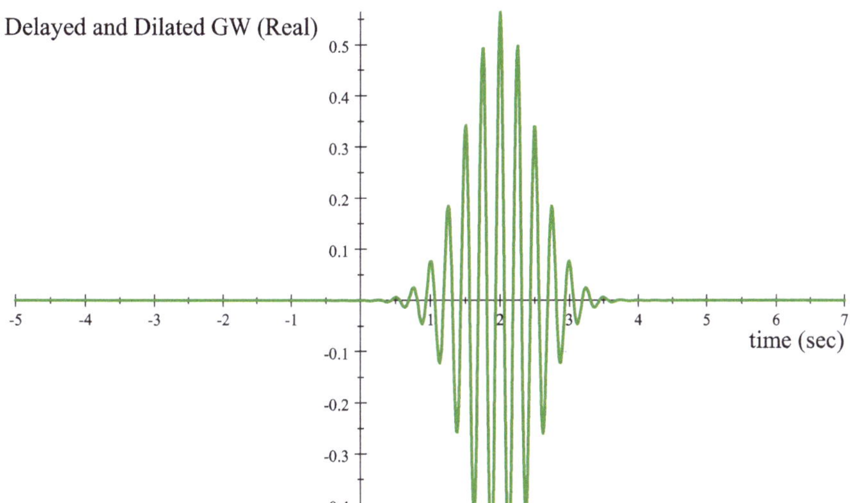

Fig. 5.27 $\frac{1}{\sqrt{0.5}} \mathrm{Re}\left\{\psi_{GW}\left(\frac{t-2}{0.5}\right)\right\} = \frac{1}{\sqrt{\pi}}\left(\cos\left(4\pi\frac{(t-2)}{0.5}\right) - e^{-8\pi^2}\right) e^{-\frac{(t-2)^2}{0.5}}$ ($f_0 = 2$, $s = 0.5$, $\tau = 2$)

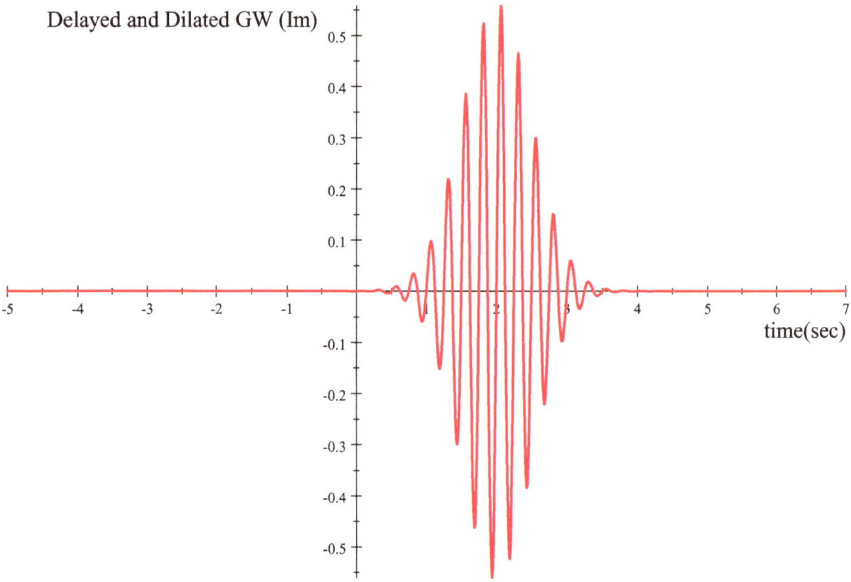

Fig. 5.28 $\text{Im}\left\{\psi_{GW}\left(\frac{t-2}{0.5}\right)\right\} = \frac{1}{\sqrt{\pi}}\left(\sin\left(4\pi\frac{(t-2)}{0.5}\right) - e^{-8\pi^2}\right)\exp\left\{-\frac{(t-2)^2}{0.5}\right\}$ ($f_0 = 2$, $s = 0.5$, $\tau = 2$)

Note that the impact of dilation/contraction is to cause spectral shift (higher frequency) while delay results in temporal selection. To elaborate, for any mother wavelet, we have

$$\mathcal{F}\left\{\frac{1}{\sqrt{s}}\psi\left(\frac{t-\tau}{s}\right)\right\} = \sqrt{s}e^{-j\Omega\tau}\Psi\left(s\Omega\right) \tag{5.159}$$

Hence,

$$\mathcal{F}\left\{\frac{1}{\sqrt{s}}\psi_{GW}\left(\frac{t-\tau}{s}\right)\right\} = \sqrt{s}e^{-j\Omega\tau}\left\{e^{-(s\Omega-2\pi f_0)^2/2} - e^{-s^2\Omega^2/2-2\pi^2 f_0^2}\right\}. \tag{5.160}$$

The temporal shift is rather obvious from these figures. However, in order to appreciate the frequency selectivity of GW, the magnitude of the FT of the shifted and contracted GW is depicted in Fig. 5.29. Note that the contraction ($s = 0.5$) resulted in shifting the center of $\Psi(\Omega)$ to $2\pi\frac{f_0}{0.5} = 8\pi$, while the "bandwidth" of the GW is doubled ($\frac{1}{s} = 2$). Hence, τ combined with the time-limited attribute of $\psi(t)$ allow for temporal selectivity, while s and bandlimited attribute of this wavelet allow for frequency selectivity. However, note that the impact of s is somewhat complicated. That is, s causes spectrum shift and dilation for $s < 1$. The reverse is also true; namely, a dilation in time ($s > 1$) results in both spectral downshift and contraction. This is somewhat different from what we observed from the concept of STFT.

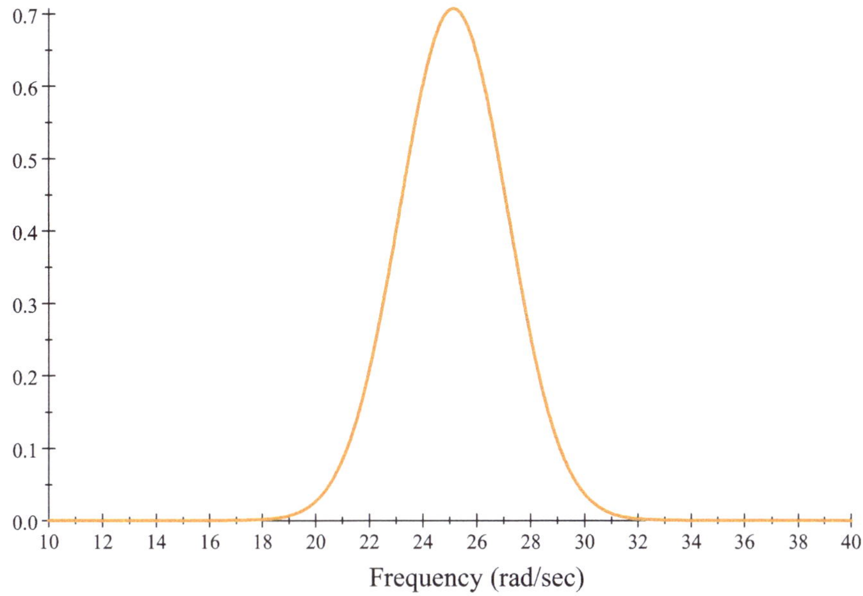

Fig. 5.29 $\sqrt{0.25}\left|\left(e^{-(0.5t-4\pi)^2/2} - e^{-0.5^2t^2/2-8\pi^2}\right)\right|$

5.3.2.3 Relationship Between Gabor Transform and STFT

The reader is cautioned not to confuse Gabor (Morlet) wavelet describe above with what is known as Gabor transform (GT). That is, if we re-examine STFT and let $g(t)$ in (5.120) be

$$g(t) = \frac{1}{\sqrt{2\pi\sigma^2}} e^{-\frac{t^2}{2\sigma^2}}, \tag{5.161}$$

then the STFT can be expressed as

$$X_{GT}(\tau, \Omega_0) = \frac{e^{-j\Omega_0\tau}}{\sqrt{2\pi\sigma^2}} \int_{-\infty}^{\infty} x(t)\, e^{-\frac{(t-\tau)^2}{2\sigma^2}} e^{-j\Omega_0(t-\tau)} dt. \tag{5.162}$$

The above is known as the Gabor transform and it is similar (but not identical) to Gabor (Morlet) wavelet transform, given by

$$w_{GW}(\tau, s) = \frac{1}{\sqrt{2\pi\,|s|}} \int_{-\infty}^{\infty} x(t) \times \left[e^{-2\pi j f_0 \frac{t-\tau}{s}} - e^{-2\pi^2 f_0^2}\right] e^{-\frac{(t-\tau)^2}{2s^2}} dt. \tag{5.163}$$

The difference is primarily due to the additional parameter f_0 in GW and the fact that s along with f_0 play a dual role in GW; $|s|$ controls the frequency resolution while f_0 along with s perform frequency translation. In the case of GT, σ is the only parameter that is used to control frequency resolution while Ω_0 may be viewed as the frequency translation component.

In other words, GT may be viewed in the frequency domain as the multiplication of $|X(\Omega)|$ by a Gaussian filter centered at $2\pi\Omega_0$ rad/s with a bandwidth that is proportional to σ^{-1} (frequency resolution). For GW, the same is true where $|X(\Omega)|$ is multiplied by a filter centered at $\frac{2\pi f_0}{|s|}$ rad/s and a bandwidth that is proportional to $|s|^{-1}$(resolution). Note that s impacts translation ($\frac{2\pi f_0}{|s|}$) and frequency resolution ($|s|^{-1}$) at the same time.

GT has one advantage over the standard STFT; it uses a window function with the smallest time-bandwidth product. In this case, τ and Ω have the same definitions as those in the case of STFT.

Hence, one can consider GT and GW as special cases of STFT with a window function that has the lowest time-bandwidth product.

5.3.2.4 Marr (Mexican Hat) Wavelet

This wavelet is defined as

$$\psi_{MARR}(t) = \frac{1}{\sqrt{2\pi}}\left(1 - t^2\right)e^{\frac{-t^2}{2}} \tag{5.164}$$

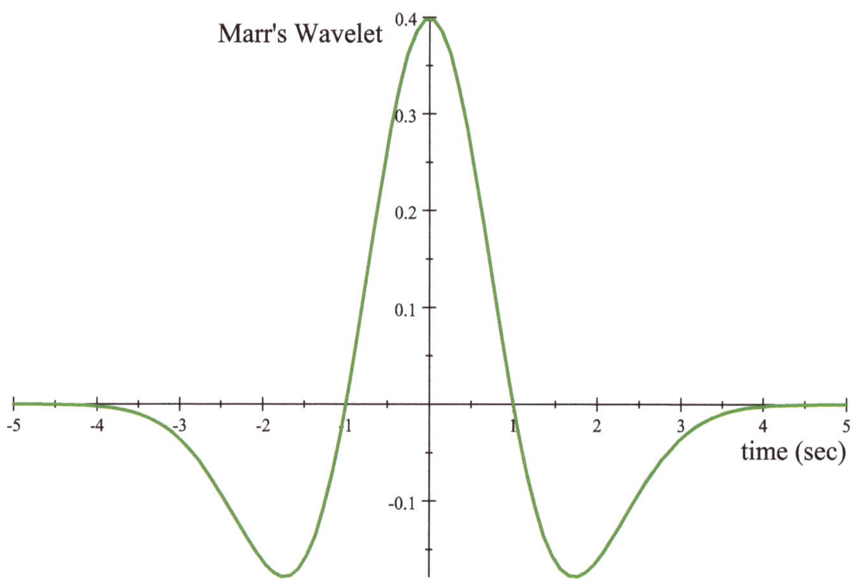

Fig. 5.30 $\psi_{MARR}(t) = \frac{1}{\sqrt{2\pi}}\left(1 - t^2\right)e^{\frac{-t^2}{2}}$

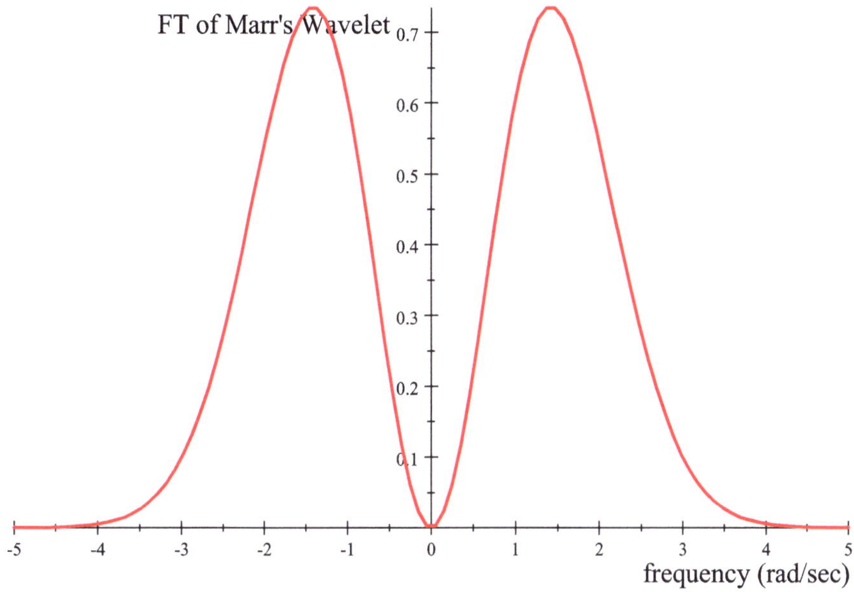

Fig. 5.31 $\Psi_{MARR}(\Omega) = \Omega^2 e^{-\frac{\Omega^2}{2}}$

and shown in Fig. 5.30 with its FT given by

$$\Psi_{MARR}(\Omega) = \Omega^2 e^{-\frac{\Omega^2}{2}}$$

and shown in Fig. 5.31. Note that the absence of f_0 makes this wavelet suitable for studying BB signals. The *admissibility* condition is satisfied here as $\int_{-\infty}^{\infty} \psi(t)\, dt = 0$. For this wavelet, time and bandwidth isolation is rather obvious as the mother wavelet exhibits isolation in both time and frequency.

Let us consider the impact of delay and dilation. In Fig. 5.32 a contracted ($s = 0.5$) and delayed version of this wavelet is depicted. Note that the temporal isolation is insured by the time-limited attribute of this wavelet. In Fig. 5.33, the absolute value of the FT of the signal in Fig. 5.32 is shown. Note that the impact of the contraction in time domain results in an expansion in the frequency domain. Also, note that unlike Gabor wavelet or Gabor transform, which allowed for frequency *selectivity* through f_0, this wavelet may be considered as a "baseband" wavelet; hence it may be inappropriate for signals with high frequency selectivity. That is, if $x(t)$ is a bandlimited signal (centered about some frequency) with limited bandwidth, this wavelet will not be an appropriate tool for identifying the significant components of the signal, specially in the presence of additive noise. Note that the inclusion of τ allows for temporal selectivity as $\psi_{MARR}(t)$ exhibits a time-limited attribute. Further, the impact of s is simply to broaden or narrow the bandwidth of the wavelet by modifying the duration of $\psi_{MARR}(t)$ in time. We also note the spectral symmetry of this wavelet Note that the inclusion of τ allows for temporal selectivity as $\psi_{MARR}(t)$

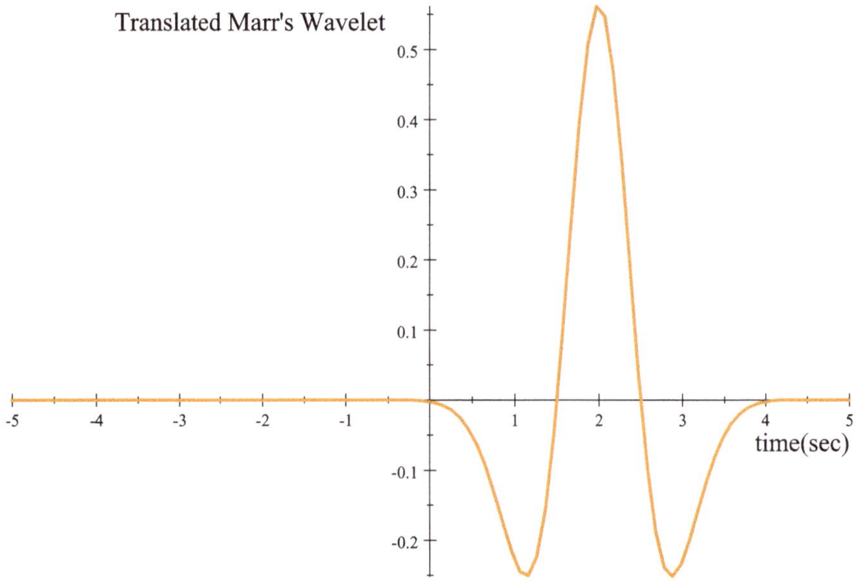

Fig. 5.32 $\frac{1}{\sqrt{s}}\psi_{MARR}\left(\frac{t-\tau}{s}\right)$ for $\tau = 2$ and $s = 0.5$

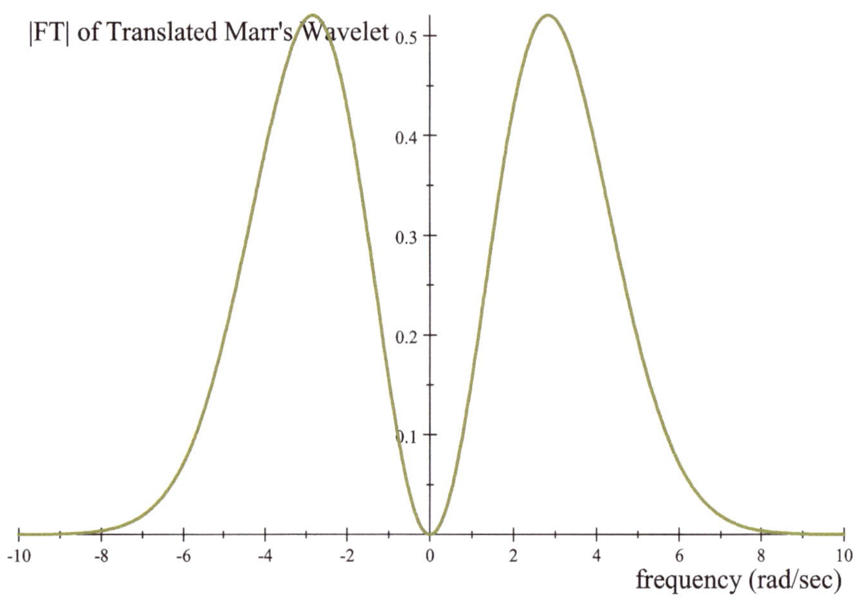

Fig. 5.33 $\sqrt{s}\,|\Psi_{MAR}(s\Omega)|$ for $\tau = 2$ and $s = 0.5$

exhibits a time-limited attribute. Further, the impact of s is simply to broaden or narrow the bandwidth of the wavelet by modifying the duration of $\psi_{MARR}(t)$ in time. We also note the spectral symmetry of this wavelet as compared with the FT of GW.

5.3.2.5 Practical Consideration for the Computation of Wavelet Transforms

CWT, as defined above, is computed for all τ and s. This approach, in practice, leads to computationally inefficient realization, unless a closed form solution for computing the CWT can be found. For biomedical signal processing, where the signal of interest does not lend itself to mathematical representation, computation of CWT is a daunting and impractical task.

Similar to STFT, we may define delay and dilation in a discrete manner. Note that, unlike FT, continuous wavelet transforms yield 2-dimensional functions (of s and τ) in response to a 1-dimensional signal, suggesting that CWT may harbor redundancies that need to be removed or minimized. More specifically, the CWT provides information for all τ and s, which may be cumbersome and/or unnecessary to compute for irregular signals or time-series where a closed-form expression is not available. Furthermore, the properties of CWT for discrete values of τ and s may only be of main interest in some problems. Note that the impacts of s and τ on the resulting WT may not be entirely independent as a change in s will alter the duration (in time) of the wavelet, which, in turn, will impact the actual interval of time over which the WT integration is carried out. Note that τ is designated to alter the integration interval of WT for a time-limited mother wavelet. Hence, the resulting integral is impacted by both s and τ.

From the sampling theorem, we understand that sampling deterministic signals (one may view the computation of CWT as a form of sampling signal for a given τ and s) beyond *Nyquist* sampling rate will not yield any additional information. Even for random, but WSS signals, samples of the signal which are obtained at the *Nyquist* sampling rate (where this rate is determined by the PSD of the random process) can be used to form an approximation of the signal that yields asymptotically zero mean-square sense error. In such a case, a larger sampling rate may not yield any additional information.

It is then of interest to consider a scenario where s defines the resolution in time (and hence frequency) while τ defines the actual integration interval (time slot). Due to the prevalent use of digital signal processing, it is then of interest to consider the discrete versions of these parameters.

To that end, let

$$s = \frac{1}{a^n} \tag{5.165}$$

while

$$\tau = kbs = \frac{kb}{a^n} \tag{5.166}$$

where n and k are integer dilation and translation indices, respectively, while a and b denote the dilation and translation increments, respectively. This is known as sampling the wavelet on a *hyperbolic lattice*. The theory for the special case of $a = 2$ and $b = 1$ has been well-developed. Hence, we focus on this special case that is most suited for digital signal processing. In that case,

$$s = \frac{1}{2^n} \tag{5.167}$$

and

$$\tau = \frac{k}{2^n}. \tag{5.168}$$

Note that s^{-1} is a power of 2 and, hence, it is suited for digital signal processing, which is based on binary computations; as one varies n, we dilate/contract by a power of 2, leading to a scenario where DSP tools can be utilized to process the wavelet transform. Also, note that the selection of τ, as shown above, allows for non-overlapping integration intervals for a mother wavelet which is defined over $[0, 1]$. To elaborate, the wavelets can be described in terms of the mother wavelet as

$$\psi_{n,k}(t) = 2^{\frac{n}{2}} \psi\left(2^n t - k\right). \tag{5.169}$$

The above selection of s and τ leads to what is known as the '*dyadic*' wavelet.

5.3.3 Father Wavelet

In conjunction with mother wavelet, one can define what is known as the *scaling function* $\phi(t)$, which is also known as the "father" wavelet. The intent here is to generate *all* wavelets, including mother wavelet, from a single function. Once again, as will be shown shortly, the introduction of father wavelet allows for an efficient computation of wavelets. The concept of scaling function will be explained shortly.

5.3.4 Orthogonal Wavelet

So far, we have described a number of wavelet functions that are known in practice without dwelling on how one can compute (efficiently) the inverse wavelet transform (or synthesize the signal from its wavelet transform). The integral that describes the WT and its inverse offer little insight into a practical implementation of WT and its inverse. Note that the WT and its inverse do not enjoy the duality that a signal and its FT offer. There is a class of wavelets that are orthogonality, and hence offer a measure of duality. If the choice of mother wavelet is such that

$$\int_{-\infty}^{\infty} \psi_{n_1,k_1}(t)\,\psi_{n_2,k_2}(t)\,dt = \delta[n_1 - n_2]\,\delta[k_1 - k_2], \tag{5.170}$$

then we have orthogonal wavelets. Note that this implies that the wavelets which are considered here are *orthonormal*. That is,

$$\int_{-\infty}^{\infty} \psi_{n,k}^2(t)\,dt = 1 \text{ for } \forall \text{ integer } n \text{ and } k. \tag{5.171}$$

These types of wavelets are of particular interest in signal processing as they provide decomposition of signals into their orthogonal (in terms of dilation and translation) components. Also, from signaling point of view, one can view a scheme where signals are generated (synthesized) by the orthogonal wavelets and then passed on through a linear channel to be recovered at another end of the link. The "information" (or the wavelet coefficients) can then be recovered via a correlation operation at the other end.

A dyadic, orthonormal wavelet is Haar wavelet, which will be discussed next.

5.3.4.1 Haar Wavelet

A popular wavelet that is used in conjunction with discrete values of τ and s is the Haar Wavelet. This wavelet offers ease of implementation and lends itself to digital representation. In the case of Haar wavelet, the mother wavelet (defined over [0, 1]) is given by

$$\psi(t) = \begin{cases} 1 & 0 \le t \le \frac{1}{2} \\ -1 & \frac{1}{2} < t \le \frac{1}{2} \\ 0 & \text{otherwise} \end{cases}. \tag{5.172}$$

In conjunction with this wavelet, the *scaling function*, which is also known as the "father" wavelet, for this wavelet is given by

$$\phi(t) = \begin{cases} 1 & 0 \le t \le 1 \\ 0 & \text{otherwise} \end{cases}. \tag{5.173}$$

The need for the scaling function, which can be used to generate all wavelets, including the mother wavelet, will be explained later. Note that we have assumed an observation interval of 1 s for the mother wavelet, but with the proper selection of k any time interval can be examined using this wavelet. Also, note that the restriction of the mother wavelet to [0, 1] allows for a dyadic orthonormal wavelet for an integer k. The wavelets for $n = 0$, $k = 0$ and 1 are depicted in Fig. 5.34 along with its corresponding scaling function in Fig. 5.35. Note the temporal isolation that the two wavelets and the scaling functions offer. This implies that no redundant information is gathered here and that there is no overlap in integrations involved for a given s (or n).

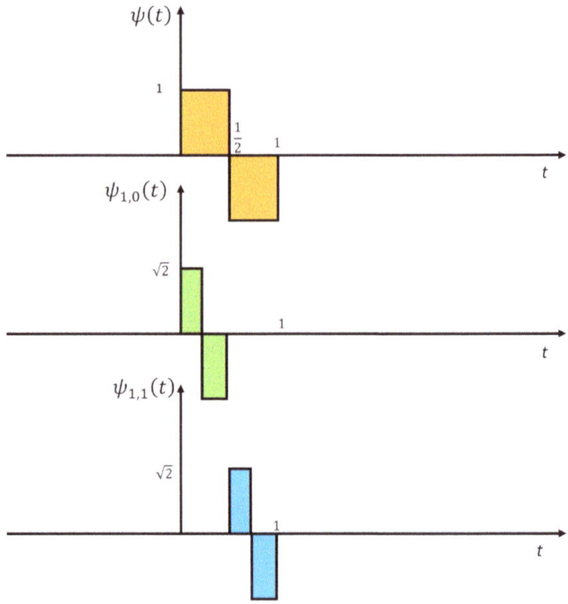

Fig. 5.34 Haar mother wavelet and its corresponding wavelets for $n = 1$ and $k = 0$ and 1

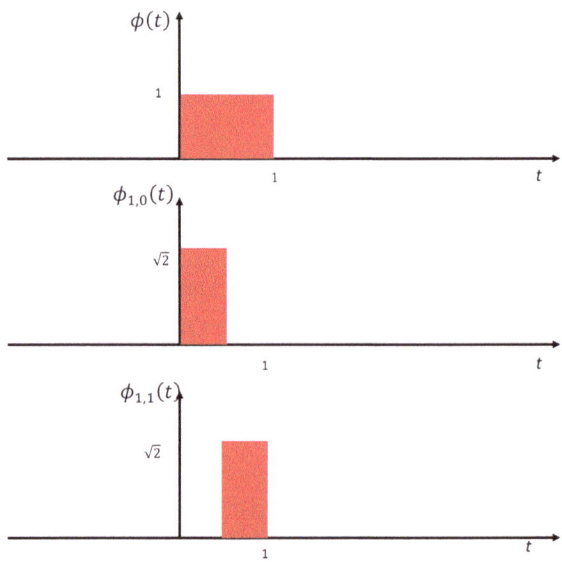

Fig. 5.35 Haar scaling function and its corresponding functions for $j_0 = 1$ and $k = 0$ and 1

The above clearly demonstrates the orthogonality of wavelets and scaling functions for all indices. Furthermore, the choice of $\phi(t)$ allows one to numerically scale the signal over a desired time slot defined by j_0 and k (allowing for the recovery of the dc level over a given time interval described by j_0 and k). For $j_0 = 0$, we see that father wavelet or $\phi(t)$ measures the dc level of the signal. We emphasize that j_0

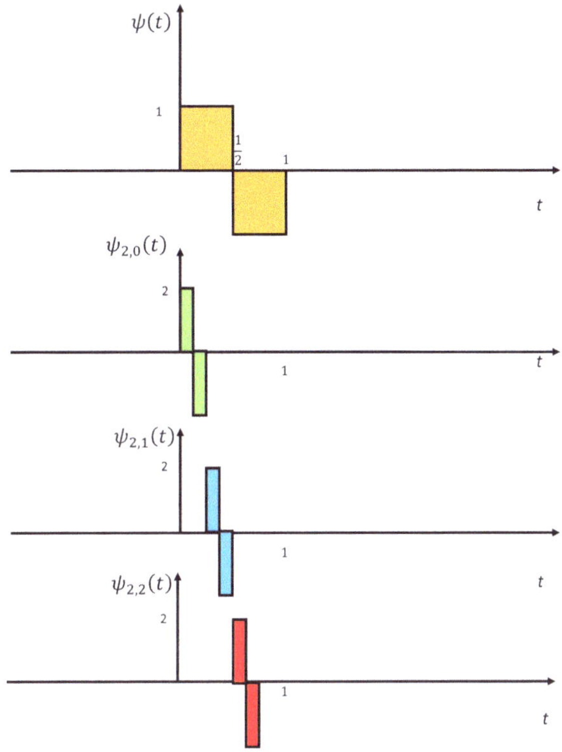

Fig. 5.36 Harr mother wavelet and its corresponding wavelets for $n = 2$ and $k = 0, 1, 2$, and 3

is typically kept fixed as the scaling is typically frequency independent, but may be transient (vary with k). That is, a signal may have a mean that is constant over a short period of time (1/2 s in the above case) but is changing with time (from $\left[0, \frac{1}{2}\right]$ to $\left[\frac{1}{2}, 1\right]$). In that case, the scaling function defined in the Haar wavelet may be used to capture this effect. In Figs. 5.36 and 5.37, we examine the Haar wavelets and scaling functions for $n = 2$ and $j_0 = 2$, respectively. Furthermore, we can validate that these wavelets are mutually orthonormal for all levels (all n). These figures are instructive in that they enable one to establish a relationship between these functions, which is critical to the practical implementation of WT. It is important to note that this wavelet has an excellent temporal isolation, while in the frequency domain provides poor isolation. Note that the selection of the above dilation and translation indices has led to a countably infinite wavelet coefficients. By comparison, this is similar to the Fourier series expansion (computed for countably infinite number of coefficients) as it relates to the Fourier transform (computed for all frequencies, which is uncountably infinite). This leads to a what is known as the wavelet series expansion, which will be discussed next.

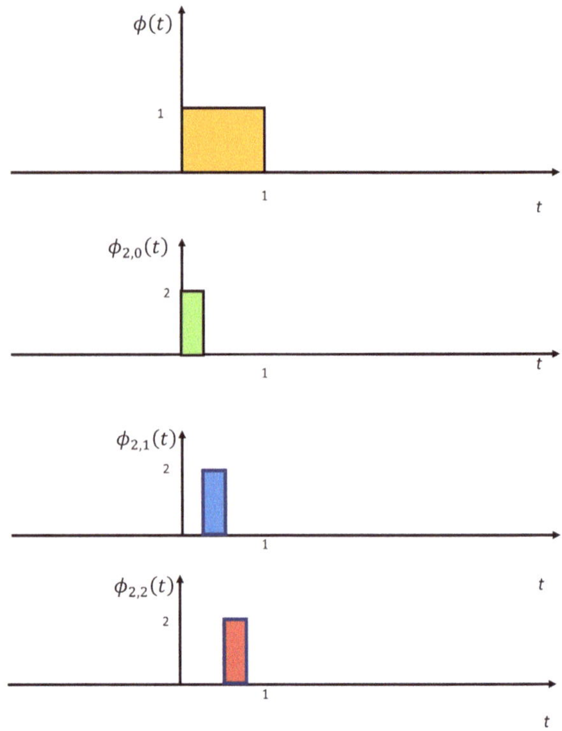

Fig. 5.37 Haar scaling function and its corresponding functions for $j_0 = 2$ and $k = 0, 1, 2,$ and 3

5.3.5 *Wavelet Series Expansion (WSE)*

CWT focused on signals with infinite durations. This consideration is, in practice, not realistic as most signals that we encounter are observed over a finite length in time. As we have seen in the case of the Fourier series expansion (FSE) where a discrete (but infinite) set of frequencies are only considered, we can consider a wavelet transform which focuses on a series expansion rather than an integral representation of $x(t)$. Similar to the FSE case, the signal does not necessarily need to be periodic as long as our interest in the signal representation is limited to a finite time interval T seconds. In that case, we can define a fundamental frequency

$$f_0 = \frac{1}{T} \text{ Hz} \tag{5.174}$$

with

$$nf_0 \tag{5.175}$$

denoting for the nth harmonic of the signal. A similar concept can be developed for the wavelet representation. That is, we only concern ourselves with a known interval of T seconds. Extending the concept of FSE to the wavelet representation when discrete s and τ, as shown above, are used, we arrive at the following representation of $x(t)$:

$$x(t) = \sum_{k=-\infty}^{\infty} a[j_0, k]\,\phi_{j_0,k}(t) + \sum_{n=j_0}^{\infty} \sum_{k=-\infty}^{\infty} w[n, k]\,\psi_{n,k}(t) \tag{5.176}$$

for some j_0 where $\phi_{j_0k}(t)$ is given by

$$\phi_{j_0,k}(t) = 2^{\frac{j_0}{2}} \phi\left(2^{j_0}t - k\right). \tag{5.177}$$

The scaling function, as noted earlier, is known as the *father* wavelet. This name is motivated by the fact that the j_0th level of the father wavelet can be used to generate the mother wavelet (and, consequently, all wavelet functions). Furthermore, the j_0th level of the father wavelet can generate the father wavelet (and, consequently, all scaling functions) as well. Note that the minimum value for n is set to j_0, which is typically 1.

Given the above formulation, one needs to compute $a[j_0, k]$ and $w[n, k]$ for all n and k.

For orthogonal wavelets, we have

$$\boxed{\int_{-\infty}^{\infty} \phi_{j_0,l}(t)\,\phi_{j_0,m}(t)\,dt = \delta[l - m],} \tag{5.178}$$

$$\boxed{\int_{-\infty}^{\infty} \psi_{n_1,k_1}(t)\,\psi_{n_2,k_2}(t)\,dt = \delta[n_1 - n_2]\,\delta[k_1 - k_2]}, \tag{5.179}$$

and

$$\int_{-\infty}^{\infty} \phi^2(t)\,dt = \int_{-\infty}^{\infty} \psi^2(t)\,dt = 1. \tag{5.180}$$

Furthermore, the scaling function and all other wavelets satisfy

$$\boxed{\int_{-\infty}^{\infty} \phi_{j_0,l}(t)\,\psi_{n,k}(t)\,dt = 0 \text{ for all } l, n, \text{ and } k.} \tag{5.181}$$

The wavelet coefficients are then given by

$$\boxed{a[j_0, k] = \int_{-\infty}^{\infty} x(t)\,\phi_{j_0,k}(t)\,dt} \tag{5.182}$$

$$\boxed{w[n, k] = \int_{-\infty}^{\infty} x(t)\,\psi_{n,k}(t)\,dt}. \tag{5.183}$$

The above transformation parallels FSE in that the transformation is carried out using orthonormal basis functions (wavelets and scaling functions). However, the above transformation is carried out in two dimensions (n and k).

For the wavelets, such as Haar, which are defined over [0, 1],

$$a[j_0, k] = \int_{\frac{j_0}{2^n}}^{\frac{j_0}{2^n}+1} x(t)\,\phi\left(2^{j_0}t - k\right) dt \tag{5.184}$$

$$w[n, k] = \int_{\frac{k}{2^n}}^{\frac{k}{2^n}+1} x(t)\,\psi\left(2^n t - k\right) dt. \tag{5.185}$$

Furthermore, if we interpret the above integrals as filtering operations, we have (we have replaced t with t' to emphasize the dummy variable nature of this parameter in the above integrals)

$$\begin{aligned} w[n, k] &= \int_{\frac{k}{2^n}}^{\frac{k}{2^n}+1} x(t')\,\psi\left(2^n\left(t' - 2^{-n}k\right)\right) dt' \\ &= \int_{\frac{k}{2^n}}^{\frac{k}{2^n}+1} x(t')\,\psi\left(-2^n\left(2^{-n}k - t'\right)\right) dt'. \end{aligned} \tag{5.186}$$

This in turn leads to

$$w[n, k] = x(t) \circ \psi\left(-2^n t\right)|_{t=k2^{-n}} \tag{5.187}$$

and, similarly,

$$a[j_0, k] = x(t) \circ \phi\left(-2^{j_0}t\right)|. \tag{5.188}$$

The above suggests that the wavelet coefficients are computed by using filter banks with the impulse responses $\psi(-2^n t)$ and $\phi\left(-2^{j_0}t\right)$ for all n and a given j_0, and sampling the output of the filters at $k2^{-n}$. Although this appears to be a simplified model, it does not lend itself to an efficient implementation as the above operation requires infinite number of filter banks.

5.3.5.1 A Filtering Perspective

One can define a pair of functions $h_0[l]$ and $h_1[l]$ which facilitate the computation of the wavelet coefficients. In particular, we claim the following:

$$\phi(t) = \sum_l h_0[l]\,\phi_{j_0,l}(t) \tag{5.189}$$

and

$$\psi(t) = \sum_l h_1[l]\,\phi_{j_0,l}(t) \tag{5.190}$$

for some $h_0[l], h_1[l]$ and j_0. Note that the limits of l are not defined as the assumption os causality is not enforced here. Now, if the wavelets and scaling functions are mutually orthonormal,

$$h_1[l] = \int_{-\infty}^{\infty} \phi_{j_0,l}(t)\,\psi(t)\,dt \tag{5.191}$$

and

$$h_0[l] = \int_{-\infty}^{\infty} \phi_{j_0,l}(t)\,\phi(t)\,dt. \tag{5.192}$$

Since all wavelets and scaling functions can be generated using $\psi(t)$ and $\phi(t)$, respectively, $h_0[l]$ and $h_1[l]$ can, in turn, be used to generate all wavelets and scaling functions. This is a useful property which will be exploited in the next section for the efficient computation of the discrete wavelet transform. Notice that the index l indicates time l and not the lth sample.

It is important to note that the selection of j_0 depends on the wavelet and on the fact that the resulting $\phi_{j_0,l}(t)'s$ must enable one to compute all other wavelets and scaling functions. In order to fully understand the implication of this selection, let us consider an example.

Example 6 For the Haar wavelet defined previously, find j_0, $h_0[l]$, *and* $h_1[l]$.

Solution: *From the definition of the Haar wavelet and with some cursory inspection, one can argue that* $j_0 = 1$ *will enable one to express* $\phi(t)$ *and* $\psi(t)$ *in terms of* $\phi_{1,l}(t)$ *and* $\psi_{1,l}(t)$, *respectively. That is, we can see that for*

$$\phi_{1,0}(t) = \sqrt{2}\{U(t) - U(t-0.5)\} \tag{5.193}$$

and

$$\phi_{1,1}(t) = \sqrt{2}\{U(t-0.5) - U(t-1)\}, \tag{5.194}$$

we have

$$\begin{aligned} h_0[l] &= \int_{-\infty}^{\infty} \phi_{1,l}(t)\,\phi(t)\,dt \\ &= \frac{1}{\sqrt{2}}\{\delta[l] + \delta[l-1]\} \end{aligned} \tag{5.195}$$

and

$$\begin{aligned} h_1[l] &= \int_{-\infty}^{\infty} \phi_{1,l}(t)\,\psi(t)\,dt \\ &= \frac{1}{\sqrt{2}}\{\delta[l] - \delta[l-1]\}. \end{aligned} \tag{5.196}$$

$h_0[l]$ and $h_1[l]$ can express $\phi(t)$ and $\psi(t)$ in terms of $h_1[l]$ and $h_0[l]$. Note that these operations are finite in length. If they are to be represented as filtering operations, then $h_1[l]$ and $h_0[l]$ may be viewed as finite impulse response, causal ($h_{0,1}[l] = 0$ for $l < 0$) filtering operations.

Finally, the discrete selection of s and τ shown above will insure that the expansion in (5.176) is unique. Note that, if one considers the Fourier transform as a counterpart to the CWT, then the above representation is similar to the Fourier series expansion, which can be used to express aperiodic signals if the expansion is used to represent the signal over a fixed interval T seconds (T may be used as the period of the signal for the Fourier series expansion). We are now ready to discuss the concept of the discrete wavelet transform that benefits from the concept of digital signal processing.

5.3.6 Discrete Wavelet Transform (DWT)

We pointed out the similarities between the WSE and its counterpart FSE. When τ and s along with time t take on a discrete nature, we arrive at what is known as the DWT. In general, the concept of the discrete wavelet transform may be viewed as the wavelet series expansion of a signal for the discrete time. For this reason, the discrete wavelet transform may be viewed as the counterpart to the discrete Fourier transform (note that the discrete Fourier transform is the discrete version of the Fourier series expansion). Since the majority of signals today are digitized in order to take advantage of DSP techniques, the discrete wavelet transform is the practical realization of the wavelet transform of a signal, which will be the focus of the remainder of this chapter.

To formulate this problem, we assume that the input signal is sampled at a rate high enough to satisfy the Nyquist sampling rate. Furthermore, we argue that a finite period of T seconds is of interest as most practical signals are observed over a finite period of time. Let the signal be

$$\mathbf{x} = [x[0], x[1], \ldots, x[N-1]]^{\mathsf{T}} \tag{5.197}$$

where

$$x[q] = x(qT_s) \tag{5.198}$$

with the sampling rate defined as

$$N_s = \frac{1}{T_s}, \tag{5.199}$$

which implies that we have assumed integer number of samples per second. Although this appears rather restrictive, in practice when DSP tools are used this assumption is readily satisfied. For instance, N_s is assume to be 128, 256, 512, 1024, etc. That is, N_s is assumed to be a power of 2. For the problem at hand, we also assume that N is a power of 2 as well. Namely,

$$N = 2^L \tag{5.200}$$

for some L. This further implies that the observation interval is $T > T_s$ given by

$$T = NT_s = \frac{N}{N_s}. \tag{5.201}$$

Note that this condition can be enforced in practice as the observation interval is often decided by the person collecting the desired data. Since both N and N_s are assumed to be powers of 2 and $N > N_s$, T becomes an integer. Hence,

$$n \in \left[0, \log_2 N\right] \tag{5.202}$$

since the number of samples cannot be divided by 2 for more than $\log_2 N$ times.

We also have to define the *mother* wavelet and its corresponding *father* wavelet for the discrete time domain. To that end, let us define the discrete version of the *mother* wavelet as

$$\psi\left[q\right] = \psi\left(qT_s\right); q \in [0, N-1] \tag{5.203}$$

and

$$\psi_{n,k}\left[q\right] = \psi_{n,k}\left(qT_s\right) = 2^{\frac{n}{2}}\psi\left(2^n qT_s - k\right). \tag{5.204}$$

Hence, $\psi_{n,k}\left[q\right]$ may now be defined as

$$\psi_{n,k}\left[q\right] = 2^{\frac{n}{2}}\psi\left[2^n q - kN_s\right]. \tag{5.205}$$

Note that the above equation has made a critical simplification; the discrete version leads to the mapping of the translation parameter k (which is in seconds) to kN_s sample in the sample space. For instance, if a signal is sampled 10 times/s, then a delay of $k = 1$ (1 s) implies a delay of 10 samples.

Similarly,

$$\phi_{n,k}\left[q\right] = 2^{\frac{n}{2}}\phi\left[2^n q - kN_s\right]. \tag{5.206}$$

To better understand these functions, let us resort to an example.

Example 7 Describe the Discrete time Haar wavelet functions for the first and second levels. Assume that the period of interest has a length of 2 s and that sampling is performed at the rate of 8 samples/s.

Solution: *In this case,* $N = 2 \times 8 = 16$. *Since the sampling rate is 8 samples/s, the wavelet is defined over 1 s, resulting in 8 non-zero samples. We then have the following mother wavelet and scaling functions*

$$\psi[q] = \frac{1}{\sqrt{8}} \{U[q] - 2U[q-4] + U[q-8]\} \tag{5.207}$$

$$\phi[q] = \frac{1}{\sqrt{8}} \{U[q] - U[q-8]\} \tag{5.208}$$

where $U[q]$ denotes the unit step function. For the first level, we have

$$\psi_{1,k}[q] = \frac{1}{\sqrt{4}} \{U[2q-8k] - 2U[2q-4-8k] + U[2q-8-8k]\} \tag{5.209}$$

and

$$\phi_{1,k}[q] = \frac{1}{\sqrt{4}} \{U[2q-8k] - U[2q-8-8k]\} \tag{5.210}$$

For translations, we have

$$\psi_{1,0}[q] = \begin{cases} \frac{1}{\sqrt{4}} & q \in [0,1] \\ -\frac{1}{\sqrt{4}} & q \in [2,3] \\ 0 & \text{otherwise} \end{cases} \tag{5.211}$$

$$\psi_{1,1}[q] = \begin{cases} \frac{1}{\sqrt{4}} & q \in [4,5] \\ -\frac{1}{\sqrt{4}} & q \in [6,7] \\ 0 & \text{otherwise} \end{cases} \tag{5.212}$$

$$\phi_{1,0}[q] = \begin{cases} \frac{1}{\sqrt{4}} & q \in [0,3] \\ 0 & \text{otherwise} \end{cases} \tag{5.213}$$

$$\phi_{1,1}[q] = \begin{cases} \frac{1}{\sqrt{4}} & q \in [4,7] \\ 0 & \text{otherwise} \end{cases}. \tag{5.214}$$

The above functions are depicted in Figs. 5.38, 5.39, 5.40, 5.41. *Note the orthonormal behavior of the scaling and wavelet functions. Also, $k = 0$ and 1 can only be accommodated as any further translation will put the wavelet outside the observation interval. Furthermore, for level 2 ($n = 2$), we have*

$$\psi_{2,k}[q] = \frac{1}{\sqrt{2}} \{U[4q-8k] - 2U[4q-4-8k] + U[4q-8-8k]\} \tag{5.215}$$

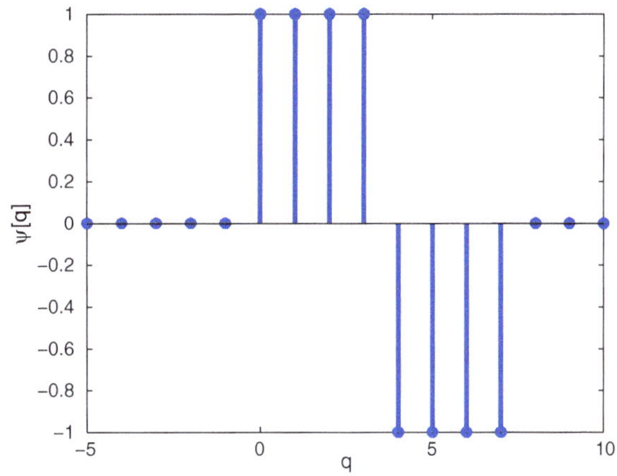

Fig. 5.38 Discrete Haar wavelet. N=16

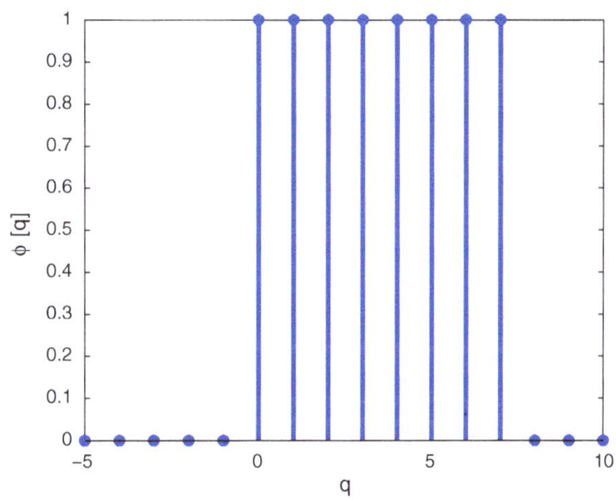

Fig. 5.39 Discrete Scaling function for Haar wavelet. N=16

and

$$\phi_{2,k}[q] = \frac{1}{\sqrt{2}} \{U[4q - 8k] - U[4q - 8 - 8k]\} \tag{5.216}$$

For translations, we have

$$\psi_{2,0}[q] = \begin{cases} \frac{1}{\sqrt{2}} & q = 0 \\ -\frac{1}{\sqrt{2}} & q = 1 \\ 0 & \text{otherwise} \end{cases} \tag{5.217}$$

$$\psi_{2,1}[q] = \begin{cases} \frac{1}{\sqrt{2}} & q = 2 \\ -\frac{1}{\sqrt{2}} & q = 3 \\ 0 & \text{otherwise} \end{cases} \tag{5.218}$$

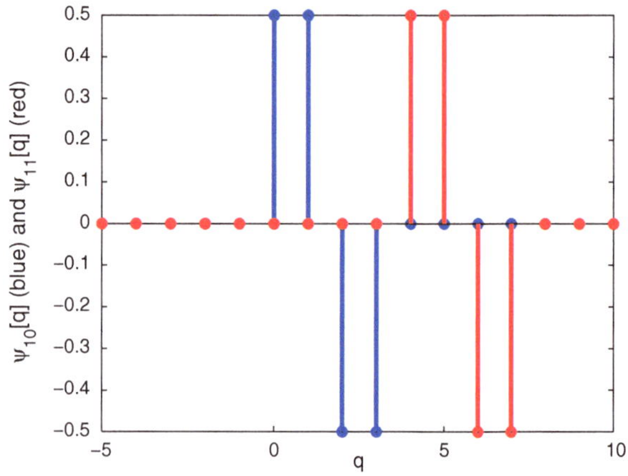

Fig. 5.40 ψ_{10} and ψ_{11} for discrete Haar wavelet. N=16

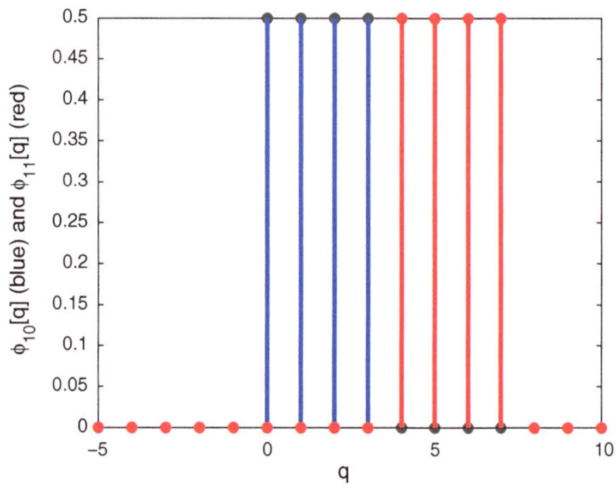

Fig. 5.41 ϕ_{10} and ϕ_{11} for discrete Haar wavelet. N=16

$$\psi_{2,2}[q] = \begin{cases} \frac{1}{\sqrt{2}} & q = 4 \\ -\frac{1}{\sqrt{2}} & q = 5 \\ 0 & \text{otherwise} \end{cases} \tag{5.219}$$

$$\psi_{2,3}[q] = \begin{cases} \frac{1}{\sqrt{2}} & q = 6 \\ -\frac{1}{\sqrt{2}} & q = 7 \\ 0 & \text{otherwise} \end{cases} \tag{5.220}$$

$$\phi_{2,0}[q] = \begin{cases} \frac{1}{\sqrt{2}} & q = 0, 1 \\ 0 & \text{otherwise} \end{cases} \tag{5.221}$$

$$\phi_{2,1}[q] = \begin{cases} \frac{1}{\sqrt{2}} & q = 2, 3 \\ 0 & \text{otherwise} \end{cases} \tag{5.222}$$

$$\phi_{2,2}[q] = \begin{cases} \frac{1}{\sqrt{2}} & q = 4, 5 \\ 0 & \text{otherwise} \end{cases} \tag{5.223}$$

$$\phi_{2,3}[q] = \begin{cases} \frac{1}{\sqrt{2}} & q = 6, 7 \\ 0 & \text{otherwise} \end{cases}. \tag{5.224}$$

The wavelet and scaling functions for $n = 2$ are shown in Figs. 5.42 and 5.43. For the case of $n = 2$, translations $0, 1, 2, 3$ (4 translations) can be accommodated for the wavelets and scaling function. While 8 translations can be realized for the

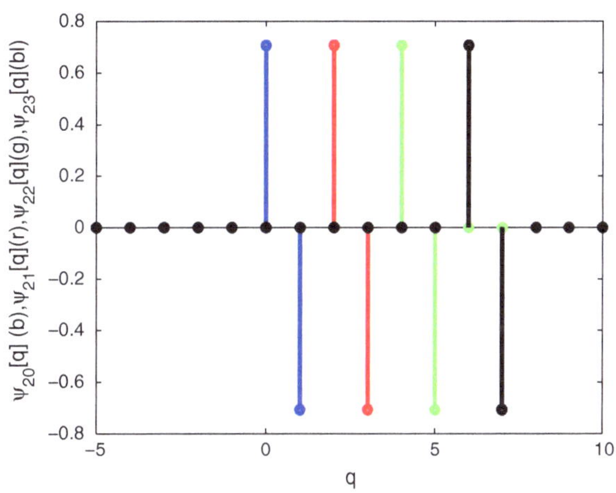

Fig. 5.42 $\psi_{20}, \psi_{21}, \psi_{22}$, and ψ_{23} for discrete Haar wavelet. N=16

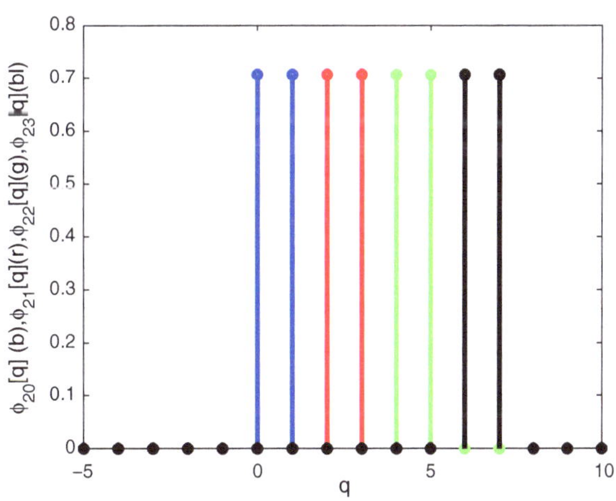

Fig. 5.43 $\phi_{20}, \phi_{21}, \phi_{22}$, and ϕ_{23} for discrete Haar wavelet. N=16

scaling function. Also, note that n must remain less than 3$(\log_2 N - 1 = L - 1 = 3)$ *for the wavelet functions while* $n = 3$ *can be considered for the scaling function, which leads to*

$$\phi_{\log_2 N,k}[q] = \phi_{3,k}[q] = \delta[q-k]. \tag{5.225}$$

The above is an important observation that will aid us in computing the wavelet transform efficiently. To elaborate, we will discuss an iterative technique that starts with the signal itself using $\phi_{\log_2 N,k}[q]$ (the highest order characteristic function) and then, in an iterative manner, builds lower order coefficients. This will be discussed shortly.

The above example puts in perspective how the discrete wavelet transform decomposes a sequence into its wavelet coefficients. A more interesting observation can also be made for this example; that is, one can express the wavelet at level j in terms of wavelets at level $j+1$ (the reader is encouraged to make an attempt to verify this assertion). We will elaborate on this property and its impact on the efficient implementation of wavelet in later section.

Now, let us consider the wavelet coefficients for a discrete time operation, which are

$$w[n,k] = \sum_{q=0}^{N-1} x[q]\,\psi_{n,k}[q] \tag{5.226}$$

$$a[j_0,k] = \sum_{q=0}^{N-1} x[q]\,\phi_{j_0,k}[q]. \tag{5.227}$$

It is important to see that for the Haar wavelet in the previous example,

$$a[3,k] = x[q]\,\delta[q-k] = x[k]. \tag{5.228}$$

Hence, we immediately have access to the scaling coefficients at the highest level when $x[q]$ is observed. In the next section, we develop a technique for computing the wavelet and scaling coefficients efficiently.

Furthermore, in the above formulation, we have to observe the following constraints as the length of the observed vector is finite:

$$0 \le n \le L-1, \tag{5.229}$$

$$0 \le j_0 \le n, \tag{5.230}$$

and

$$0 \le k \le 2^n - 1. \tag{5.231}$$

We then have

$$x[q] = \sum_k a[j_0, k]\,\phi_{j_0,k}[q] + \sum_{n=j_0}^{L-1} \sum_k w[n, k]\,\psi_{n,k}[q]. \tag{5.232}$$

Note that we have not specified the range for k as it will depend upon the duration in time of the mother and father wavelets. This completes the definition of the DWT.

As seen above, this requires computing the vector products for all k and n for a given L. Since $N = 2^L$, once N is determined (total number of samples available in time), then L can be set.

One can see that the computational complexity grows exponentially with n as k grows exponentially (2^n) with n. This demands for a further investigation of how to reduce this computational complexity. For practical reasons, which will be discussed next, $j_0 = 1$, and hence, we have

$$x[q] = \sum_k a[1, k]\,\phi_{0,k}[q] + \sum_{n=0}^{L-1} \sum_k w[n, k]\,\psi_{n,k}[q]. \tag{5.233}$$

We are now in a position to discuss the efficient implementation of the above decomposition.

5.3.7 *Efficient Realization of DWT*

As we observed above, computing the wavelet coefficients requires multiplication/integration (or filtering) operation to compute each coefficient. However, this poses a serious problem; that is, one requires a unique filtering operation for each coefficient, resulting in a large computational complexity. This is, for obvious reasons, an impractical solution. We will show in this section that one can use only a pair of filtering operations followed by an iterative operation to compute the wavelet coefficients for a time-limited signal. This is quite similar to the FFT computation of DFT, which operates on N samples of a signal obtained at the sampling rate N_s. Now, since $\phi(t)$ is known as the father wavelet, we claimed earlier that we can find a pair of functions $h_0[l]$ and $h_1[l]$ (which was observed in the previous section). Now, let us consider the digital version of the previously shown results. That is, let us define $h_0[l]$ and $h_1[l]$ as

$$h_0[l] = \sum_{q=0}^{N-1} \phi_{j_0,l}[q]\,\phi[q] \tag{5.234}$$

and

$$h_1[l] = \sum_{q=0}^{N-1} \phi_{j_0,l}[q]\,\psi[q], \tag{5.235}$$

which results in (orthonormal scaling and wavelet functions)

$$\phi[q] = \sum_{l} h_0[l]\,\phi_{j_0,l}[q] \tag{5.236}$$

and

$$\psi[q] = \sum_{l} h_1[l]\,\phi_{j_0,l}[q]. \tag{5.237}$$

We need to point out that the index l in the above equations refers to a somewhat different point in time. That is, while q points to qth sample in time, l refers to lN_sth samples in time.

Using (5.234)–(5.237), and assuming that the wavelet and the scaling functions are all real, we can show from the above that

$$\Psi[\omega] = \sqrt{2}\Phi\left[\frac{\omega}{2}\right] H_1\left[\frac{\omega}{2}\right] \tag{5.238}$$

and

$$\Phi[\omega] = \sqrt{2}\Phi\left[\frac{\omega}{2}\right] H_0\left[\frac{\omega}{2}\right], \tag{5.239}$$

where

$$H_0[\omega] = \sum_{l=0}^{L_h} h_0[l]\,e^{-j\omega l N_s} \tag{5.240}$$

and

$$H_1[\omega] = \sum_{l=0}^{L_h} h_1[l]\,e^{-j\omega l N_s}. \tag{5.241}$$

To arrive at the above, it is assumed that the length of the filters is L_h. This parameter is different for different wavelets and hence will be assumed to be a variable when dealing with different wavelets.

It is noteworthy that L_h denotes the number of samples of the filter response in 1 s interval. A direct consequence of this assumption is the following:

$$\boxed{\Phi[0] = \sqrt{2}\Phi[0]\,H_0[0] \rightarrow H_0[0] = 1 \rightarrow \textstyle\sum_l h_0[l] = \sqrt{2}.} \tag{5.242}$$

We also have for most wavelets, $\Psi[0] = 0$ ($\int \psi(t)\,dt = 0$), which leads to

$$\boxed{\Psi[0] = \sqrt{2}\Phi[0]\,H_1[0] \rightarrow H_0[0] = 0 \rightarrow \sum_l h_1[l] = 0.} \tag{5.243}$$

One can also recognize the following important facts (we assume unit energy mother and father wavelets) for the class of orthogonal wavelets:

$$\int_{-\infty}^{\infty} \phi_{1,l}(t)\,\phi_{1,m}(t)\,dt = \delta[l-m], \tag{5.244}$$

$$\int_{-\infty}^{\infty} \phi^2(t)\,dt = \int_{-\infty}^{\infty} \psi^2(t)\,dt = 1, \tag{5.245}$$

and

$$\int_{-\infty}^{\infty} \phi(t)\,\psi(t)\,dt = 0. \tag{5.246}$$

Note that the above implies

$$\int_{-\infty}^{\infty} \phi_{1,l}^2(t)\,dt = 1 \text{ for all } l.$$

Using the above identities, we arrive at the following important results:

$$\boxed{\sum_l h_0^2[l] = \sum_l h_1^2[l] = 1} \Longrightarrow \tag{5.247}$$

(for a real wavelet $H[-\omega] = H^*[\omega]$)

$$\boxed{\frac{1}{2\pi}\int_{-\pi}^{\pi} |H_0[\omega]|^2\,d\omega = \frac{1}{2\pi}\int_{-\pi}^{\pi} |H_1[\omega]|^2\,d\omega = 1} \tag{5.248}$$

and

$$\boxed{\sum_l h_0[l]\,h_1[l] = 0.} \tag{5.249}$$

To see this the above identities, consider

$$\begin{aligned}
\int_{-\infty}^{\infty} \phi^2(t)\,dt &= \sum\sum h_0[l_1]\,h_0[l_2] \int_{-\infty}^{\infty} \phi_{1,l_1}(t)\,\phi_{1,l_2}(t)\,dt \\
1 &= \sum_l h_0^2[l] \int_{-\infty}^{\infty} \phi_{1,l}^2(t)\,dt \\
1 &= \sum_l h_0^2[l].
\end{aligned} \tag{5.250}$$

The other results can be found using $\int_{-\infty}^{\infty} \phi(t)\,\psi(t)\,dt = 0$ and $\int_{-\infty}^{\infty} \psi^2(t)\,dt = 1$.

This approach suggests that the first level scaling function (i.e., $\phi_{1,0}(t)$) can be used to generate all wavelets and scaling functions. It is interesting to note that $\sum_l h_0[l]\,h_1[l] = 0$ leads to

$$
\begin{aligned}
\sum_l h_0[l]\,h_1[l] &= \frac{1}{(2\pi)^2}\int_{-\pi}^{\pi}\int_{-\pi}^{\pi}\sum_l H_1[\omega_1]\,H_0[\omega_2]\,e^{j(\omega_1+\omega_2)lN_s}\,d\omega_1 d\omega_2 \\
0 &= \frac{1}{2\pi}\int_{-\pi}^{\pi}\int_{-\pi}^{\pi} H_1[\omega_1]\,H_0[\omega_2]\,\delta[\omega_1+\omega_2]\,d\omega_1 d\omega_2 \\
0 &= \frac{1}{2\pi}\int_{-\pi}^{\pi} H_1[\omega]\,H_0[-\omega]\,d\omega. \qquad (5.251)
\end{aligned}
$$

For a real wavelet where $H[-\omega] = H^*[\omega]$, we arrive at

$$\frac{1}{2\pi}\int_{-\pi}^{\pi} H_1[\omega]\,H_0^*[\omega]\,d\omega = 0. \qquad (5.252)$$

One can insure this condition by choosing $H_0[\omega]$ as a low pass filter while forcing $H_1[\omega]$ to behave as a high pass filter.

We also notice that considering

$$\frac{1}{2\pi}\int_{-\pi}^{\pi} |H_0[\omega]|^2\,d\omega = \frac{1}{2\pi}\int_{-\pi}^{\pi} |H_1[\omega]|^2\,d\omega = 1,$$

we have

$$\frac{1}{2\pi}\int_{-\pi}^{\pi}\left\{|H_1[\omega]|^2 + |H_0[\omega]|^2\right\}d\omega = 2. \qquad (5.253)$$

Furthermore, using (5.252), we arrive at

$$\frac{1}{2\pi}\int_{-\pi}^{\pi} |H_1[\omega] + H_0[\omega]|^2\,d\omega = 2. \qquad (5.254)$$

Example 8 For the Haar wavelet, demonstrate that the properties stated above hold.

Solution: *For the Haar wavelet ($L_h = 2$), we have shown that*

$$h_0[l] = \frac{1}{\sqrt{2}}\{\delta[l] + \delta[l-1]\} \qquad (5.255)$$

and

$$h_1[l] = \frac{1}{\sqrt{2}}\{\delta[l] - \delta[l-1]\}. \qquad (5.256)$$

It is easy to verify that

$$\sum h_0^2[l] = \sum h_1^2[l] = 1, \ \sum h_0[l] = \sqrt{2},$$
$$\sum h_1[l] = 0, \sum h_1[l]\, h_0[l] = 0.$$

Also, note that (for N_s samples per second) $H_0[\omega] = \frac{1}{\sqrt{2}}\left[1 + e^{-j\omega N_s}\right]$ and $H_1[\omega] = \frac{1}{\sqrt{2}}\left[1 - e^{-j\omega N_s}\right]$ and $\frac{1}{2\pi}\int_{-\pi}^{\pi} H_1[\omega]\, H_0[-\omega]\, d\omega = \frac{1}{2\pi}\int_{-\pi}^{\pi} \sin(N_s\omega)\, d\omega = 0$ ($N_s \geq 1$ is integer).

Next, let us consider the following:

$$\psi[q] = \sum_l h_1[l]\, \phi_{1,l}[q] \tag{5.257}$$

and

$$\phi[q] = \sum_l h_0[l]\, \phi_{1,l}[q]. \tag{5.258}$$

Now, let $q \to 2^n q - kN_s$, which leads to

$$\psi\left[2^n q - kN_s\right] = \sum_l h_1[l]\, \phi_{1,l}\left[2^n q - kN_s\right].$$

Note that $\psi\left[2^n q - kN_s\right] = \psi_{n,k}[q]$. Hence,

$$\begin{aligned}\psi_{n,k}[q] &= \sum_l h_1[l]\, \phi\left[2\left(2^n q - kN_s\right) - lN_s\right] \\ &= \sum_l h_1[l]\, \phi\left[2^{n+1} q - 2kN_s - lN_s\right] \\ &= \sum_l h_1[l]\, \phi_{n+1,l+2k}[q].\end{aligned} \tag{5.259}$$

Now, let $l + 2k = m$,

$$\begin{aligned}\psi_{n,k}[q] &= \sum_m h_1[m-2k]\, \phi_{n+1,m}[q] \\ &= \sum_m h_1[m-2k]\, \phi_{n+1,m}[q].\end{aligned} \tag{5.260}$$

The same is also true for the scaling function. That is,

$$\phi_{n,k}[q] = \sum_m h_0[m-2k]\, \phi_{n+1,m}[q]. \tag{5.261}$$

Note that we need to obtain $h_0[l]$ and $h_1[l]$ for all l. Now, using (5.260) and (5.261), we have

$$\begin{aligned} w[n,k] &= \sum_{q=0}^{N-1} x[q]\,\psi_{n,k}[q] \\ &= \sum_{q=0}^{N-1} x[q] \sum_m h_1[m-2k]\,\phi_{n+1,m}[q] \\ &= \sum_m h_1[m-2k] \sum_{q=0}^{N-1} x[q]\,\phi_{n+1,m}[q] \\ &= \sum_m h_1[m-2k]\,a[n+1,m]. \end{aligned} \tag{5.262}$$

This leads to

$$\boxed{w[n,k] = \textstyle\sum_m h_1[m-2k]\,a[n+1,m].} \tag{5.263}$$

Similarly,

$$\begin{aligned} a[j_0,k] &= \sum_{q=0}^{N-1} x[q]\,\phi_{j_0,k}[q] \\ &= \sum_{q=0}^{N-1} x[q] \sum_m h_0[m-2k]\,\phi_{j_0+1,m}[q] \\ &= \sum_m h_0[m-2k] \sum_{q=0}^{N-1} x[q]\,\phi_{j_0+1,m}[q] \\ &= \sum_m h_0[m-2k]\,a[j_0+1,m]. \end{aligned} \tag{5.264}$$

Hence,

$$\boxed{a[j_0,k] = \textstyle\sum_m h_0[m-2k]\,a[j_0+1,m].} \tag{5.265}$$

The last two equations in brackets give us a method for generating the wavelet coefficient and scaling functions using a filtering operation in an efficient manner. More specifically, it appears that filters with impulse responses $h_0[-l]$ and $h_1[-l]$ are indeed needed in order to decompose a signal. Furthermore, the above operations (note $2k$ in the arguments of $h_0[-l]$ and $h_1[-l]$) points to a decimation of the output of the filters by 2. That is, let us consider $w[n,k]$

$$w[n,k] = \sum_m h_1[m-2k]\,a[n+1,m]. \tag{5.266}$$

Now, let us consider the following:

$$r[k] = \sum_m h_1[-(k-m)]\zeta[m], \tag{5.267}$$

which shows that $r[k]$ is the convolution of $\zeta[m] = a[n+1, m]$ with $h_0[-m]$ (since $h_0[m]$ is causal, $h_0[-m]$ is anti-causal) at time instance k. Now, it is easy to see that $w[n, k]$ is the decimated version of $r[k]$. That is,

$$\begin{aligned} w[n,k] &= r[2k] \\ &= r[k]\downarrow_2 . \end{aligned} \tag{5.268}$$

We then have

$$w[n,k] = a[n+1,k] \circ h_1[-k]\downarrow_2 . \tag{5.269}$$

Similarly,

$$a[n,k] = a[n+1,k] \circ h_0[-k]\downarrow_2 . \tag{5.270}$$

For the Haar wavelet,

$$a\left[\log_2 N, k\right] = x[q]\,\delta[q-k]. \tag{5.271}$$

This in turn suggests an iterative method to compute the wavelet coefficients. That is, starting with $a\left[\log_2 N, k\right]$, one can compute the lower scaling coefficients and, subsequently, wavelet coefficients using a pair of filtering followed by down sampling by a factor of 2. This is quite similar to the FFT algorithm for computing DFT and is depicted in Fig. 5.44. The above recursive operation is repeated until all coefficients are synthesized. Note that there are L stages in this operation. Given the hierarchal nature of this operation, one can compute the wavelet in an efficient manner.

The next question that can be asked here is how to recreate the original signal using the wavelet and scaling coefficients in a recursive (and hence) efficient manner. Such an operation is referred to as *synthesis*, whereas the operation of obtaining the wavelet coefficients may be viewed as *analysis*. Note that, since the filters shown above are FIR (although they are anti-causal), it can be shown with relative ease that one can arrive at the input of the filter shown above by upsampling by a factor of 2 and filtering the resulting signal using filters whose impulse responses are $g_0[-l]$

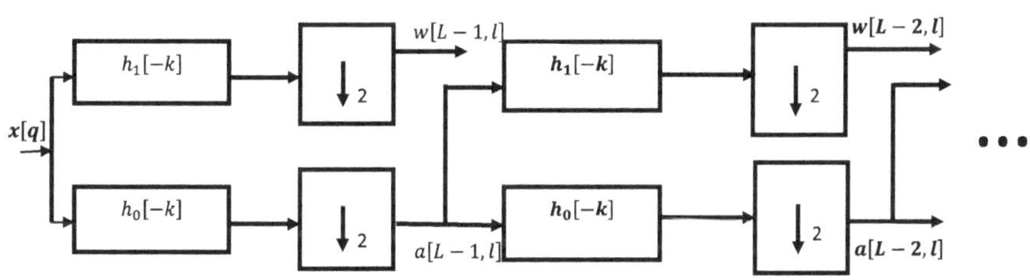

Fig. 5.44 DWT analysis block

(corresponding to $h_0[-l]$) and $g_1[-l]$ (corresponding to $h_1[-l]$). Next, we address the reconstruction of the synthesis filters.

5.3.8 Signal Synthesis Using DWT

In order to implement the synthesis operation, which for all intends and purposes is the inverse of the operation shown in Fig. 5.44, we need to consider the concept of *perfect reconstruction*, shown in Fig. 5.45. Note that the first two blocks on both branches produce the wavelet and scaling coefficients. It is only appropriate to perform an upsampling operation (which may not correct the aliasing which may have occurred due to downsampling) and then process the signals using the "inverse" filters. If this operation is done perfectly, the output will be a "perfect" delayed and scaled version of the original signal. The delay p corresponds to the total delay introduced by each branch of the system shown and c denotes the constant attenuation/amplification of the signal which has no bearing on the structure and shape of the reconstructed signal. This operation, quite appropriately, may be viewed as a *perfect reconstruction.* From a mathematical point of view and if we take the process into the Fourier domain, the perfect reconstruction implies:

$$X[\omega]\{H_1[\omega]G_1[\omega]+H_0[\omega]G_0[\omega]\}=2X[\omega]e^{-j\omega p} \tag{5.272}$$

which implies

$$H_1[\omega]G_1[\omega]+H_0[\omega]G_0[\omega]=2e^{-j\omega p}. \tag{5.273}$$

The factor of 2 is added to account for the summation of the signal over 2 separate branches. Furthermore, the downsampling followed by upsampling operations may result in aliasing. The total effect of aliasing may be viewed as $H_1[\omega]G_1[\omega+\pi]+H_0[\omega]G_0[\omega+\pi]$, and hence, in order to avoid aliasing,

$$H_1[\omega]G_1[\omega+\pi]+H_0[\omega]G_0[\omega+\pi]=0. \tag{5.274}$$

This concept is shown in Fig. 5.46, where the top figure depicts the system function of the filter that will not cause aliasing once its output is down sampled by a factor

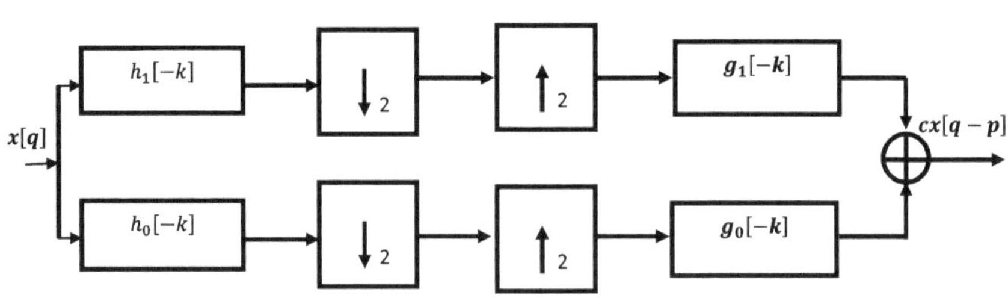

Fig. 5.45 Perfect reconstruction algorithm

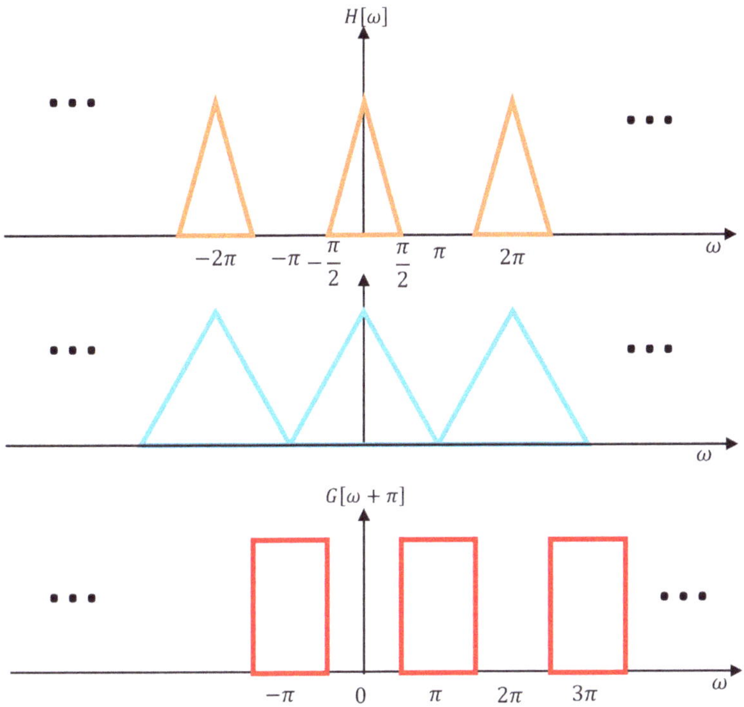

Fig. 5.46 Perfect reconstruction in the frequency domain

of 2 (the spectrum is limited to $\left[-\frac{\pi}{2}, \frac{\pi}{2}\right]$). The middle plot depicts the output of the filter when it is downsampled, which shows, rather obviously, no aliasing. The filter $g\,[]$ will operate on the upsampled version of the signal, and hence must also be confined to $\left[-\frac{\pi}{2}, \frac{\pi}{2}\right]$. The plot at the bottom shows $G\,[\omega+\pi]$. It can readily be seen that $H\,[\omega]\,G\,[\omega+\pi]=0$. Note that this condition is sufficient, but not necessary as we are concerned with the total contribution of aliasing and not that of the individual contribution of each branch. That is, although $H_1\,[\omega]\,G_1\,[\omega+\pi]=0$ and $H_0\,[\omega]\,G_0\,[\omega+\pi]=0$ lead to $H_1\,[\omega]\,G_1\,[\omega+\pi]+H_0\,[\omega]\,G_0\,[\omega+\pi]=0$, the actual condition for not having an impact due to aliasing is $H_1\,[\omega]\,G_1\,[\omega+\pi]+H_0\,[\omega]\,G_0\,[\omega+\pi]=0$, which implies

$$H_1\,[\omega]\,G_1\,[\omega+\pi]=-H_0\,[\omega]\,G_0\,[\omega+\pi]\,. \tag{5.275}$$

This will give us some degree of freedom in selecting $h\,[n]$ and $g\,[n]$.

There are two methods to satisfy these conditions. One method leads to what is known as the *orthogonal wavelet* and the other to the bi-orthogonal *wavelet*. Before discussing those cases, based upon the above observation/discussions, one may implement an iterative method to generated (synthesize) the original signal using DWT scaling and wavelet coefficients. This method is depicted in Fig. 5.47.

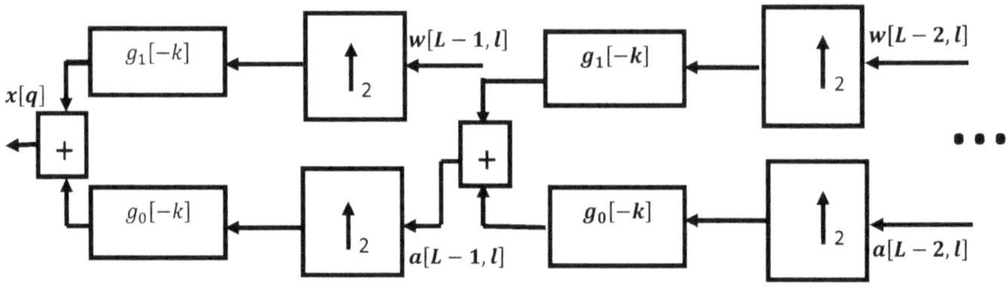

Fig. 5.47 DWT synthesis block

This completes the discussion on the efficient analysis and synthesis of signals using DWT. Next, we introduce the concept of orthogonal and bi-orthogonal DWT.

Orthogonal Case: In the orthogonal case, we assume that

$$G_0[\omega] = H_0[-\omega]\, e^{-j\omega N} = H_0^*[\omega]\, e^{-j\omega N} \tag{5.276}$$

and

$$G_1[\omega] = H_1[-\omega]\, e^{-j\omega N} = H_1^*[\omega]\, e^{-j\omega N} \tag{5.277}$$

where it is assumed that $p = N$. This leads to (see (5.272) and (5.277))

$$|G_0[\omega]|^2 + |G_1[\omega]|^2 = |H_0[\omega]|^2 + |H_1[\omega]|^2 = 2. \tag{5.278}$$

The above condition is known as the *Smith-Barnwell* property. One immediate observation that can be made is that since the sum of the two system functions (i.e., $H_0[\omega]$ and $H_1[\omega]$) over the entire band is constant, the two filters complement one another. That is, if $h_0[l]$ is an ideal low pass filter whose system function is limited to $[0, \pi/2]$, then $h_1[l]$ will be an ideal high pass filter with system function that is confined to $[\pi/2, \pi]$. The last condition also satisfies

$$\frac{1}{2\pi}\int_{-\pi}^{\pi}\left\{|H_0[\omega]|^2 + |H_1[\omega]|^2\right\} d\omega = 2$$

which was discussed earlier.

Let the length of the impulse response for $h_0[l]$ and $h_1[l]$ be L_h (even). Then, conditions (5.276) and (5.277) are satisfied when

$$\boxed{g_0[l] = h_0[L_h - l]} \tag{5.279}$$

and

$$\boxed{g_1[l] = h_1[L_h - l].} \tag{5.280}$$

This implies that the inverse operation involves filtering using filters with time-reversed impulse responses. Furthermore, the second condition for avoiding aliasing results in

$$\begin{aligned} &H_1[\omega]\,G_1[\omega+\pi]+H_0[\omega]\,G_0[\omega+\pi] \\ &= H_1[\omega]\,H_1^*[\omega+\pi]\,e^{-jN\omega}+H_0[\omega]\,H_0^*[\omega+\pi\,]\,e^{-jN\omega}=0 \\ &H_1[\omega]\,H_1^*[\omega+\pi]+H_0[\omega]\,H_0^*[\omega+\pi]=0. \end{aligned} \tag{5.281}$$

This condition is satisfied when (assuming real $h_0[l]$ and $h_1[l]$ and L_h is even)

$$H_1[\omega]=-e^{-j\omega(N-1)}H_0^*[\omega-\pi], \tag{5.282}$$

which results in

$$\begin{aligned} &H_1[\omega]\,H_1^*[\omega+\pi]+H_0[\omega]\,H_0^*[\omega+\pi] \\ &= e^{j\pi}H_0[\omega]\,H_0^*[\omega-\pi]+H_0[\omega]\,H_0^*[\omega+\pi] \\ &= -H_0[\omega]\,H_0^*[\omega+\pi]+H_0[\omega]\,H_0^*[\omega+\pi]=0. \end{aligned} \tag{5.283}$$

The last step is justified by the fact that $H_0^*[\omega-\pi] = H_0^*[\omega-\pi+2\pi] = H_0^*[\omega+\pi]$. Therefore, the impact of aliasing has been removed. Using the inverse FT,

$$\boxed{h_1[l]=(-1)^{l+1}\,h_0[L_h-1-l]\,;\, l\in[0,L_h-1]} \tag{5.284}$$

which insures that $\sum_l h_0[l]\,h_1[l]=0$ for an even L_h. It is assumed here that $h_0[l]$ is non-zero in the interval $[0, N-1]$, where L_h is determined by the shape and duration of the mother and father wavelets (see (5.234) and (5.235)). Furthermore, if $\sum_l h_0^2[l]=1$, then it is easy to see that $\sum_l h_1^2[l]=1$ as well. It is important to note that index l refers the sample in time of lN_s. Since $\sum_l h_0[l]\,h_1[l]=0$, this is referred to as the orthogonal DWT.

Finally, one can readily verify that

$$\boxed{g_1[l]=(-1)^l\,g_0[L_h-1-l]\,;\, l\in[0,L_h-1].} \tag{5.285}$$

This completes our reconstruction step. There is, however, one issue with this design; that is, the filter functions are all derived from a single filter impulse response $h_0[l]$, which results in the above solution for $h_1[l]$ with nonlinear phase response. This is an unattractive attribute of this approach. To overcome this shortcoming, one can resort to the idea of bi-orthogonal DWT, which will be discussed next.

Bi-orthogonal Case: For the bi-orthogonal wavelet the "perfect reconstruction" requirements must also be satisfied. The main difference is that the inverse DWT is reconstructed assuming a given $h_0[l]$ and $h_1[l]$. It is important to note that if the wavelet and its scaling functions are know a priori, then $h_0[l]$ and $h_1[l]$ will also be know ahead of time. Hence, the task of finding the reconstruction functions becomes one of finding $g_0[l]$ and $g_1[l]$ only. Now, let

$$\boxed{g_1[l] = (-1)^l h_0[l]} \tag{5.286}$$

and

$$\boxed{g_0[l] = (-1)^{l+1} h_1[l].} \tag{5.287}$$

Note that these selections result in

$$G_1[\omega] = H_0[\omega - \pi] \tag{5.288}$$

$$G_0[\omega] = -H_1[\omega - \pi]. \tag{5.289}$$

Now, let us look at

$$\begin{aligned} &H_1[\omega] G_1[\omega] + H_0[\omega] G_0[\omega] \\ &= H_1[\omega] H_0[\omega - \pi] - H_0[\omega] H_1[\omega - \pi]. \end{aligned} \tag{5.290}$$

Since $H_1[\omega]$ and $H_0[\omega]$ form a pair of system functions that complement each other in the frequency domain, the above expression is a constant with the exception of a phase shift, which is exactly what is needed here. Furthermore, as far as aliasing effect is concerned,

$$\begin{aligned} &H_1[\omega] G_1[\omega + \pi] + H_0[\omega] G_0[\omega + \pi] \\ &= H_1[\omega] H_0[\omega] - H_0[\omega] H_1[\omega] = 0. \end{aligned} \tag{5.291}$$

Therefore, the above selections of $g_0[l]$ and $g_1[l]$ meet the requirements for a perfect reconstruction. Also, since we have not imposed any relationship between $h_0[l]$ and $h_1[l]$, as is done in the case of orthogonal DWT, phase linearity can be maintained. In order to avoid aliasing due to down sampling by a factor of 2 (see Fig. 5.46), we need to have $H_1[\omega - \pi] H_1[\omega] = H_0[\omega] H_0[\omega - \pi] = 0$ (see Fig. 5.46). Hence,

$$\begin{aligned} \frac{1}{2\pi}\int G_0[\omega] H_1[\omega] d\omega &= -\frac{1}{2\pi}\int H_1[\omega - \pi] H_1[\omega] d\omega = 0 \\ \frac{1}{2\pi}\int G_1[\omega] H_0[\omega] d\omega &= \frac{1}{2\pi}\int H_0[\omega - \pi] H_0[\omega] d\omega = 0 \\ \Rightarrow \sum_{l=0}^{L_h-1} g_0[l] h_1[l] &= \sum_{l=0}^{L_h-1} g_1[l] h_0[l] = 0. \end{aligned} \tag{5.292}$$

This leads to the name *bi-orthogonal* as the sequence $g_0[l]$ is "orthogonal" to the sequence $h_1[l]$ and the sequence $g_1[l]$ is "orthogonal" to $h_0[l]$. Also, note that

$$G_0[0] = H_0[0] = \sum_{l=0}^{L_h-1} g_0[l] = \sum_{l=0}^{L_h-1} h_0[l] = \sqrt{2}. \tag{5.293}$$

As noted earlier, this approach allows one to work with a pair of pre-defined wavelet decomposition filter functions $h_1[l]$ and $h_0[l]$, which can be designed to provide phase linearity. For such scenarios, this method can be used to construct the reconstruction filters $g_0[l]$ and $g_1[l]$.

Next, we provide some well-known DWT and follow that up with a series of examples aimed at gaining a deeper insight into the potential of DWT in analyzing biomedical signals. Before we do so, let us summarize the properties of orthogonal and bi-orthogonal DWT we have discussed above.

Wavelet type	Orthogonal
(1)	L_h even
(2)	$g_0[l] = h_0[-l]$
(3)	$g_1[l] = h_1[-l]$
(4)	$h_1[l] = (-1)^{l+1} h_0[L_h - 1 - l]$
(5)	$g_1[l] = (-1)^l g_0[L_h - 1 - l]$
(6)	$\sum_l h_0[l] = \sum_l g_0[l] = \sqrt{2}; \sum_l h_1[l] = \sum_l g_1[l] = 0$
(7)	$\sum_l h_0^2[l] = \sum_l g_0^2[l] = \sum_l h_1^2[l] = \sum_l g_1^2[l] = 1; \sum_l h_0[l] h_1[l] = 0$

Wavelet type	Bi-orthogonal
(1)	L_h even
(2)	filters have linear phase responses
(3)	$g_0[l] = (-1)^{l+1} h_1[l]$
(4)	$g_1[l] = (-1)^l h_0[l]$
(5)	$\sum_l g_0[l] h_1[l] = \sum_l g_1[l] h_0[l] = 0$
(6)	$\sum_l h_0[l] = \sum_l g_0[l] = \sqrt{2}; \sum_l h_1[l] = \sum_l g_1[l] = 0$
(7)	$\sum_l h_0^2[l] = \sum_l g_0^2[l] = \sum_l h_1^2[l] = \sum_l g_1^2[l] = 1$

The above set of equations may be used to construct the filter functions (decomposition and reconstruction). We now introduce a number of useful DWT which are widely known.

5.3.8.1 Orthogonal Wavelets-Daubechies (Db1) or Haar DWT

The class of Daubechies (Db) wavelets, attributed to *Ingrid Daubechies*, are used frequently for image analysis and is one of the well-known DWT tools. Db wavelets fall into the category of *orthogonal* wavelets. Here, we present Db1–Db8, which are commonly used. At its fundamental level ($L_h = 2$), Db1 is identical to the *Haar* wavelet, discussed previously. For this wavelet, one has to only define $h_0[l]$ as all other filter functions can be defined in terms of $h_0[l]$ (see the above table for orthogonal wavelets). That is,

$$h_1[l] = (-1)^{l+1} h_0[L_h - 1 - l], \tag{5.294}$$

$$g_1[l] = h_1[-l], \tag{5.295}$$

and

$$g_0[l] = h_0[-l]. \tag{5.296}$$

For $L_h = 2$,

$$h_0[l] = \{0.7071, 0.7071\} \tag{5.297}$$

Note that Db1 (or *Haar*) wavelet is an orthogonal wavelet as $\sum h_0[l]\, h_1[l] = 0$. In the above, the first element corresponds to $l = 0$ and the last to $l = L_h - 1$.

Finally, it is important to note that the index l in defining the wavelet filters refers to lN_s samples in time and not to the actual sample in time. This fact must be taken into account when computing DWT of various sequences.

Db2 DWT For this wavelet ($L_h = 4$), which is also an orthogonal wavelet,

$$h_0[l] = \{-0.1294, 0.2241, 0.8365, 0.4830\}$$

Db3 DWT $L_h = 6$

$$h_0[l] = \{0.0352, -0.0854, -0.1350, 0.4599, 0.8069, 0.3327\} \tag{5.298}$$

Db4 DWT $L_h = 8$

$$h_0[l] = \{-0.0106, 0.0329, 0.0308, -0.1870, -0.0280, 0.6309, 0.7148, 0.2304\}$$

Db5 DWT $L_h = 0$

$$h_0[l] = \{0.0033, -0.0126, -0.0062, 0.0776, -0.0322, -0.2423, 0.1384, 0.7243, 0.6038, 0.1601\} \tag{5.299}$$

Db6 DWT $L_h = 12$

$$h_0[l] = \{-0.0011, 0.0048, 0.0006, -0.0316, 0.0275, 0.0975, -0.1298, -0.2263, 0.3153, 0.7511, 0.4946, 0.1115\} \tag{5.300}$$

Db7 DWT $L_h = 14$

$$h_0[l] = \{0.0004, -0.0018, 0.0004, 0.0126, -0.0166, -0.0380, 0.0806, 0.0713, -0.2240, -0.1439, 0.4698, 0.7291, 0.3965, 0.0779\} \tag{5.301}$$

Finally,

Db8 DWT $L_h = 16$

$$h_0[l] = \{-0.0001, 0.0007, -0.0004, -0.0049, 0.0087, 0.0140, -0.0441, -0.0174, 0.1287, 0.0005, -0.2840, -0.0158, 0.5854, 0.6756, 0.3129, 0.0544\} \tag{5.302}$$

Symlet DWT

Symlet is yet another DWT that falls into the category of orthogonal wavelets. Similarly, we present Symlet2-8 (corresponding to $L_h = 4$ to 16).

Symlet2 DWT For this wavelet when $L_h = 4$,

$$h_0[l] = \{-0.1294, 0.2241, 0.8365, 0.4830\} \tag{5.303}$$

Symlet3 DWT $L_h = 6$

$$h_0[l] = \{0.0352, -0.0854, -0.1350, 0.4599, 0.8069, 0.3327\} \tag{5.304}$$

Symlet4 DWT $L_h = 8$

$$h_0[l] = \{-0.0758, -0.0296, 0.4976, 0.8037, 0.2979, -0.0992, -0.0126, 0.0322\} \tag{5.305}$$

Symlet5 DWT $L_h = 10$

$$\begin{aligned} h_0[l] = \{&0.0273, 0.0295, -0.0391, 0.1994, 0.7234, 0.6340, 0.0166, -0.1753, \\ &-0.0211, 0.0195\} \end{aligned} \tag{5.306}$$

Symlet6 DWT $L_h = 12$

$$\begin{aligned} h_0[l] = \{&0.0154, 0.0035, -0.1180, -0.0483, 0.4911, 0.7876, 0.3379, \\ &-0.0726, -0.0211, 0.0447, 0.0018, -0.0078\} \end{aligned} \tag{5.307}$$

Symlet7 DWT $L_h = 14$

$$\begin{aligned} h_0[l] = \{&0.0027, -0.0010, -0.0126, 0.0305, 0.0679, -0.0496, 0.0174, 0.5361, \\ &0.7678, 0.2886, -0.1400, -0.1078, 0.0040, 0.0103\} \end{aligned} \tag{5.308}$$

Symlet8 DWT $L_h = 16$

$$\begin{aligned} h_0[l] = \{&-0.0034, -0.0005, 0.0317, 0.0076, -0.1433, -0.0613, 0.4814, 0.7772, \\ &0.3644, -0.0519, -0.0272, 0.0491, 0.0038, -0.0150, -0.0003, 0.0019\}. \end{aligned} \tag{5.309}$$

The next class of DWT that we will consider is the Coiflets DWT, which is also designed by *Ingrid Daubechies*. This wavelet falls under the general category of orthogonal wavelet. We will consider Coiflets1-4 as $N = 6n$ where n denotes the wavelet number.

Coiflets1 DWT For this wavelet ($L_h = 6$),

$$h_0[l] = \{-0.0157, -0.0727, 0.3849, 0.8526, 0.3379, -0.0727\} \tag{5.310}$$

Coiflets2 DWT $L_h = 12$

$$h_0[l] = \{-0.0007, -0.0018, 0.0056, 0.0237, -0.0594, -0.0765, 0.4170, 0.8127, 0.3861, -0.0674, -0.0415, 0.0164\} \tag{5.311}$$

Coiflets3 DWT $L_h = 18$

$$h_0[l] = \{-0.0000, -0.0001, 0.0005, 0.0011, -0.0026, -0.0090, 0.0159, 0.0346, -0.0823, -0.0718, 0.4285, 0.7938, 0.4052, -0.0611, -0.0658, 0.0235, 0.0078, -0.0038\} \tag{5.312}$$

Coiflets4 DWT $L_h = 24$

$$h_0[l] = \{-0.0000, -0.0000, 0.0000, 0.0001, -0.0003, -0.0006, 0.0013, 0.0038, -0.0057, -0.0152, 0.0251, 0.0393, -0.0962, -0.0666, 0.4344, 0.7822, 0.4153, -0.0561, -0.0813, 0.0267, 0.0161, -0.0073, -0.0016, 0.0009\} \tag{5.313}$$

5.3.8.2 Bi-orthogonal Wavelets

As noted above, for the bi-orthogonal wavelet, we need to define $h_0[l]$ as well as $h_1[l]$. Once again, we focus on some examples of well-known DWT and make no attempt to be exhaustive. *Haar* wavelet is known as the Bi-orthogonal 1.1 wavelet. Hence, *Haar* is not only an orthogonal wavelet, but also satisfies the bi-orthogonal condition. In general, for Bi-orthogonal $m.n$ wavelet,

$$L_h = \begin{cases} 2n + m - 1 & m \text{ odd} \\ 2n + m & m \text{ even} \end{cases}. \tag{5.314}$$

Below are some examples of Bi-orthogonal $m.n$ wavelets.

Bi-orthogonal 1.3 For this wavelet, $L_h = 6$.

$$h_0[l] = \{-0.0884, 0.0884, 0.7071, 0.7071, 0.0884, -0.0884\} \tag{5.315}$$

$$h_1[l] = \{0, 0, -0.7071, 0.7071, 0, 0\} \tag{5.316}$$

Bi-orthogonal 1.5 For this wavelet, $L_h = 10$.

$$h_0[l] = \{0.0166, -0.0166, -0.1215, 0.1215, 0.7071, 0.7071, 0.1215, -0.1215, -0.0166, 0.0166\} \quad (5.317)$$

$$h_1[l] = \{0, 0, 0, 0, -0.7071, 0.7071, 0, 0, 0, 0\} \quad (5.318)$$

Bi-orthogonal 2.2 For this wavelet, $L_h = 6$.

$$h_0[l] = \{0, -0.1768, 0.3536, 1.0607, 0.3536, -0.1768\} \quad (5.319)$$

$$h_1[l] = \{0, 0.3536, -0.7071, 0.3536, 0, 0\} \quad (5.320)$$

Bi-orthogonal 2.4 For this wavelet, $L_h = 10$.

$$h_0[l] = \{0, 0.0331, -0.0663, -0.1768, 0.4198, 0.9944, 0.4198, -0.1768, -0.0663, 0.0331\} \quad (5.321)$$

$$h_1[l] = \{0, 0, 0, 0.3536, -0.7071, 0.3536, 0, 0, 0, 0\} \quad (5.322)$$

Bi-orthogonal 2.6 For this wavelet, $L_h = 14$.

$$h_0[l] = \{0, -0.0069, 0.0138, 0.0470, -0.1077, -0.1699, 0.4475, 0.9667, 0.4475, -0.1699, -0.1077, 0.0470, 0.0138, -0.0069\} \quad (5.323)$$

$$h_1[l] = \{0, 0, 0, 0, 0, 0.3536, -0.7071, 0.3536, 0, 0, 0, 0, 0, 0\} \quad (5.324)$$

Bi-orthogonal 2.8 For this wavelet, $L_h = 18$.

$$h_0[l] = \{0, 0.0015, -0.0030, -0.0129, 0.0289, 0.0530, -0.1349, -0.1638, 0.4626, 0.9516, 0.4626, -0.1638, -0.1349, 0.0530, 0.0289, -0.0129, -0.0030, 0.0015\} \quad (5.325)$$

$$h_1[l] = \{0, 0, 0, 0, 0, 0, 0, 0.3536, -0.7071, 0.3536, 0, 0, 0, 0, 0, 0, 0, 0\} \quad (5.326)$$

Bi-orthogonal 3.1 For this wavelet, $L_h = 4$.

$$h_0[l] = \{-0.3536, 1.0607, 1.0607, -0.3536 \quad (5.327)$$

$$h_1[l] = \{-0.1768, 0.5303, -0.5303, 0.1768\} \quad (5.328)$$

Bi-orthogonal 3.3 For this wavelet, $L_h = 8$.

$$h_0[l] = \{0.0663, -0.1989, -0.1547, 0.9944, 0.9944, -0.1547, -0.1989, 0.0663\} \quad (5.329)$$

$$h_1[l] = \{0, 0, -0.1768, 0.5303, -0.5303, 0.1768, 0, 0\} \quad (5.330)$$

Bi-orthogonal 3.5 For this wavelet, $L_h = 12$.

$$h_0[l] = \begin{matrix}\{-0.0138, 0.0414, 0.0525, -0.2679, -0.0718, 0.9667, 0.9667, \\ -0.0718, -0.2679, 0.0525, 0.0414, -0.0138\}\end{matrix} \quad (5.331)$$

$$h_1[l] = \{0, 0, 0, 0, -0.1768, 0.5303, -0.5303, 0.1768, 0, 0, 0, 0\} \quad (5.332)$$

Bi-orthogonal 3.7 For this wavelet, $L_h = 16$.

$$h_0[l] = \begin{matrix}\{0.0030, -0.0091, -0.0168, 0.0747, 0.0313, -0.3012, -0.0265, 0.9516, \\ 0.9516, -0.0265, -0.3012, 0.0313, 0.0747, -0.0168, -0.0091, 0.0030\}\end{matrix} \quad (5.333)$$

$$h_1[l] = \{0, 0, 0, 0, 0, 0, -0.1768, 0.5303, -0.5303, 0.1768, 0, 0, 0, 0, 0, 0\} \quad (5.334)$$

Bi-orthogonal 3.9 For this wavelet, $L_h = 20$.

$$\begin{aligned} h_0[l] = \{&-0.0007, 0.0020, 0.0051, -0.0206, -0.0141, 0.0991, 0.0123, -0.3202, \\ &0.0021, 0.9421, 0.9421, 0.0021, -0.3202, 0.0123, 0.0991, -0.0141 \\ &-0.0206, 0.0051, 0.0020 - 0.0007\} \end{aligned} \quad (5.335)$$

$$h_1[l] = \{0, 0, 0, 0, 0, 0, 0, 0, -0.1768, 0.5303, -0.5303, 0.1768, 0, 0, 0, 0, 0, 0, 0, 0\} \quad (5.336)$$

Example 9 In this example, we examine the property of DWT of known signals. To that end, let us consider sinusoids with frequencies of 5 and 300 Hz, when observed over a period of 1 s. It is assumed that the sampling rate is 1024, which for 1 s observation yields 1024 samples ($L = \log_2 1024 = 10$). Compute the DWT and scaling coefficients for the first 4 highest levels (i.e., $n = 9, 8, 7, 6$) when Db1 (Haar) wavelet is used.

Solution: *For a 5 Hz sinusoid, we observe Figs.* 5.48, 5.49, 5.50 and 5.51. *Each figure consists blue (scaling coefficients) and red (DWT coefficients) plots. Given the low-pass nature of* $h_0[l]$, *we observe that the scaling coefficients are dominant, somewhat resembling the* 5 *Hz sinusoid. This is expected from the scaling function for the Haar wavelet. The high-pass filtering attribute of* $h_1[l]$ *is also apparent from these figures, as the 5 Hz signal is almost entirely suppressed by the filter, underscoring the small contribution of the wavelet coefficients in the reconstruction of the low frequency sinusoid. Also, it is noteworthy that the scaling and wavelet coefficients form a pair*

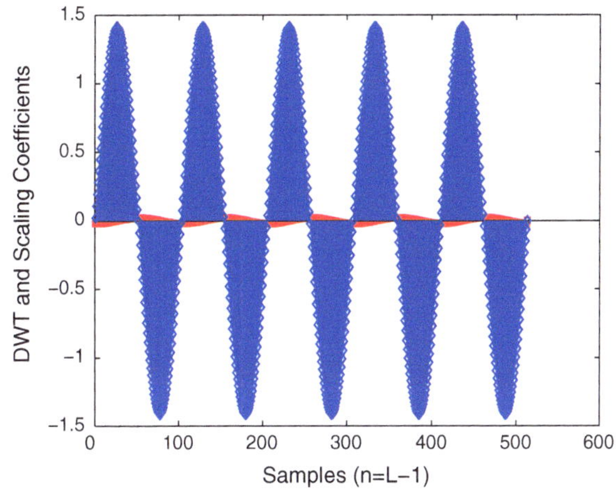

Fig. 5.48 DWT and the scaling coefficients of a sinusoidal signal

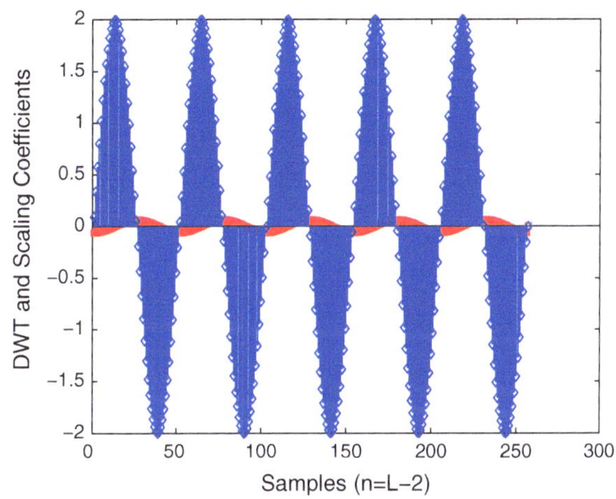

Fig. 5.49 DWT and the scaling coefficients of a 5 Hz sinusoidal signal

of orthogonal signals. That is, the wavelet coefficients in time for a given level of dilation for a sinusoid appears to be a cosine signal. This is not surprising as the Haar DWT belongs to the family of orthogonal wavelets.

For the case of the 300 *Hz sinusoid, we observe Figs.* 5.52, 5.53, 5.54 and 5.55. *Contrary to the previous case, the wavelet coefficients become dominant as the frequency of the sinusoid qualifies the signal as a "high" frequency signal. Note that the frequency of the sinusoid is well below the Nyquist sampling rate, and yet an "aliasing-like" effect emerges in these plots. This underscores the fundamental difference between FT and WT. Furthermore, a surprising result presents itself; that is, not only the scaling and wavelet coefficients become dependent on n, but also the dominance of scaling coefficient over wavelet coefficients becomes inconsistent. This may be attributed to the fact that signal presents a high frequency characteristics, which yields widely different set of responses for different levels of decimation (n).*

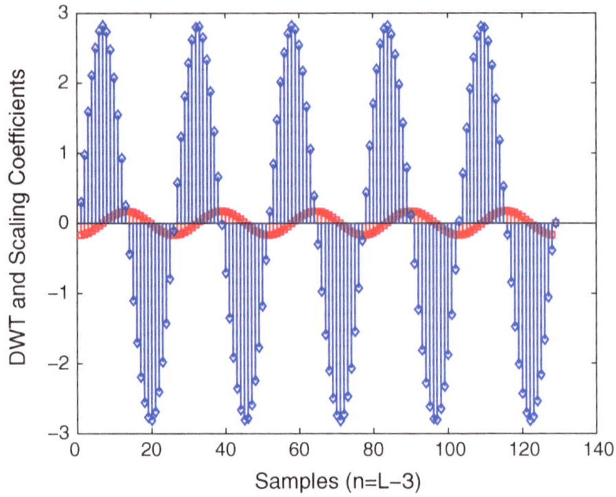

Fig. 5.50 DWT and the scaling coefficients of a 5 Hz sinusoidal signal

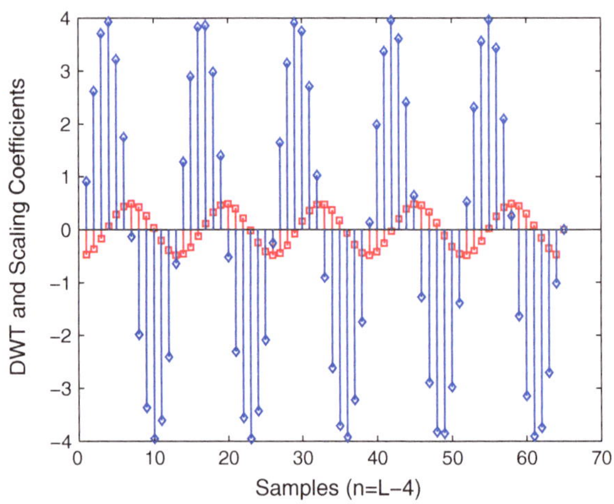

Fig. 5.51 DWT and the scaling coefficients of a 5 Hz sinusoidal signal

However, the orthogonality property of the scaling and wavelet coefficients holds. Since wavelets and scaling coefficients are computed using integrations over an interval of time which may contain several periods, as well as partial periods, of a high-frequency sinusoid, these coefficient may be impacted by the choice of n in a drastic manner. This result is somewhat disappointing as it does not allow the reader to gain much insight into the fundamental properties of this signal. This is expected as WT is suitable in analyzing signals with temporal selectivity. Signals such as a sinusoid are void of such selectivity. Now, let us divert our attention to signals that exhibit temporal selectivity. To that end, let us consider the following example.

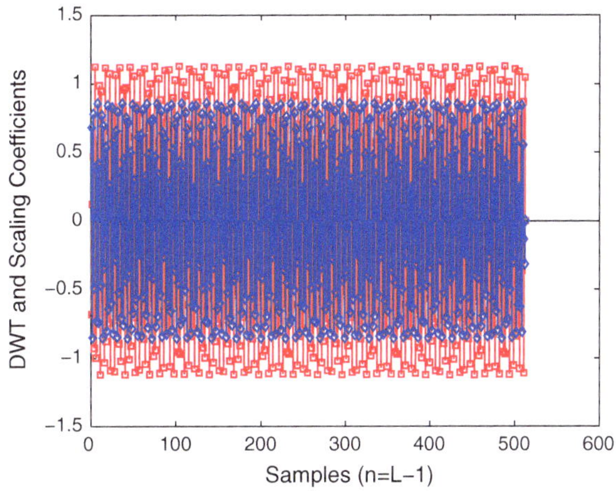

Fig. 5.52 DWT and the scaling coefficients of a 300 Hz sinusoidal signal

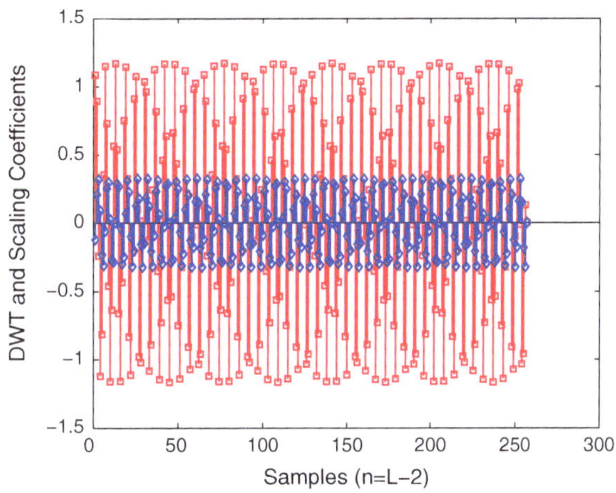

Fig. 5.53 DWT and the scaling coefficients of a 300 Hz sinusoidal signal

Example 10 In this example, we examine a pulse-shape signal that is given below:

$$x(t) = \frac{1}{\sqrt{2\pi 10^{-3}}} \times \left\{ \exp\left[-\frac{\left(t - 0.5 + 10^{-5}\right)}{2 \times 10^{-3}} \right] - \exp\left[-\frac{\left(t - 0.55 - 10^{-5}\right)}{2 \times 10^{-3}} \right] \right\} \quad (5.337)$$

The signal is defined over $t \in [0, 1]$. It is assumed that 1024 samples are taken. Compute the DWT and scaling coefficients for the first 4 highest levels (i.e., $n = 9, 7, 4, 0$) when dB1 (Haar) wavelet is used.

Solution: *The signal is depicted in Fig. 5.56. This signal, unlike a sinusoid, exhibits a significant level of temporal selectivity. In Figs. 5.57, 5.58, 5.59 and 5.60, the scaling and wavelet coefficients are shown for* $n = 9, 7, 4,$ *and* 0, *respectively.*

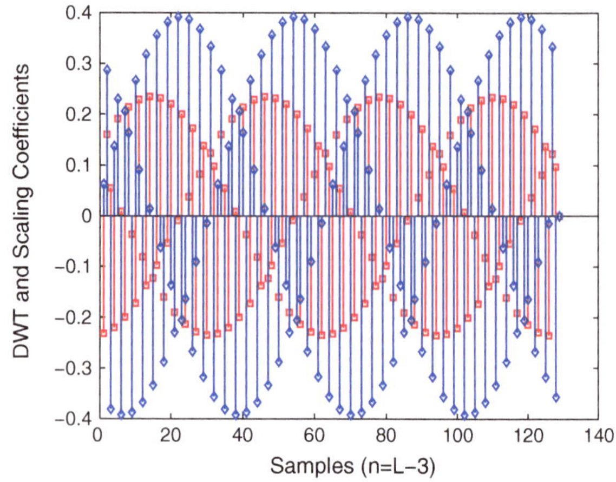

Fig. 5.54 DWT and the scaling coefficients of a 300 Hz sinusoidal signal

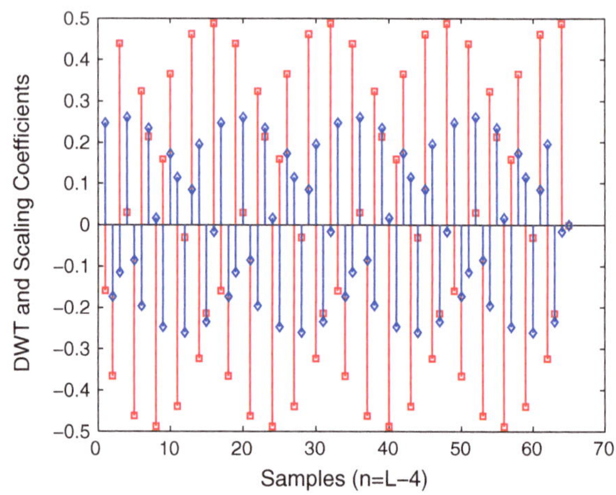

Fig. 5.55 DWT and the scaling coefficients of a 300 Hz sinusoidal signal

Quite clearly, these coefficients underscore the presence of a major "transition" in the signal behavior about the center of the pulse. For $n = 4$*, we see the presence of a dominant coefficient at the center sample. Hence, these coefficients may be used to render a decision regarding the presence of this pulse in time. In fact, if the objective is to detect "when" this pulse will occur in time, wavelet and scaling coefficients at all levels collectively offer a means of achieving that goal. Furthermore, the wavelet and scaling coefficients unravel the significant duration of the pulse, which will be important in biomedical signal processing as the duration of an "event" is often of significant importance. In all figures, we can see that the pulse duration is about* 320 *samples (in real time). This is apparent form Fig.* 5.57*. However, other figures suggest the same conclusion. For instance, from Fig.* 5.59*, we see that the wavelet and scaling coefficients for the pulse at level* $n = 4$ *are approximately* 5 *samples long,*

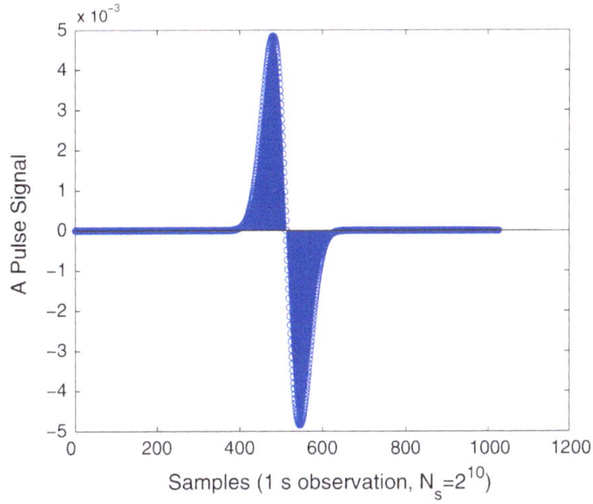

Fig. 5.56 A pulse signal

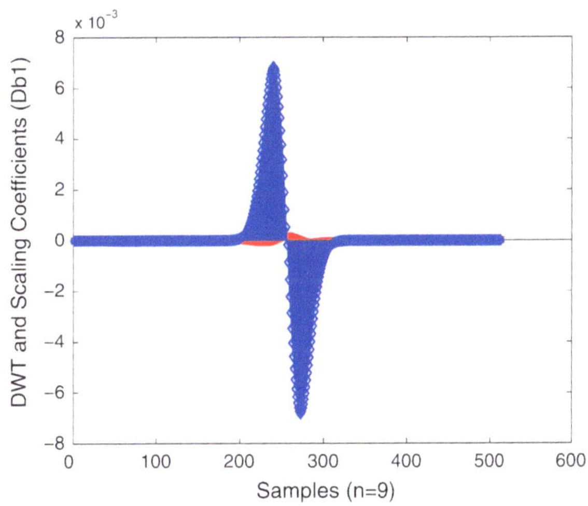

Fig. 5.57 DWT and the scaling coefficients of a pulse signal

which, when upsampled to the proper level, results in $5 \times \frac{1024}{2^n} = 5 \times \frac{1024}{16} = 320$ *samples as the duration of the original pulse.*

The above example underscored the utility of DWT in identifying the presence of signals and their respective temporal selectivities. It is important to note that the shape of the signal played in our favor as the *Haar* wavelet offers temporal characteristics that match those of the signal under investigation. In some instances, this approach may lead to a less obvious result. That is, let us consider the above example when Db8 (instead of Db1) wavelet is used. In that case, we arrive at Figs. 5.61, 5.62, 5.63 and 5.64, which although identify the onset of the pulse in time, they do not offer an accurate measure of the pulse duration (compare Figs. 5.59, 5.60, 5.61, 5.62 and 5.63 as well as Figs. 5.60, 5.61, 5.62, 5.63 and 5.64). This is due to the fact that the impulse

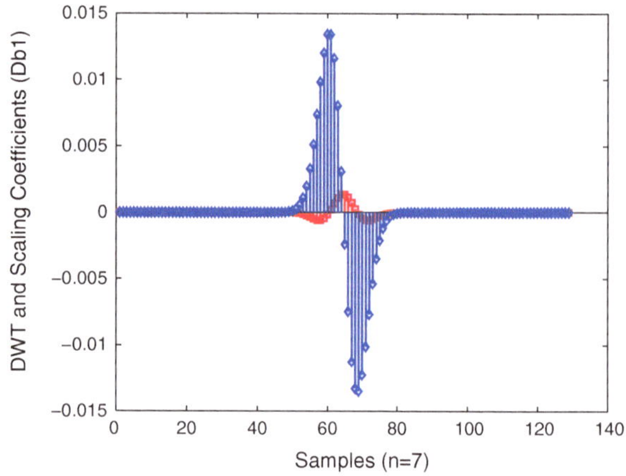

Fig. 5.58 DWT and the scaling coefficients of a pulse signal

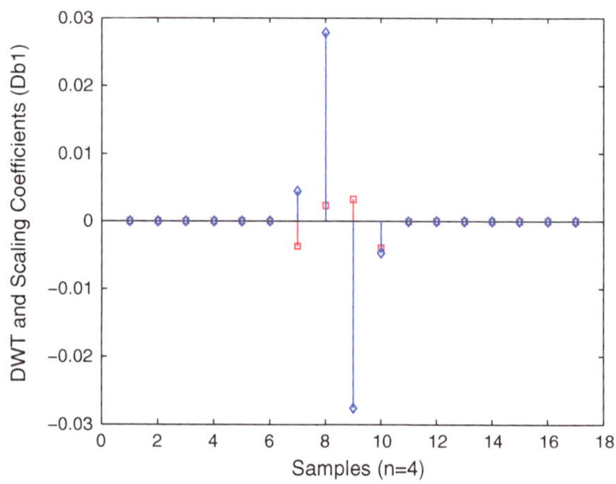

Fig. 5.59 DWT and the scaling coefficients of a pulse signal

response of the decomposition filters for Db8 is longer than that of Db1, leading to unintended ripples in the response, which, in turn, hinder the determination of the duration of the pulse in question. Next, we investigate the performance of DWT in identifying signal characteristics in the presence of noise. To that end, let us consider the following examples.

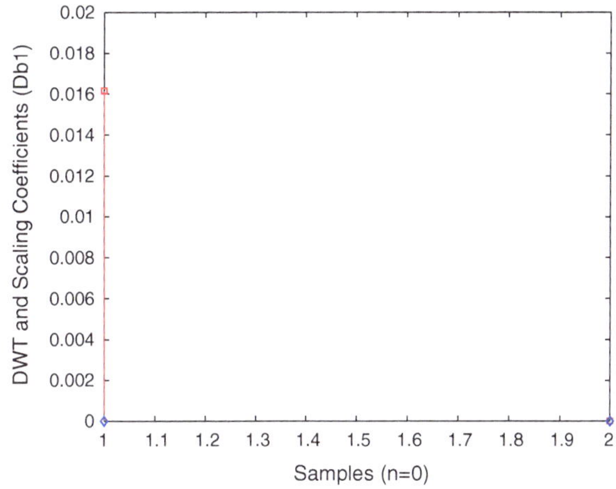

Fig. 5.60 DWT and the scaling coefficients of a pulse signal

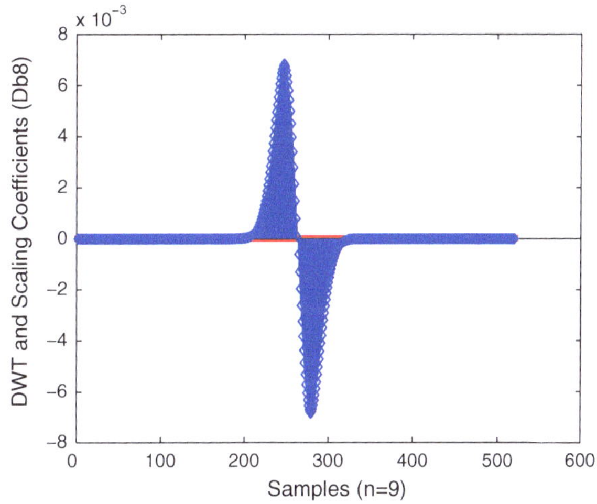

Fig. 5.61 DWT and the scaling coefficients of a pulse signal

Example 11 In the previous example, use Db1 and Db8 to study the same signal when it is corrupted by AWGN. The noise adds an i.i.d Gaussian noise with a standard deviation of 10^{-3} to the samples of the signal.

Solution: *The signal along with the Db1 wavelet and scaling coefficients are depicted in Figs.* 5.65, 5.66, 5.67, 5.68 and 5.69. *Interestingly enough, for lower values of* n, *the presence of pulse is correctly identified. In particular, for* $n = 4$ *(Fig.* 5.68*), the presence and the location of the two pulses with opposite polarities are identified. This is due to the fact that lower values of* n *result in wavelet and scaling coefficient computations that allow integrations over large number of samples, and provided*

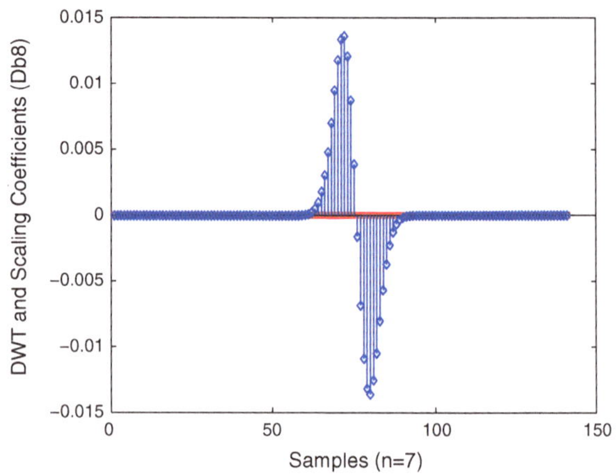

Fig. 5.62 DWT and the scaling coefficients of a pulse signal

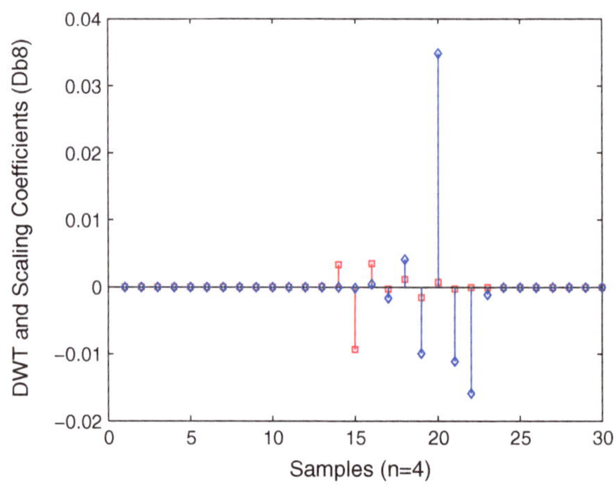

Fig. 5.63 DWT and the scaling coefficients of a pulse signal

that AWGN has a 0 mean, such integration removes the impact of AWGN. When Db8 is used (see Figs. 5.70, 5.71, 5.72 and 5.73*), as expected, similar results emerge, except that the long impulse response of Db8 filter results in distortion in time that hinders the estimation of the original pulse duration at low values of n. Nonetheless, the presence and its the location in time in AWGN was verified using these methods.*

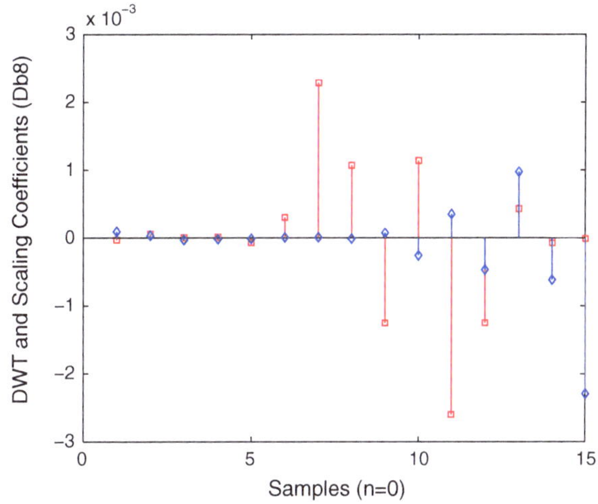

Fig. 5.64 DWT and the scaling coefficients of a pulse signal

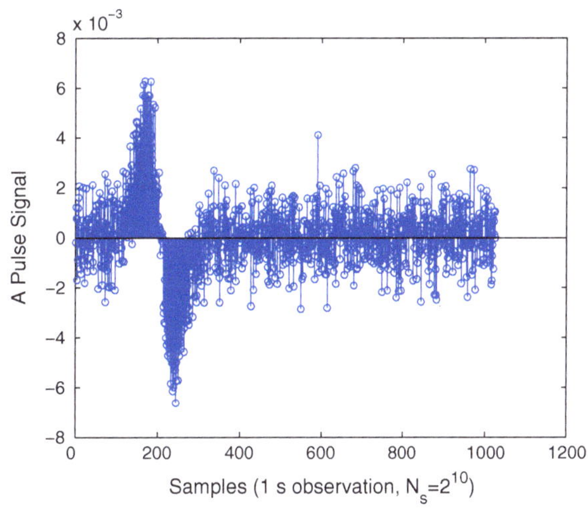

Fig. 5.65 A noisy pulse signal

Example 12 So far, we have used deterministic as well as AWGN signals to study DWT and its potential as a signal processing tool. Now, let us consider DWT as a tool for studying biomedical signals, such as EEG. To that end, examine a 4-s observation of an EEG signal from MIT-BIH using Db1.

Solution: *From MIT-BIH, we choose chb01_01_edfm record (approximately first hour of child 1 EEG). The sampling rate is* 256 *Hz and there are* 23 *rows in the record, corresponding to* 23 *probes. We use the first probe and consider the record for the period of* $t \in [4, 8]$ *s. It is assumed that the length of observation remains*

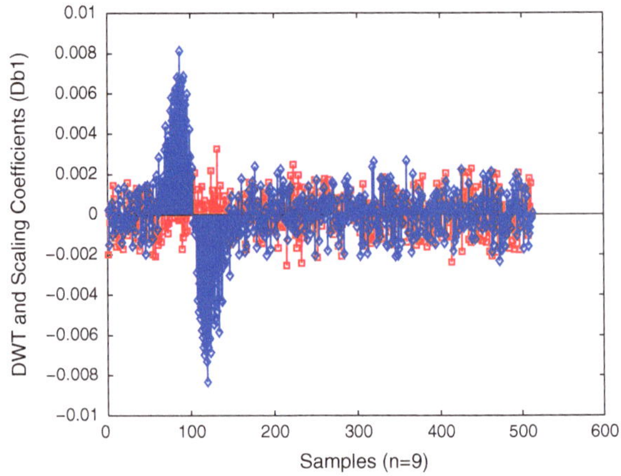

Fig. 5.66 DWT and the scaling coefficients of a noisy pulse signal

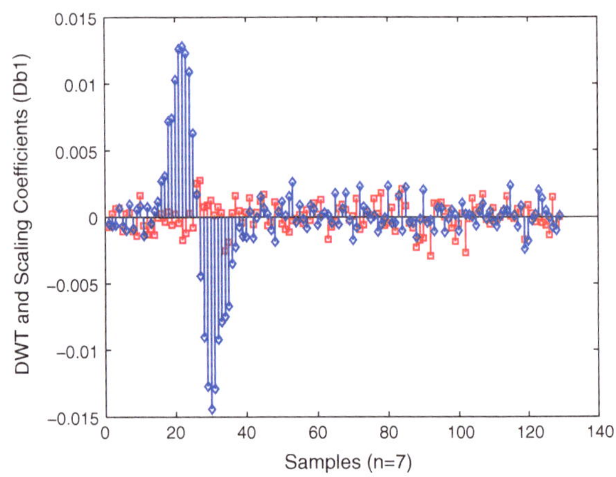

Fig. 5.67 DWT and the scaling coefficients of a noisy pulse signal

as 1024 *samples* (4×256). *Therefore, the wavelet coefficients are computed for* $n = 0, 1, \ldots, 9$ ($N = 1024$, $n_{\max} = \log_2(N) - 1$). *The resulting Db1 wavelet and scaling coefficients are depicted in Figs.* 5.74, 5.75, 5.76, 5.77 and 5.78. *Note that the signal exhibits two short term peaks at approximately 180 and 700 samples. Although such high peaks may be of little clinical significance, they may indicate the onset of brain's anomalous function in the ensuing minutes or hours. For Db1, it is clear that these peaks pose a significant contribution to scaling and wavelet coefficients at all levels of* n. *In particular, for* $n = 4$, *we clearly identify the location*

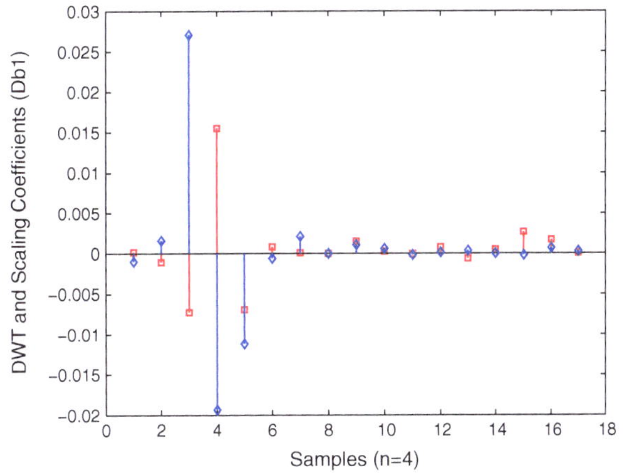

Fig. 5.68 DWT and the scaling coefficients of a noisy pulse signal

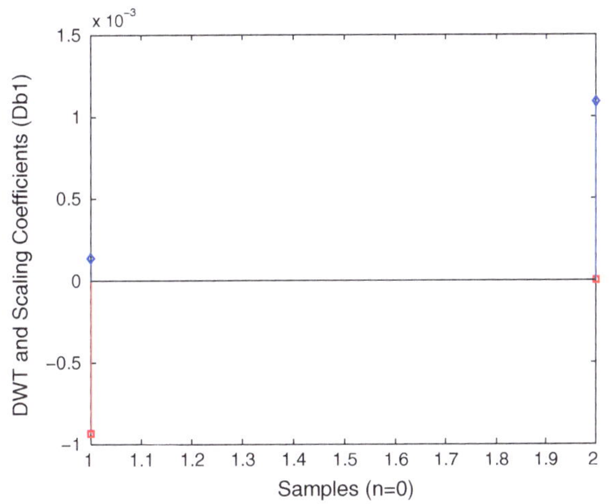

Fig. 5.69 DWT and the scaling coefficients of a noisy pulse signal

of the two peaks ($3\frac{1024}{2} = 192$ *and* $11\frac{1024}{2^7} = 704$, *which agree with the original signal's attributes). Also, note that the scaling coefficients remain dominant. The reason for this observation is that the signal is relatively slow varying for a* 256 *samples/s scenario. That is, in general, for EEG signals with most significant signal components limited to well below* 100 *Hz, the wavelet coefficients are less dominant than the scaling coefficients (which are the result of filtering with the low-pass* $H_0[\omega]$ *filter). In fact, the wavelet coefficients remain negligible for all levels of decimation except for* $n = 4$. *Note that at* $192th$ *sample (3rd sample for* $n = 4$) *and* $704th$ *sample* ($11th$ *sample of* $n = 4$), *we have rather fast variations with large magnitudes in*

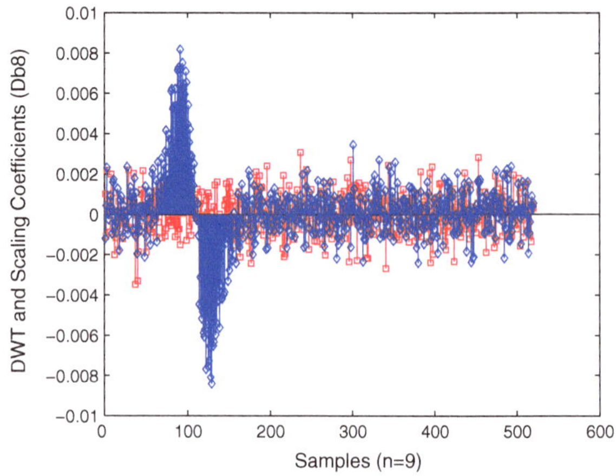

Fig. 5.70 DWT and the scaling coefficients of a noisy pulse signal

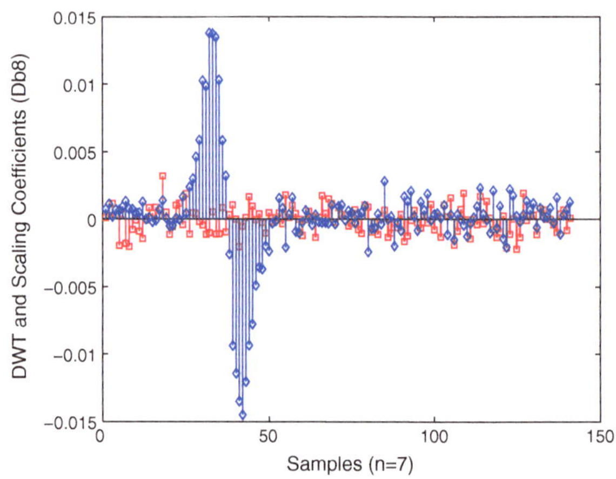

Fig. 5.71 DWT and the scaling coefficients of a noisy pulse signal

the signal. For this reason, we have large values for both the wavelet and scaling coefficients at those points, identifying the onset of a transition. Per Haar (Db1) wavelet structure, the scaling coefficient is proportional to the signal amplitude while wavelet coefficients underscore the presence of a fast varying signal. Hence, the presence of strong scaling and wavelet coefficients at a point in time indicate the onset of a fast varying signal with large magnitude. As noted earlier, this observation in and of itself may not be of clinical value, but it allows for diagnosis of changes in the behavior of EEG. In general, clinicians are interested in sustained transitions in the behavior of EEG. To note that, let us consider the same EEG signal during a period where the signal experiences a sustained change in its behavior.

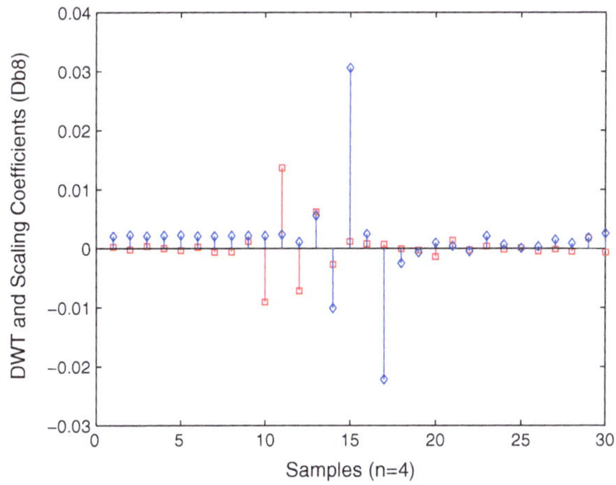

Fig. 5.72 DWT and the scaling coefficients of a noisy pulse signal

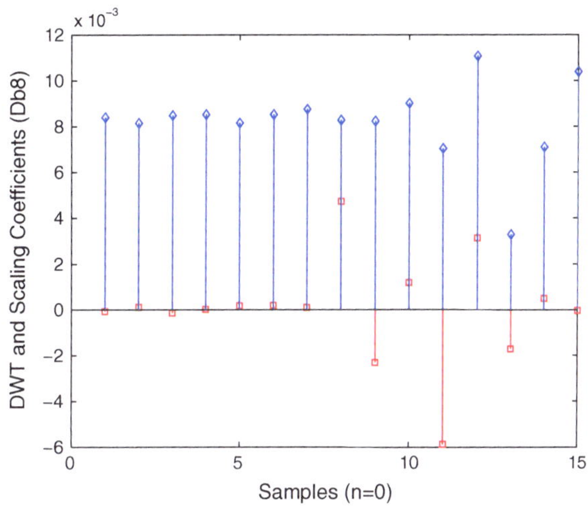

Fig. 5.73 DWT and the scaling coefficients of a noisy pulse signal

Example 13 In the previous example, we observed that DWT may be used to identify rapid, large variations in the signal and their corresponding periods. In this example, we look at EEG signal for the same patient, but consider an interval identified by samples ranging from 14112 to 15136. This corresponds to the time interval $\left[\frac{14112}{256}, \frac{15136}{256}\right] = [55.12s, 59.12s]$. Use DWT to examine the signal and identify the transition point(s) in the signal behavior. Use Db1 to perform the analysis.

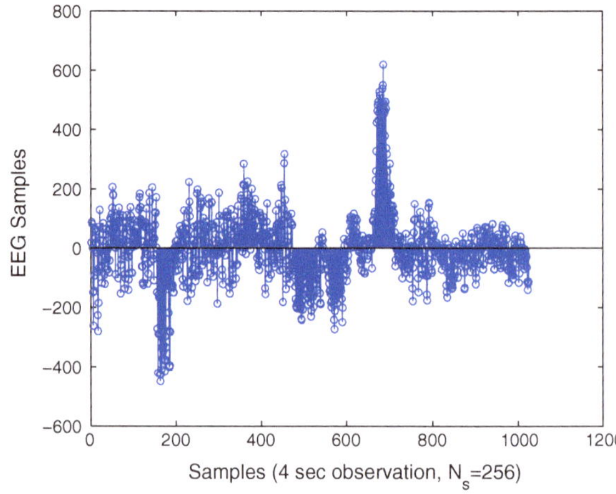

Fig. 5.74 An EEG signal

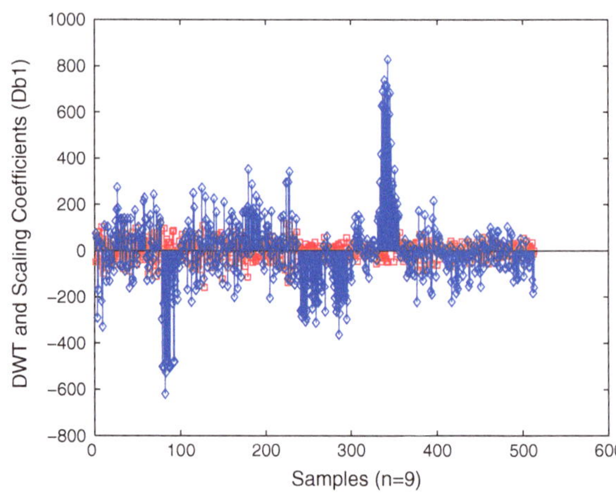

Fig. 5.75 DWT and the scaling coefficients of an EEG signal

Solution: *As we have learned, DWT may be used to identify the onset of large amplitude variations (scaling coefficient) as well as fast variations (wavelet coefficients) in the signal. The signal is shown in Fig.* 5.79. *Note that the signal around sample number* 200 *experiences a rapid, large amplitude change followed by a sustained, large-amplitude fast variations. To see whether DWT can identify these changes, in Figs.* 5.80, 5.81, 5.82 and 5.83 *we are depicting the scaling and wavelet coefficients*

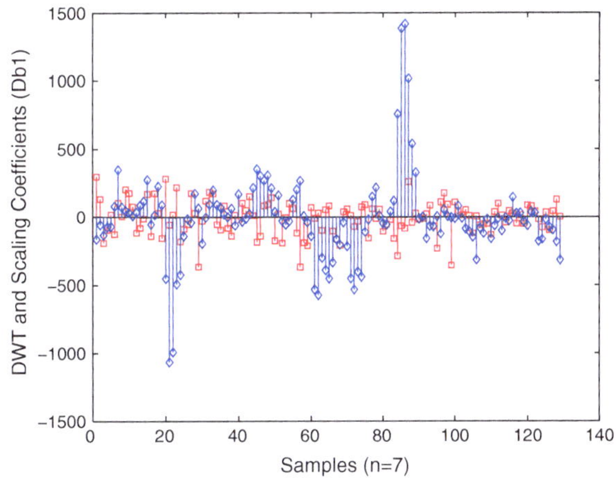

Fig. 5.76 DWT and the scaling coefficients of an EEG signal

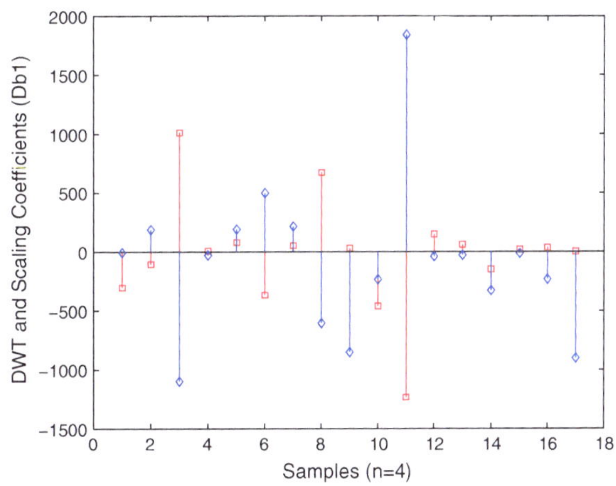

Fig. 5.77 DWT and the scaling coefficients of an EEG signal

for $n_9 - n_0$. For $n = 4$, at sample 4 (corresponding to the sample $4\frac{1024}{16} = 256$) we observe a transition in amplitude (scaling coefficient) and frequency (wavelet coefficients). This transition is followed by a sequence of high amplitude (large scaling coefficient) and high speed changes. Although we do not attribute this to a physiologically anomaly, with the aid of DWT, a physician can identify the onset of a period of time for which signal experiences large amplitudes and frequency changes.

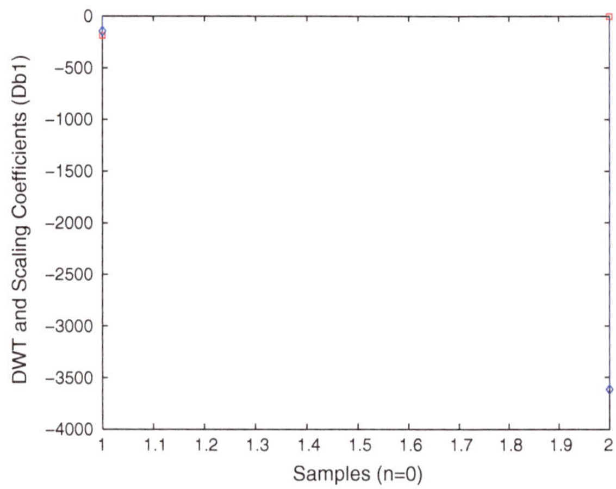

Fig. 5.78 DWT and the scaling coefficients of an EEG signal

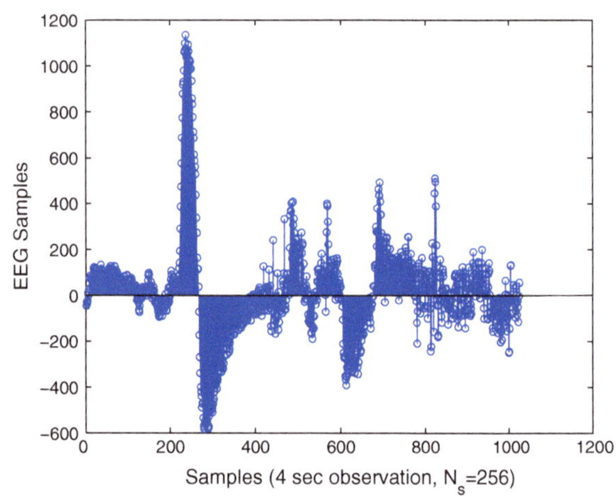

Fig. 5.79 An EEG signal

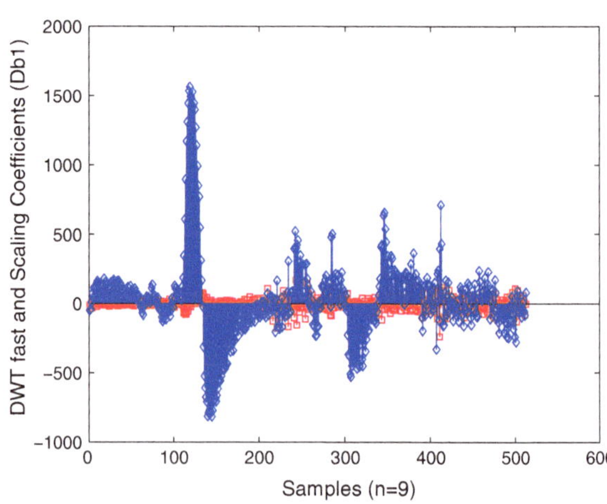

Fig. 5.80 DWT and the scaling coefficients of an EEG signal

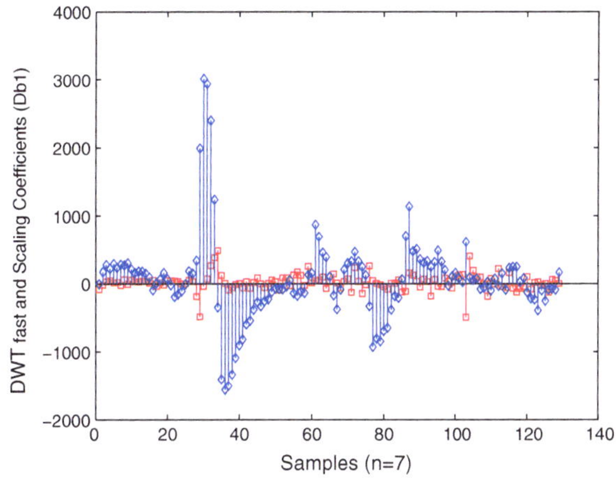

Fig. 5.81 DWT and the scaling coefficients of an EEG signal

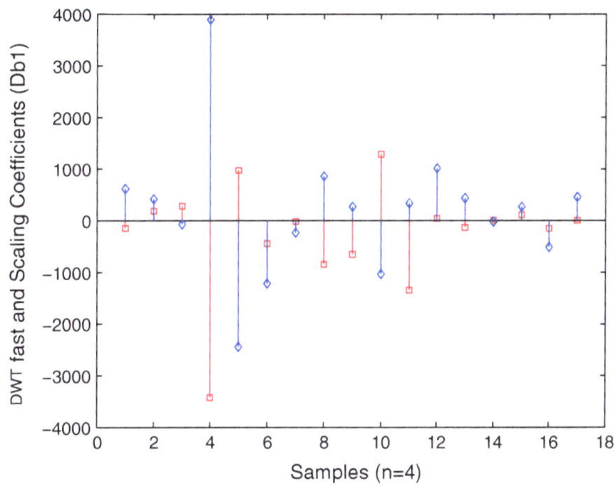

Fig. 5.82 DWT and the scaling coefficients of an EEG signal

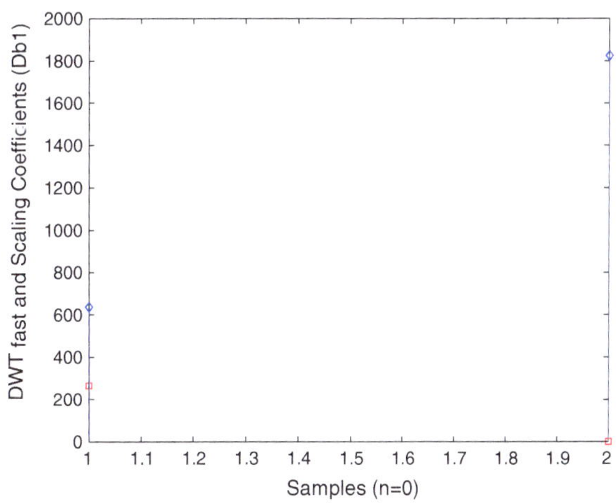

Fig. 5.83 DWT and the scaling coefficients of an EEG signal

In conclusion, we observe that the discrete wavelet transform has some unique features that make it suitable for biomedical signal processing. As we observed at the beginning of this chapter, standard signal processing tools, such as the Fourier transform, do not adequately identify the characteristics of non-stationary signals, such as the onset of transitions in the state of the signal and the duration of such transitions. Biomedical signals often present time-varying characteristics, which can only be identified by signal processing tools that allow for temporal as well as spectral selectivity. DWT, as was seen above, enables one to identify periods of time where signal experiences changes not only in its the magnitude but also in its spectral content. This feature of DWT makes it suitable for studying signals such as EEG, which are non-stationary and are prone to sudden changes in their amplitudes and frequencies. Although we have not been exhaustive in our treatment of EEG signals, the discussion on DWT and its property which are provided in this chapter have been exhaustive. The reader should be adequately equipped at this point to bring to bear signals processing tools, such as Matlab, to analyze biomedical signals of interest, such as EEG, with a high level of confidence.

Chapter 6
References and Concluding Remarks

With the advent of Internet and the proliferation of wireless and other communication devices, it is rather easy for someone armed with a set of "keywords" and access to google/google scholar, or other search engines, to launch a search of any of the topics tackled in this text from almost anywhere in the world and come up with an impressive list of references, as is done routinely in some textbooks today. Such a list is of little comfort to a novice in the field and will not be didactical for such an individual. Often, the references cited include a long list of scholarly publications. Many of those articles, which appear in the refereed international journals and conferences, are presented in a manner that demands a high level of sophistication on the part of the reader in understanding the fundamental principles tackled here, hence making such references unsuitable as an educational tool for a novice in the field. Hence, we avoid such an approach and assume that the reader is well-equipped to conduct further Internet searches if need arises. Instead, we take a didactical approach here and provide a list of further readings that complement discussions in this textbook and/or provide an alternate view of the principles presented here. This leads us to an approach that is informative rather than exhaustive. To that end, we tackle each subject separately to allow the reader an opportunity to address reader's deficiencies and/or in-depth questions in an effective manner. Finally, it is important to note that many fundamental concepts tackled here exhibit a time-less attribute; hence, some of the references cited here, although aged, provide excellent references for the concepts presented in this textbook.

6.1 Signals, Systems (Linear), Digital Signal Processing

There are many excellent references that address the problem of signals and systems. The main focus of these texts are linear systems and means to analyze such systems using standard signal processing tools. A good treatment of FT can typically be found in such texts as a tool for studying the spectral properties of signals and the response

K. Kiasaleh, *Biological Signals Classification and Analysis*,
Lecture Notes in Bioengineering, DOI 10.1007/978-3-642-54879-6_6

of linear systems to signals. References [2–4] present a solid treatment of continuous and discrete signals, followed by the response of linear systems, both in continuous and discrete times, to signals. There are also numerous exercises in these texts to enhance understanding of the key concepts. We will be remiss if we do not mention Ref. [5], which has been a classic textbook dealing with digital signal processing and the implementation of efficient tools for computing DFT. A careful examination of these references will provide the reader with a deep understanding of linear system theory and deterministic signal analysis using the Fourier transform in continuous and discrete time domains.

6.2 Random Signals, System Response to Random Signals, and Detection/Estimation Theory

Much has been written on the subject of random variables, processes, and how the "noisy" observations can be classified as random processes. Further, for obvious reasons, the response of systems to noisy inputs has been subject of many textbooks. The interest in this subjects stems from the fact that almost all processes observed in natural as well as those generated by artificial means subscribe to some class of stochastic processes. This coupled with the need to make informed "decisions" in the face of uncertainty presented by noisy observations makes the subject of random processes of paramount importance to the world of signal processing. References [6–8] are excellent references for the subject of random processes. In particular, Ref. [7] builds on top of the basic understanding of random variables and introduces reader to the concepts of random vectors, random sequences, and finally random processes. References [8] provides an excellent reference for stochastic calculus, Markov chains, and other useful techniques in the world of random signal processing. Finally, if nonlinear processing of random processes is a subject of interest, in addition to the material presented in this textbook, Ref. [9] provides a unique perspective on the subject.

On the subject of detection and estimation, Refs. [10–13] provide a comprehensive treatment of the detection and estimation of parameters of noisy signals. References [10] focuses on statistical signal processing in the discrete time domain, while the rest present the general concepts of detection and estimation for AWGN and other types of communications channels. Much of the concepts presented in these texts can be extended to scenarios that do not lend themselves to the AWGN model case. In particular, the concept of hypothesis testing is covered in some detail in Refs. [11, 13].

If adaptive filtering in the face of noisy observation is of interest, in addition to Ref. [10], the reader will benefit greatly by examining Ref. [14].

6.3 Biological Signals

For the most part, signals presented in this text are obtained from MIT-BIH PhysioNet ATM [1], which offers the end user access to variety of biological data, collected for healthy as well as for patients suffering from clinically significant conditions. The data can be downloaded in a number of formats, of which *.mat files are used here to input the desired data into matlab for further signal processing. The data can also be plotted using plotting tools provided in the web site (http://www.physionet.org/cgi-bin/atm/ATM?tool=\&database=aami-ec13\&rbase=\&srecord=\&annotator=\&signal=\&sfreq=\&tstart=\&tdur=\&tfinal=\&action=\&tfmt=\&dfmt=\&nbwidth=\#tools).

6.4 Principle and Independent Component Analysis

Principle component analysis has its roots in SVD principle, and hence, it is recommended that any reader interested in this topic examine the early work on the subject of SVD. There are numerous sources for this subject. Among the possible references, we refer the reader to Ref. [15], where the subject is treated in a tutorial manner. This paper addresses not only the principle of SVD but also describes the computation and the applications of this principle. On the subject of ICA, one needs to examine Ref. [16], where the general concept of blind signal separation is addressed from a statistical point of view. Naturally, the concept of ICA falls under the general umbrella of blind signal separation, and the reader will benefit greatly from examining this reference. Some potential application of ICA are also addressed in Refs. [17, 18]. Once again, the aim of citing these references is to provide the reader with articles which are tutorial in nature. The aim of these articles is an important application of ICA; namely, ICA as a tool for studying brain function. Similar applications of PCA/ICA for ECG can also be found in the literature.

6.5 Wavelet Transform

On the subject of WT, we refer the reader to Refs. [19, 22] for further reading on the mathematical foundation and properties of wavelets along with their applications in signal processing. Additionally, Refs. [20, 21] by Daubechies offer valuable insights into wavelet generation and application. References [23, 24] provide a tutorial approach to the application of wavelets in biomedical field, and hence, can be used as relevant references for the materials presented in this textbook.

References

1. PhysioNet ATM, PhysioNet, MIT Room E25–505A, 77 Massachusetts Avenue Cambridge, MA 02139 USA
2. A.V. Oppenheim, A.S. Willsky, S.H. Nawab, *Signals and Systems* (Pearson Education Limited, Edinburgh Gate, 2014)
3. M.J. Roberts, *Signals and Systems: Analysis using Transform Methods and Matlab* (McGraw Hill, New York, 2004)
4. B.P. Lathi, *Linear Systems and Signals* (Berkeley-Cambridge Press, Carmichael, 1992)
5. A.V. Oppenheim, R.W. Schafer, *Digital Signal Processing* (Prentice-Hall, Englewood Cliffs, 1975)
6. A. Papoulis, *Probability, Random Variables, and Stochastic Processes*, 2nd edn. (McGraw-Hill, Now York, 1984)
7. H. Stark, J.W. Woods, *Probability, Statistics, and Random Processes for Engineers*, 4th edn. (Pearson Education Inc., New York, 2012)
8. H.J. Larson, B.O. Schubert, *Probabilistic Models in Engineering Sciences, Volume II: Random Noise, Signals, and Dynamic Systems* (Wiley, New York, 1979)
9. W.B. Davenport, W.L. Root, *An Introduction to the Theory of Random Signals and Noise* (McGraw-Hill, New York, 1958)
10. C.W. Therrien, *Discrete Random Signals and Statistical Signal Processing* (Prentice Hall, Englewood Cliffs, 1992)
11. H.L. Van Trees, *Detection, Estimation, and Modulation Theory Part I* (Wiley, Now York, 1968)
12. R.N. McDonough, A.D. Whalen, *Detection of Signals in Noise*, 2nd edn. (Academic Press, New York, 1995)
13. C.W. Helstrom, *Elements of Signal Detection & Estimation* (Prentice Hall, Englewood Cliffs, 1995)
14. S. Haykin, *Adaptive Filter Theory*, 2nd edn. (Prentice Hall, Englewood Cliffs, 1991)
15. V.C. Klema, A.J. Laub, The singular value decomposition: its computation and some applications. IEEE Trans. Autom. Control **25**(2), 164–176 (1980)
16. J.F. Cardoso, Blind signal separation: statistical principles. Proc. IEEE **86**(10), 2009–2025 (1998)
17. A. Kachenoura, L. Albert, L. Secnhadji, P. Comon, ICA: a potential tool for BCI systems. IEEE Signal Process. Mag. 57–68 (2008)
18. T.P. Jung, S. Makeig, M.J. Mckeown, A.J. Bell, T.W. Lee, T.J. Sejnowski, Imaging brain dynamics using independent component analysis. Proc. IEEE **89**(7), 1107–1122 (2001)
19. G.W. Wornell, *Signal Processing with Fractals: A Wavelet-Based Approach* (Prentice Hall, Upper Saddle River, 1996)
20. I. Daubechies, Orthogonal bases of compactly supported wavelets. Comm. Pure Appl. Math. **41**, 909–996 (1988)
21. I. Daubechies, The wavelet transform, time-frequency localization and signal analysis. IEEE Trans. Inf. Theory **36**(5), 961–1005 (1990)
22. A. Cohen, J. Kovacevic, Wavelets: the mathematical background. Proc. IEEE **84**(4), 514–522 (1996)
23. M. Unser, A. Aldroubi, A review of wavelets in biomedical applications. Proc. IEEE **84**(4), 626–638 (1996)
24. M. Akay, *Wavelets, Wavelet Packets, and Matching Pursuits with Biomedical Applications* (Wiley-IEEE Press Ebook, 1998)

MIX
Papier aus verantwortungsvollen Quellen
Paper from responsible sources
FSC® C105338

If you have any concerns about our products,
you can contact us on
ProductSafety@springernature.com

In case Publisher is established outside the EU,
the EU authorized representative is:
Springer Nature Customer Service Center GmbH
Europaplatz 3, 69115 Heidelberg, Germany

Printed by Libri Plureos GmbH
in Hamburg, Germany